Archaeology from Historical Aerial and Satellite Archives

William S. Hanson • Ioana A. Oltean
Editors

Archaeology from Historical Aerial and Satellite Archives

Springer

ArcLand
HERITAGE · DATA · KNOWLEDGE

University of Glasgow | Centre for Aerial Archaeology

Editors
William S. Hanson
Department of Archaeology
Centre for Aerial Archaeology
University of Glasgow
Glasgow, United Kingdom

Ioana A. Oltean
Department of Archaeology
University of Exeter
Exeter, United Kingdom

ISBN 978-1-4899-8548-4 ISBN 978-1-4614-4505-0 (eBook)
DOI 10.1007/978-1-4614-4505-0
Springer NewYork Heidelberg Dordrecht London

Printed on acid-free paper

Springer is part of Springer Science+Business Media (www.springer.com)

Contents

List of Figures

List of Tables

Contributors

Peter Barton The Tunnellers Memorial Fund, Faversham, Kent, UK

Natal'ya S. Batanina Historical-Cultural Reserve Arkaim, Chelyabinsk, Russia

Anthony R. Beck School of Computing, University of Leeds, Leeds, UK

Robert Bewley Heritage Lottery Fund, London, UK

Jean Bourgeois Department of Archaeology, University of Ghent, Ghent, Belgium

David Cowley Royal Commission on the Ancient and Historic Monuments of Scotland, Edinburgh, UK

Wim De Clercq Department of Archaeology, University of Ghent, Ghent, Belgium

Damian Evans Department of Archaeology, University of Sydney, Sydney, Australia

Lesley Ferguson Royal Commission on the Ancient and Historic Monuments of Scotland, Edinburgh, UK

Martin J.F. Fowler Les Rocquettes, Winchester, UK

Bryan K. Hanks Department of Anthropology, University of Pittsburgh, Pittsburgh, PA, USA

William S. Hanson Department of Archaeology, Centre for Aerial Archaeology, University of Glasgow, Glasgow, UK

Davy Herremans Department of Archaeology, University of Ghent, Ghent, Belgium

José Iriarte Department of Archaeology, University of Exeter, Exeter, UK

Rebecca H. Jones Royal Commission on the Ancient and Historic Monuments of Scotland, Edinburgh, UK

David Kennedy Classics and Ancient History, The University of Western Australia, Crawley, WA, Australia

Peter McKeague Royal Commission on the Ancient and Historic Monuments of Scotland, Edinburgh, UK

Elizabeth Moylan School of Geosciences, University of Sydney, Sydney, Australia

Ioana A. Oltean Department of Archaeology, University of Exeter, Exeter, UK

Iván Fumadó Ortega La Escuela Española de Historia y Arqueología en Roma – Consejo Superior de Investigaciones Científicas, Rome, Italy

Rog Palmer Air Photo Services, Cambridge, UK

Graham Philip Department of Archaeology, University of Durham, Durham, UK

Tony Pollard Centre for Battlefield Archaeology, University of Glasgow, Glasgow, UK

José Carlos Sánchez-Pardo Fundación Española para la Ciencia y Tecnología, Ministerio de Ciencia e Innovación, Madrid, Spain

Birger Stichelbaut Department of Archaeology, University of Ghent, Ghent, Belgium

Patrizia Tartara Consiglio Nazionale delle Ricerche – Istituto Beni Archeologici e Monumentali, Campus Universitario, Lecce, Italy

Zsolt Visy Department of Archaeology, University of Pécs, Pécs, Hungary

Allan Williams Royal Commission on the Ancient and Historic Monuments of Scotland, Edinburgh, UK

Andrew Young Historic Environment, Cornwall Council, Truro, Cornwall, UK

Part I
Introduction

Chapter 1
A Spy in the Sky: The Potential of Historical Aerial and Satellite Photography for Archaeological Research

William S. Hanson and Ioana A. Oltean

Abstract Aerial photography has facilitated recognition of the density, diversity and complexity of human settlement activity across the fertile lowlands of Europe over millennia, but application of the standard technique of observer-directed archaeological aerial reconnaissance is not universal for a variety of reasons. This introductory chapter highlights the considerable and largely untapped potential of historical aerial and satellite photography for archaeological area survey and landscape analysis, contextualising the examples contained in the volume, which range widely both geographically and chronologically. It draws attention to the range of archival sources available and to the additional benefits of using them, including visualisation of the landscape as it was half a century or more ago before the destructive impact of late twentieth-century development; time-change analysis of the condition of known archaeological monuments; and the discovery of archaeological sites now destroyed.

The impact of aerial photographic discoveries on British and European archaeology has already been immense. In particular, it has facilitated recognition of the density, diversity and complexity of settlement activity across the fertile lowlands over millennia, greatly extending the distribution of many site types. It has been variously estimated that something in excess of 50% of all archaeological sites in Britain have

W.S. Hanson (✉)
Department of Archaeology, Centre for Aerial Archaeology, University of Glasgow, Glasgow G12 8QQ, UK
e-mail: william.hanson@glasgow.ac.uk

I.A. Oltean
Department of Archaeology, University of Exeter, Laver Building, North Park Road, Exeter EX4 4QE, UK
e-mail: I.A.Oltean@exeter.ac.uk

W.S. Hanson and I.A. Oltean (eds.), *Archaeology from Historical Aerial and Satellite Archives*, DOI 10.1007/978-1-4614-4505-0_1,

been discovered from the air (British Academy 2001), primarily as a result of archaeological aerial reconnaissance which has been applied both extensively and intensively since the end of the Second World War. Traditional archaeological aerial reconnaissance, now often referred to as observer-directed reconnaissance, usually involves selective oblique photography of sites identified by observation from light aircraft flying at a height of around 500 m. In that way, it provides a filtered set of glimpses of the archaeological heritage which, given the large number of variable conditions that need to be in place to reveal the remains, are made at what are often unique or unrepeatable moments in time. But some countries, even in Europe, have operated a closed-skies policy until recently or, indeed, continue to do so (such as Greece, Bulgaria and Turkey), while others impose severe bureaucratic restraints on the use of light aircraft for such photo reconnaissance. As a result, many areas of Europe, particularly in the east (see Chap. 10 by Visy, this volume), have been able to adopt this methodology fully only in recent years, and for much of the Middle East, with the notable exception of Jordan (Chap. 13 by Bewley and Kennedy, this volume), such reconnaissance still remains impossible, as for example in Armenia and Syria (Chap. 16 by Palmer and Chap. 15 by Beck and Philip, this volume).

These difficulties, or this late start, can never be entirely overcome by newly acquired imagery, even current high-resolution satellite imagery. However, historical archives of vertical photographs and satellite photography obtained for other purposes offer considerable and largely untapped potential for archaeological research. These may involve sources external to the area concerned, such as declassified military reconnaissance acquired for intelligence-gathering purposes, or those acquired internally, by state authorities or commercial companies, for mapping or other landscape survey and monitoring activities.

The comprehensive survey of Britain by the RAF at a scale of c. 1:10000 made immediately after Second World War (c. 1945–1950) is comparatively well known in the Britain, reasonably readily accessible through the relevant National Monuments Records for England, Scotland or Wales, and provides a good example of the value of such imagery (Chap. 7 by Young, this volume). Though it often goes strangely unacknowledged, this photography has been used consistently as a starting point for the assessment of the archaeological landscape in many areas of Britain. However, although numerous countries possess various collections of vertical aerial photography taken for a wide variety of purposes, with a few notable exceptions, the level of access to and use of historical archival imagery for archaeological research seen in Britain has not been reflected elsewhere in Europe, around the Mediterranean or, indeed, further afield. In Italy, there has been a long tradition of utilising such data, partly at least because of the restrictions on observer-directed aerial reconnaissance from light aircraft which were lifted only in the last decade (Chap. 8 by Tartara, this volume). Similarly, in parts of Eastern Europe such as in Hungary or Romania, where opportunities to undertake archaeological aerial reconnaissance were more limited after the Second World War, the potential of such resources was recognised and utilised by a small number of pioneers (e.g. Visy 1997 and Chap. 10, this volume; Stefan 1986; Bogdan Cătăniciu 1996). The former Soviet Union carried out systematic black-and-white aerial photography within its territorial boundaries at regular intervals during the second half of the twentieth century. Though access to this historical imagery is still largely restricted by government agencies, several archaeological projects have been able to utilise it since the early 1990s as its classification changed from 'confidential' to 'for official use only' (Chap. 12 by Batanina and Hanks, this volume).

Wherever else the technique of aerial photography was applied, however, the primary focus has been on the acquisition of new data from observer-directed reconnaissance specifically for archaeology. This is reflected in the standard textbooks on the subject from different European countries which barely mention the use of non-archaeological archival imagery (e.g. Dassié 1978; Wilson 2000; Braasch 2005).

Apart from Britain, several countries in Europe and further afield maintain their own substantial archives of aerial photographs with national coverage, as in Belgium (Belgian Royal Army Museum – Chap. 5 by Stichelbaut et al., this volume), Italy (Istituto Geografico Militare and Aerofototeca – Chap. 8 by Tartara, this volume), Israel (Aerial Photos of Israel 1917–1919 – Chap. 13 by Bewley and Kennedy, this volume) and Uruguay (Servicio Geográfico Militar – Chap. 14 by Iriarte, this volume). Elsewhere, such material tends to be more dispersed and even less well-known, as for example in Spain and Portugal (Chap. 11 by Fumadó Ortega and Sánchez Pardo, this volume) and Cambodia (Chap. 17 by Evans and Moylan, this volume), while even in countries with more centralised archives, additional imagery, particularly that acquired by commercial companies, can be widely scattered (see, for example, the account in Chap. 13 by Bewley and Kennedy of the situation in Jordan).

However, tens of millions of mainly vertical photographs of areas of Europe and much further afield, derived primarily from military sorties (mainly RFC/RAF and other Allied air forces, Luftwaffe and USAAF) taken during both the First and Second World Wars and shortly thereafter, are potentially available for consultation. Many of these photographs are housed in three major international archives. Two are located in Britain at The National Collection of Aerial Photography (formerly The Aerial Reconnaissance Archive – TARA) in Edinburgh (http://aerial.rcahms.gov.uk; Chap. 2 by Cowley et al. this volume) and the Imperial War Museum in London (Chap. 6 by Pollard and Barton, this volume; Stichelbaut et al. 2010). The third is housed in the National Archives and Records Administration (NARA) at various locations in the Washington D.C. area in the USA (http://www.archives.gov; Going 2002; Abicht 2010). This much-underused resource has been exploited by a relatively few knowledgeable academic researchers, but the potential for further analysis is extremely high. In addition to this vast historical aerial photographic resource, the 5 years between 1995 and 2000 saw the declassification of a range of US satellite photographs taken between 1960 and 1980 primarily for the purposes of military intelligence and mapping (Chap. 4 by Fowler; Chap. 15 by Beck and Philip; Chap. 17 by Evans and Moylan, this volume). This archive, which runs to approximately 900,000 photographs with a wide geographical coverage, has been made available commercially at relatively modest cost through the United States Geological Survey and can be searched and bought online.

These various sets of data have a number of particular advantages over more recent imagery (whether aerial or satellite). First and foremost, they provide a unique insight into the character of the landscape across parts of Europe and beyond as it was approximately century or more ago before the destructive impact of later twentieth century development, whether from the increasing mechanisation of agriculture, intensive industrialisation or urban expansion. Thus, in eastern Romania, various such developments have had a major impact on the survival and current visibility of archaeological sites, for example: the expansion of urban areas such as Mangalia and Galati; the construction of massive industrial complexes covering several hectares; the construction of the navigable canal between the Danube and the Black Sea; the intensification of quarrying; and the expansion of arable agriculture (Chap. 18 by Oltean and Hanson; Chap. 9 by Oltean, this volume). Similarly, in Hungary, various elements of the Roman frontier along the Danube now lie concealed under buildings and factories or have been destroyed by intensive cultivation (Chap. 10 by Visy, this volume). In Cornwall, in south-western England, the later twentieth century witnessed the widespread breaking in of moorland, a move towards deep ploughing and a considerable expansion of towns, which had a similar impact. Agricultural improvement schemes in the basalt zone in Syria have resulted in the clearance of fields, walls and cairns by bulldozing, while enhancements to the road and rail networks, and the concomitant increase in associated settlement activity, have destroyed archaeological features there (Chap. 15 by Beck and Philip, this volume). In Jordan a combination of urban expansion and agricultural development, particularly as the tapping of deep water sources has allowed expansion into previously uncultivated areas, has damaged or destroyed numerous sites (Chap. 13 by Bewley and Kennedy, this volume), while in Armenia, villages have expanded to cover archaeological features whose only record now is that depicted on declassified satellite photographs. The area of south-eastern Uruguay has changed dramatically since the 1970s with the drainage of wetland for rice cultivation (Chap. 14 by Iriarte, this volume), while in Belgium, the photographs from early in the First World War represent the landscape as it was before the devastating transformational impact of the conflict on the Western Front (Chap. 5 by Stichelbaut et al., this volume). The second half of the twentieth century in the former Soviet Union saw a substantial increase in livestock grazing, mineral exploitation, the construction of hydroelectric dams and an emphasis on intensified agricultural production, which in some cases had substantial detrimental impact on, particularly, prehistoric sites (Chap. 12 by Batanina and Hanks, this volume). Similarly, Cambodia experienced major changes to the social, cultural and physical landscape in the later twentieth century resulting in rapid development and urbanisation in many areas. More specifically, the Khmer Rouge regime brought a radical restructuring of the agrarian landscape, involving widespread destruction of field systems and topographic features that had evolved over centuries (Chap. 17 by Evans and Moylan, this volume). In all these areas, archival aerial and/or satellite photography has allowed archaeologists to turn back the clock and identify archaeological features in the landscape that have been erased both from view and from memory.

Moreover, the historic character of the imagery means that it can often provide large-scale 'snapshots' of the landscape at various points in time. This can facilitate time-change analysis of the condition of known archaeological monuments, providing insight into any site attrition or any modifications that may have occurred within the landscape. This, in turn, can aid in any quantification of the impact of landscape change on the archaeological resource and help demonstrate the need for rigorous protection of what remains, thus making a contribution to the management of the cultural resource.

Obviously, the military reconnaissance archival material provides a photographic record of the contemporary landscape, which can be important in its own right, particularly in relation to the archaeology of twentieth century conflict whose importance has been increasingly recognised in recent decades (Chap. 6 by Pollard and Barton; Chap. 9 by Oltean, this volume; Stichelbaut et al. 2009). But in addition, it and other historical coverage offer the prospect of the better survival and visibility of archaeological remains reflecting the earlier history of that landscape, whether manifest as earthworks, cropmarks or soilmarks. This is all the more important in those areas of Eastern Europe, the Middle East or further afield with no history of sustained aerial reconnaissance for archaeology which might be drawn upon. Even in areas where archaeological aerial reconnaissance is now possible, such historical archival data can considerably augment that reconnaissance. In Romania, it has provided important coverage of areas still inaccessible for political or security reasons, such as along the Bulgarian border and north of the Danube-Black Sea Canal (Chap. 18 by Oltean and Hanson, this volume). In the former Soviet Union, Batanina and Hanks (Chap. 12, this volume) have highlighted the importance of the Soviet Era historical imagery for the detection, identification and interpretation of archaeological sites, including the discovery of completely new categories of monument, while the investigation of existing aerial photographs has been of considerable importance in increasing the number of the known or suspected archaeological sites in Hungary (Chap. 10 by Visy, this volume). In Jordan, Bewley and Kennedy (Chap. 13, this volume) have demonstrated the considerable value and importance of integrating historical imagery into their ongoing programme of aerial reconnaissance, while in Syria, historical satellite imagery was used to investigate long-term human-landscape interaction in contrasting environmental zones experiencing expansion of settlement and agriculture (Chap. 15 by Beck and Philip, this volume). In Cambodia, archival imagery has provided an opportunity to both evaluate and add archaeological understanding to oral histories of landscape change (Chap. 17 by Evans and Moylan, this volume).

The character of the historical archival photography is markedly different from that obtained through traditional archaeological aerial reconnaissance, which still depends on observer-directed flying. In the latter, the initial recognition and primary interpretation take place in the air, both in terms of whether a particular feature is recognised and recorded, and also in relation to the chosen flight path for the aircraft, thus compounding the serendipitous nature of the survey process. By contrast, historical archival imagery is primarily vertical block coverage which is more systematic, intensive and, often, quite extensive. Such coverage greatly enhances the potential for taking a more landscape-focused approach, enabling the identification and mapping of more widespread and sometimes more ephemeral remains, such as roads and trackways, field systems, cultivation traces and mining activity, and interrelationships between settlement foci and the wider man-made landscape (Chap. 18 by Oltean and Hanson; Chap. 7 by Young; Chap. 8 by Tartara; Chap. 17 by Evans and Moylan, this volume).

It is clear, therefore, that historical archival aerial and satellite photography can add considerably to the known archaeological record, a theme which runs consistently through the chapters in this volume. One of the main problems in utilising this archival photography, however, is its accessibility. The archive of declassified US satellite photography is readily accessible and can be searched online, where the approximate coordinates of the four corners of the ground footprint of each photograph can be displayed on a Google Map backdrop (Chap. 4 by Fowler, this volume). By way of contrast, however, of the four main Second World War collections held within The National Collection of Aerial Photography in Britain, that is the Allied Central Interpretation Unit (ACIU) Archive, the Mediterranean Allied Photo Reconnaissance Wing (MAPRW) Archive, the archive of GX (Luftwaffe) Reconnaissance Imagery and of the Joint Air Reconnaissance Intelligence Centre (JARIC), only the first currently has an online GIS catalogue providing easy access to the original sortie plots (http://aerial.rcahms.gov.uk; Chap. 2 by Cowley et al. this volume). This is not to say that the other archives are completely inaccessible,[1] but they require some prior knowledge of the historical context of the reconnaissance to assess the likelihood of the photographic material being found in the archive (see Chap. 9 by Oltean, this volume). It is hoped that the necessary resources, both in terms of time and manpower, can be found to allow the enhancement of the finding aids for MAPRW, GX and JARIC archives held in Edinburgh in order to make these large and important collections more readily accessible (Chap. 2 by Cowley et al.; Chap. 3 by McKeague and Jones, this volume). Similarly, though a high proportion of the aerial photographs held by the National Archives and Records Administration in the USA is geographically indexed, the finding aids are not available online (Cowley et al. 2010: 2).

The original stimulus for this volume came in November 2000 at the NATO-sponsored Advanced Research Workshop on Aerial Archaeology in Leszno, Poland, when Chris Going (2002) drew attention to the potential of Second World War aerial reconnaissance photographs for archaeology, focusing on those taken by the Luftwaffe. Though already well aware of the archaeological use of immediately post-war RAF photography of Britain (c.f. Chap. 7 by Young, this volume), we had not fully appreciated the vast scale of the historical archival material potentially available for much of Europe and further afield. Though one of us had already made some use of declassified CORONA satellite photography (Oltean 2002), this was primarily as a transcription aid. However, we subsequently made fuller interpretative use of both Allied and German photography as part of our work in Dobrogea, Romania, so we soon came to better appreciate that potential at first hand. The latter work was briefly reported to a SPIE conference on Remote Sensing for Environmental Monitoring, GIS Applications, and Geology in Florence in September 2007

[1] The two chapters by Oltean and by Oltean and Hanson (Chaps. 9 and 18) provide examples of the use of MAPRW and GX reconnaissance imagery in Romania.

(Oltean and Hanson 2007 – a much revised and expanded version of that paper is presented in this volume, Chap. 18). Finally, the fact that the topic was still seen as sufficiently current and under-examined to feature as a theme at the annual Aerial Archaeology Research Group conference, held in Copenhagen later in September 2007, prompted us to develop the proposal for this volume and seek out appropriate contributors. Though strictly speaking all aerial photographs become archival at the point at which they are indexed and potentially made accessible to the wider public, the focus of this volume, as has been made clear above, is on historical photography taken at least 40–50 years ago for purposes other than archaeological survey.

A primary aim of this volume is to draw to wider attention the existence, scope and potential access to such historical archival aerial and satellite photographs in order to encourage their use in a wider range of archaeological and landscape research. Though the theme also features in several other chapters (e.g. Chap. 8 by Tartara; Chap. 9 by Oltean; Chap. 15 by Beck and Philip), the first of the two parts – *Opening Doors: Aerial and Satellite Archives* – focuses specifically on familiarising the reader with the resources of, and access to, some of the most important historical archives of international relevance, such as TARA and the United States Geological Survey's Earth Resources Observation and Science (EROS) Center for CORONA satellite photography, both containing declassified military intelligence-gathering material (Chap. 2 by Cowley et al.; Chap. 4 by Fowler). Looking to the future, McKeague and Jones (Chap. 3) provide informative details about the latest developments in the organisation and accessibility of archival material. Though based on the archives of the Royal Commission on the Ancient and Historic Monuments of Scotland, the approaches they detail have potentially much wider applicability.

The second part – *Historical Aerial and Satellite Photographs in Archaeological Research* – is the core of the volume and presents a series of examples or case studies addressing the use of historical aerial and satellite archives for archaeological research. Most of the papers make use of aerial photography, whether derived from First World War or Second World War intelligence-gathering reconnaissance or from internally generated survey photography, either individually or in combination. Two illustrate the value of declassified Cold War satellite photography, while two demonstrate the benefit of the integration of both aerial and satellite imagery. The papers range widely both geographically and chronologically in order to demonstrate the widespread applicability of the methodological approach both to students and academic researchers. Thus, their coverage extends across Europe, both eastern (Chap. 12 by Batanina and Hanks; Chap. 10 by Visy) and western (Chap. 11 by Fumadó-Ortega and Sánchez-Pardo; Chap. 5 by Stichelbaut et al.; Chap. 6 by Pollard and Barton; Chap. 7 by Young) to the borders of the Black Sea and around the Mediterranean (Chap. 9 by Oltean; Chap. 18 by Oltean and Hanson; Chap. 8 by Tartara). Through them, we travel from the uplands of the Caucasus (Chap. 16 by Palmer) and the dryer climates of the Middle East (Chap. 15 by Beck and Philips; Chap. 13 by Bewley and Kennedy) to the marshes of South America (Chap. 14 by Iriarte) and the wetlands of Cambodia (Chap. 17 by Evans and Moylan). The chapters demonstrate that historical archival photography has proved equally useful when dealing with archaeological remains dating from the activities of some of the earliest agriculturalists to the investigation of sites associated with the First and the Second World Wars. By drawing attention to this massive archival resource, providing examples of its successful application to archaeological questions across a wide geographical and chronological range and offering advice how to access and utilise the resource, the volume seeks to bring this material to wider attention, demonstrate its huge potential for archaeology, encourage its further use and stimulate a new approach to archaeological survey and the study of landscape evolution internationally. Many of these photographs were originally acquired and subsequently maintained in secret in the context of major conflict, either open warfare or clandestine surveillance. Now, half a century or more later, they are proving to be of much wider benefit, with enormous and unanticipated value for the elucidation of our archaeological heritage.

Bibliography

Abicht, M. J. (2010). Using wartime aerial photographs to locate lost grave sites. In D. Cowley, R. A. Standring, & M. J. Abicht (Eds.), *Landscapes through the lens. Aerial photographs and historic environment* (pp. 263–265). Oxford: Oxbow.

Bogdan Cătăniciu, I. (1996). I valli di Traiano nella Dobrugia. Considerazioni sulle fotografie aeree. In M. Porumb (Ed.), *Omaggio a Dinu Adameşteanu* (pp. 201–225). Cluj-Napoca: Clusium.

Braasch, O. (2005). *Vom heiteren Himmel. Luftbildarchäologie*. Esslingen: Gesellschaft für Vor- und Frühgeschichte.

British Academy. (2001). *Aerial survey for archaeology. Report of a British Academy working party 1999*. Compiled by Robert Bewley. London: British Academy.

Cowley, D., Standring, R. A., & Abicht, M. J. (2010). Landscapes through the lens: An introduction. In D. Cowley, R. A. Standring, & M. J. Abicht (Eds.), *Landscapes through the lens. Aerial photographs and historic environment* (pp. 1–6). Oxford: Oxbow.

Dassié, J. (1978). *Manuel d'archéologie aérienne*. Paris: Technip.

Going, C. J. (2002). A neglected asset. German aerial photography of the Second World War period. In R. H. Bewley & W. Raczkowski (Eds.), *Aerial archaeology: Developing future practice* (Nato science series, pp. 23–30). Amsterdam: IOS Press.

Oltean, I. A. (2002). The use of satellite imagery for the transcription of oblique aerial photographs. In R. H. Bewley & W. Raczkowski (Eds.), *Aerial archaeology: Developing future practice* (Nato science series, pp. 224–232). Amsterdam: IOS Press.

Oltean, I. A., & Hanson, W. S. (2007). Reconstructing the archaeological landscape of Southern Dobrogea: Integrating imagery. In M. Ehlers & U. Michel (Eds.), *Remote sensing for environmental monitoring, GIS applications, and geology VII, Proceedings of SPIE* (Vol. 6749, no. 674906). Bellingham.

Stefan, A. S. (1986). Archéologie aérienne en Roumanie. *Photo-interpretation,* 25(1 and 2) (Special ed.). Paris.

Stichelbaut, B., Bourgeois, J., Saunders, N., & Chielens, P. (Eds.). (2009). *Images of conflict: Military aerial photography and archaeology*. Newcastle upon Tyne: Cambridge Scholars Publishing.

Stichelbaut, B., Gheyle, W., & Bourgeois, J. (2010). Great war aerial photographs: The imperial war museum's box collection. In D. Cowley, R. A. Standring, & M. J. Abicht (Eds.), *Landscapes through the lens. Aerial photographs and historic environment* (pp. 225–236). Oxford: Oxbow.

Visy, Z. (1997). Stand und Entwicklung der archäologischen Luftprospektion in der DDR, der Tschechoslowakei und Ungarn in den Jahren 1945 und 1990. In J. Oexle (Ed.), *Aus der Luft. Bilder unserer Geschichte. Luftbildarchäologie in Zentraleuropa* (pp. 22–27). Dresden: Landesamt für Archaeologie mit Landesmuseum für Vorgeschichte.

Wilson, D. R. (2000). *Air photo interpretation for archaeologists* (2nd ed.). Stroud: Tempus.

Part II
Opening Doors: Aerial and Satellite Archives

Chapter 2
The Aerial Reconnaissance Archives: A Global Aerial Photographic Collection

David C. Cowley, Lesley M. Ferguson, and Allan Williams

Abstract Recognition of the importance of historic aerial photographs for the primary discovery and recording of previously unrecognised archaeological sites, for the documentation of already known sites and for the characterisation of the wider cultural landscape continues to grow. This chapter describes one of the world's largest collections of historical aerial photographs, The Aerial Reconnaissance Archives (TARA), outlining its contents and potential for archaeologists. Issues of accessibility to TARA and the importance of 'best practice' in the use of historic aerial photographs are discussed.

2.1 Introduction

The last two decades has seen a growing recognition of the importance of historic aerial photographs for primary discovery and recording of previously unrecognised archaeological sites, for the documentation of already known sites and for the characterisation of the wider cultural landscape (e.g. papers in Bewley and Rączkowski 2002; Cowley and Ferguson 2010; Cowley et al. 2010a; Cowley and Stichelbaut 2012; Ferguson 2011). Despite increased awareness of potential applications, including their systematic use as one of a range of sources in England's National Mapping Programme (Horne 2009, 2011; Winton and Horne 2010), on a global level these photographs remain a largely unexplored resource for a number of reasons, including problems of access and research traditions. This chapter

D.C. Cowley (✉) • L.M. Ferguson • A. Williams
Survey and Recording, Collections Royal Commission on the Ancient and Historical Monuments of Scotland, 16 Bernard Terrace, Edinburgh EH8 9NX, UK
e-mail: Dave.Cowley@rcahms.gov.uk; Lesley.Ferguson@rcahms.gov.uk; Allan.Williams@rcahms.gov.uk

W.S. Hanson and I.A. Oltean (eds.), *Archaeology from Historical Aerial and Satellite Archives*, DOI 10.1007/978-1-4614-4505-0_2,

describes one of the world's largest collections of aerial photographs, The Aerial Reconnaissance Archives (TARA), which is curated by the Royal Commission on the Ancient and Historical Monuments of Scotland (RCAHMS) in Edinburgh. TARA holds aerial photographs of locations throughout the world, dating from 1938 to 1989, with further imagery being added as it is declassified. As a collection with a global reach, these photographs can often be a unique resource, providing the earliest aerial records of landscapes from above, which predate the major landscape changes of the mid- and late twentieth century. The imagery can be a unique source and is thus of interest to the whole range of historical and landscape studies, but this contribution emphasises its importance to archaeologists. These uses are a prime example of the 'serendipity effect' of aerial photographs collected for one reason (military intelligence) but which contain valuable information, recorded by chance, for other purposes (Brugioni 1989; Cowley et al. 2010b: 1; Fowler 2004: 118).

2.2 The Aerial Reconnaissance Archives

The origins of TARA date back to the Second World War, when millions of aerial photographs had been amassed at the Allied Central Interpretation Unit (ACIU), based at RAF (Royal Air Force) Medmenham, Buckinghamshire. The peacetime potential for the images was immediately recognised, and there was interest in the wider uses of the photographs for teaching and research purposes by university academics – many of whom had been wartime intelligence officers (e.g. Bradford 1957; Ferguson 2011). In the late 1950s, Stanley Beaver, Professor of Geography at the University College of North Staffordshire (now Keele University), headed a series of discussions with the Air Ministry, which led to the transfer of the ACIU reconnaissance photographs of western Europe to Keele in the early 1960s and the establishment of the Keele Air Photo Library (Walton 1975), later renamed The Aerial Reconnaissance Archives. With its earliest holdings dating from 1938, and a major focus of the collection on the period of the Second World War, TARA continues to grow as declassified British military intelligence photographs are added.

Until recently, the main use of TARA has been by European bomb-disposal experts, as a means of locating unexploded wartime ordnance; in addition, some limited use has also been made by historians, environmentalists and archaeologists. There are four separate and distinct collections within TARA, each with differing characteristics which have a direct bearing on how they may be used. Vitally, the majority of imagery was captured as stereoscopic pairs, providing the added value to photo-interpretation that the 3-D view allows. However, only one of the collections is readily available for research at present (2011), but developing access to the others remains a long-term goal. A summary of the collections follows, but full details, their history and access information are available on the website (aerial.rcahms.gov.uk.), and this will be updated as research progresses to make more material accessible.

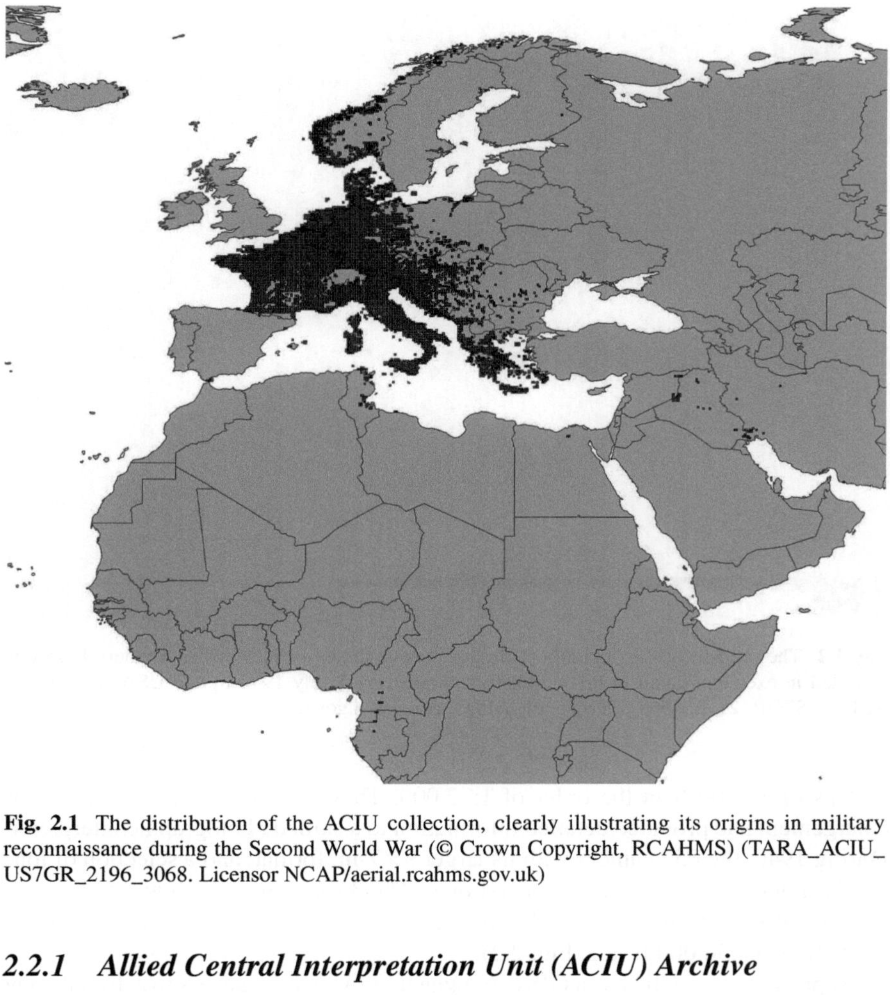

Fig. 2.1 The distribution of the ACIU collection, clearly illustrating its origins in military reconnaissance during the Second World War (© Crown Copyright, RCAHMS) (TARA_ACIU_US7GR_2196_3068. Licensor NCAP/aerial.rcahms.gov.uk)

2.2.1 Allied Central Interpretation Unit (ACIU) Archive

The ACIU Archive is the most accessible part of TARA; it comprises reconnaissance prints, mainly vertical images, of locations throughout western Europe (Fig. 2.1) taken during the Second World War by the RAF, the Royal Canadian Air Force (RCAF), the South African Air Force (SAAF) and the United States Army Air Force (USAAF). The analyses undertaken at the ACIU headquarters at RAF Medmenham informed the planning of practically every wartime operation. Over 1,700 people worked at the ACIU in 1944 and, with a daily intake of imagery in 1945 averaging 25,000 negatives and 60,000 prints, an extensive resource had been accumulated by the end of the war (Babington Smith 1957).

The earliest ACIU cover dates from 1939 with the initiation of high-level, high-speed, and unarmed reconnaissance by the RAF. These Spitfire sorties carried the F24 camera equipped with a lens of 5-in. focal length. The 5×5-in. format of the negatives produced small images, with contact-print scales from the operating

Fig. 2.2 The rampart of the oppidum at Braquemont to the east of Dieppe in northern France is recorded in excellent detail in this vertical view taken on 5 July 1944 by the USAAF (TARA_ACIU_US7GR_2196_3068. Licensor NCAP/aerial.rcahms.gov.uk)

heights of 30,000 ft in the order of 1:72,000. They were nevertheless capable of enlargement and provided detailed information to a skilled photographic interpreter. During 1940, the F52 camera with its larger 9 × 7-in. format, and a selection of longer focal length lenses, became operational. The 36-in. lens, which was to become the standard reconnaissance lens by the following year, produced negative scales of 1:10,000 from 30,000 ft (e.g. Fig. 2.2).

At present, the ACIU has a GIS catalogue to provide easy access to a digital copy of the original sortie plots, and the imagery is preserved on microfilm. In excess of 5.5 million ACIU images are accessible via either the RCAHMS search room or the Paid Search Service.

2.2.2 *Mediterranean Allied Photo Reconnaissance Wing (MAPRW) Archive*

The Mediterranean Allied Photo Reconnaissance Wing Archive comprises vertical and oblique aerial reconnaissance prints of southern and central Europe, including many classical sites, taken by the RAF, SAAF and USAAF. It contains some 150,000 images dating to between 1943 and 1945 (Fig. 2.3) (see also Chap. 9 by Oltean and

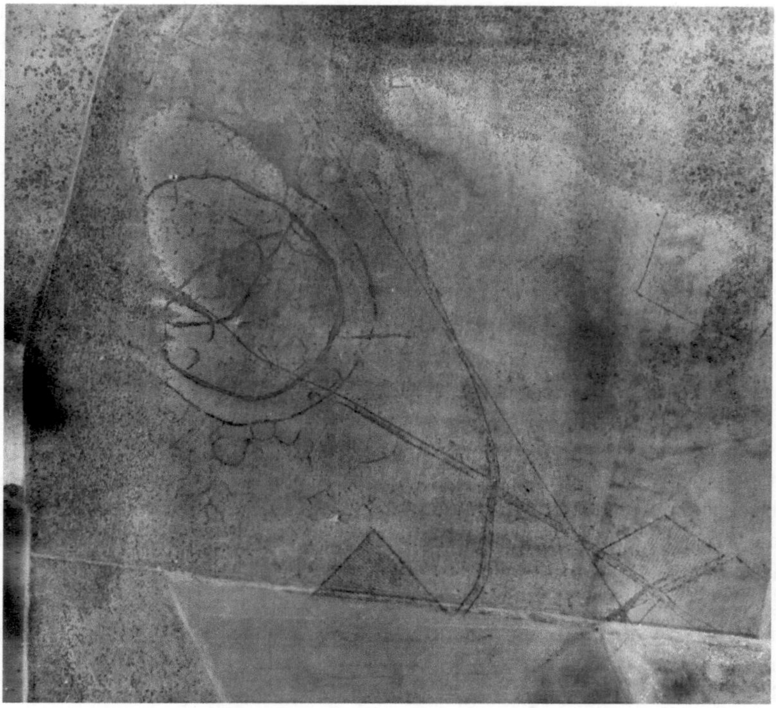

Fig. 2.3 An extract from a vertical frame taken on 21 June 1945 of the prehistoric village, field boundaries and trackways at Masseria Cascavilla, San Giovanni Rotondo in Apulia, southern Italy (Jones 1987), recorded as vegetation marks by chance on a reconnaissance photograph (TARA_SJ_682_L21_3678. Licensor NCAP/aerial.rcahms.gov.uk)

Chap. 18 by Oltean and Hanson, this volume). After the war, the photographs were initially held by the British School in Rome before being transferred to the Pitt Rivers Museum, Oxford and, in the 1980s, to Keele University.

The MAPRW archive is currently inaccessible but will be available for research in the future when the finding aids are catalogued.

2.2.3 GX (*Luftwaffe*) Reconnaissance Imagery

At the end of the Second World War, over one million German aerial photographs were discovered by the British and American forces. Taken by the Luftwaffe between 1939 and 1945, the mainly vertical aerial reconnaissance photographs are of locations mostly in Eastern Europe and Russia. Hidden by the Germans, much of this material came from Hitler's mountain retreat at Berchtesgaden in Bavaria, but other collections were discovered in Vienna, Oslo and Berlin. All the Luftwaffe material was consolidated at RAF Medmenham, where preliminary sorting continued until 1949. This provided mass aerial imagery of Eastern Europe and the Soviet Union on a scale otherwise unachievable at the time, and it remained a key intelligence resource for more than two decades. With incredible potential for both site discovery and analysis of landscape change (see Chap. 9 by Oltean and Chap. 18 by Oltean and Hanson, this volume), this imagery will also provide the earliest aerial views in some countries. After the fall of the Soviet Union, the imagery was declassified and transferred to Keele University in the early 1990s.

There are no finding aids for the Luftwaffe imagery, and access is extremely difficult with few clues to even geographical location.

2.2.4 *Joint Air Reconnaissance Intelligence Centre (JARIC)*

The final component of TARA is the archive from the Joint Air Reconnaissance Intelligence Centre, the main provider of imagery intelligence to the UK Ministry of Defence. This is by far the largest and most recent of all TARA holdings. Taken by the RAF, RCAF, Royal Australian Air Force (RAAF), Royal Navy, Royal New Zealand Air Force (RNZAF), SAAF and USAAF, and other NATO countries, there are in excess of ten million vertical and oblique aerial photographs of locations throughout the world dating from 1938 to 1989. Many of the photographs are from the Second World War (Fig. 2.4), but there are also holdings for countries where there has since been military activity, such as Suez in 1956 or the Korean War (1949–1954). The collection contains a microfilm copy of all surviving original photographs and associated sortie plots and includes many millions of unique images – the original films having been destroyed.

Key to research in the JARIC archive is knowledge of modern history, and a list of military campaigns is included in the collection description on the website (aerial.rcahms.gov.uk). With this knowledge and research into the deployment of particular squadrons in intelligence records at The National Archives at Kew (London), exact sortie references will lead to the identification of imagery.

2.3 TARA: Unique Opportunities

With many millions of aerial photographs spanning the period between the late 1930s and 1989, and with more imagery added as it is declassified, TARA is a major information source for a broad range of disciplines (Cowley and Ferguson 2010; Cowley et al. 2010a). Any treatise on the value of historic aerial photographs should stress the importance of records from the mid-twentieth century in documenting landscapes before the wholesale changes brought about by the increasing mechanisation of agriculture, urbanisation and industrialisation that have characterised the last half century (Cowley and Ferguson 2010).

Fig. 2.4 This vertical view of the southeast of Rome was taken as part of a run of images on 20 August 1944. This frame records the Baths of Caracalla and the Aurelian Walls. Comparisons with contemporary images (e.g. Google Earth™ 2002) show extensive development across the open ground outside the walls since 1944. Inside the walls, while changes have been less sweeping, there has been significant infilling of gap sites and alterations in land use and vegetation. Of particular note are material changes in the Baths of Caracalla, with, for example, the removal of tiered seating from the southwest courtyard. Such records of ancient sites are especially important where ongoing consolidation and restoration may have altered fabric (TARA_JARIC_106G_2353_3012. Licensor NCAP/aerial.rcahms.gov.uk)

Large parts of the globe have been swallowed up by expanding cities, and for these areas, early photographs are effectively the only source from which sites that are now beyond the reach of the archaeologist may be studied. Changing agricultural practices have destroyed archaeological sites wholesale and, again, the view from 50 or 60 years ago is priceless – it is the only means of recovering what has been lost to our understanding of the past. Survey is usually equated with flying in a small aircraft or walking across the ground, but probably the most effective means of survey, of discovering previously unrecorded sites and creating a basic record of them, remains existing aerial photographs. Undertaking 'survey' in boxes of aerial photographs or viewing digital images on screen may lack the appeal or glamour of aerial or field survey, but for simple value for money in discovering and recording sites, we contend that it has no equal.

Monitoring condition and material change in known archaeological monuments offers a recent historical perspective on the attrition of sites and can be a powerful means of demonstrating the need for rigorous protection of what remains. An understanding of changes in the material condition of sites also has a direct research application, for example, in identifying parts of sites that may have been disturbed in the recent past or where vegetation cover may have altered. In all these cases, and others, the textured view that aerial photographs provide has immediacy that maps and plans do not and so can help communicate change and damage to a monument in a highly effective way.

Parts of the world are fortunate in having virtually unrestricted access to the skies to conduct aerial survey and to collections of aerial images. This can be a recent phenomenon, even in parts of Europe (Braasch 2002), and, from a global perspective, the more usual state of affairs is that aerial survey is viewed with suspicion and may be highly restricted. Indeed, access to and use of aerial photographs is also frequently viewed in the same way. TARA's holdings may thus be the primary survey record for many parts of the world from the mid-twentieth century, and, notwithstanding the problems of limited finding aids, the major relatively accessible source material. In these areas, the potential for such historical photographs to aid discovery of previously unknown sites and monuments is magnified enormously as recently demonstrated by an aerial research programme in Romania (Oltean and Hanson 2007 and Chap. 18, this volume, also Chap. 9 by Oltean this volume).

2.4 TARA: A Global Collection in Context

While this chapter is focussed on TARA, in general terms, much of it is applicable to the other global collection at National Archives and Records Administration (NARA) in the USA and national collections such as the Aerofototeca in Italy (Ceraudo and Shepherd 2010). It should be noted that these collections can often be complementary, individually containing material not represented elsewhere and so of primary importance for particular issues. Thus, an understanding of patterns of acquisition, preservation and discard (i.e. destruction of images once they had no perceived value) can be vital to making effective use of the aerial photograph resources (Cowley et al. 2010b: 3–4). Beyond the need to understand the source material, such resources will be at their most effective in the context of broadly-based landscape studies, as the serendipity effect will more likely pay dividends for a range of interests rather than the narrow focus of, for example, a period-specific study.

2.5 Opening up TARA: Aspirations and Issues

The importance of TARA has been identified for some time (Going 2002; Rączkowski 2004) and, while the application of its resources by archaeologists has been limited, this and other publications (Cowley et al. 2010a) demonstrate a growing awareness of such material in a wider archaeological community, though there remain some blocks on maximising usage. This section outlines the main issues as they relate directly to TARA, though aspects of this discussion have a more general applicability.

2.5.1 Access: Website, Search Room and Finding Aids

With the move to RCAHMS, two key developments were required to instantly improve access to TARA. These were achieved in 2009, that is, the development of a new website and the creation of a search room facility for visitors. A new website (*aerial.rcahms.gov.uk*) has been launched to provide information about the archives, their historical background and content, and news about the latest discoveries and developments (Fig. 2.5). Full descriptions, broad geographical content and access arrangements are outlined in a series of summaries for all the distinct groups of material. Ongoing digitisation and preservation programmes provide a steadily growing number of images visible online. As the centre point of each image has been established, access is primarily by geographical location. Low-resolution 'thumbnail' images can be viewed free of charge and, for a modest subscription, enhanced quality images and extra features, such as mosaicked data, are available.

Through the website, it is possible to commission some basic research to identify imagery for certain areas, and guidelines describe the information required by staff to undertake the work, for which a charge is levied. But within the RCAHMS search room in Edinburgh, a whole new public service has been established for visitors to undertake their own research. By prior appointment, all the indices and flight plots to the ACIU archives are accessible, along with access to millions of good quality microfilm images. In early 2010, the RCAHMS search room facilities were extended to create additional workstations for the growing number of visitors (Fig. 2.6).

One of the main barriers and challenges to improving access is the sheer size of TARA. It might seem incomprehensible that there is not a ready-made index to the archives, but this is the case for whole collections, and there are tens of thousands of sorties which have not been geographically located – even to a specific country. Addressing this problem will remain one of the priorities for the coming years.

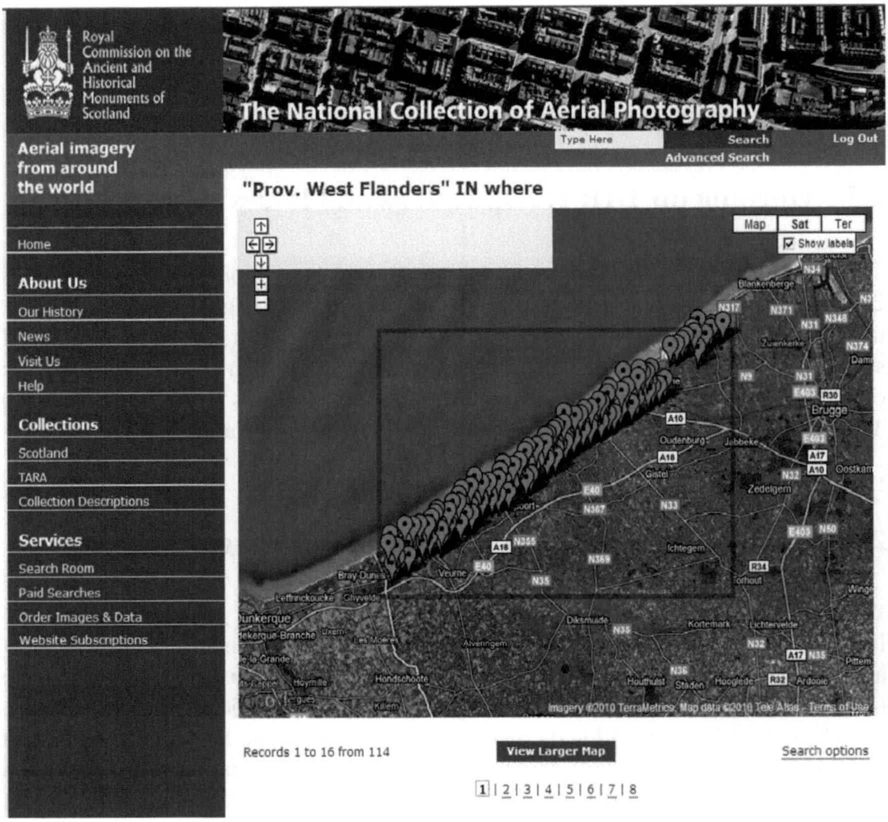

Fig. 2.5 The TARA website allows remote users to search for images against a Google Earth™ background and view low-resolution versions of digital holdings, a facility that continues to expand as digitisation continues

Much has been achieved already for the ACIU imagery with the digitisation of all the flight plots and, although these are currently only accessible in the RCAHMS search room in Edinburgh, there is considerable potential for online delivery, as witnessed already through the launch of the *evidence in camera* website in 2004, when TARA was still at Keele.

In the medium and long term, work will continue on rationalising and digitising existing but less-coherent finding aids. All will require major investments in time and resource to even allow for basic location identification, and finding external finance for such exercises will be absolutely essential. TARA is not alone in experiencing access difficulties. Archive material held elsewhere, for example, in The National Archives at Kew or in the USA, may offer routes into some areas of the collection, and working in partnership may bring mutually beneficial results.

Fig. 2.6 The recently expanded public search facilities at RCAHMS allow visitors to undertake their own research on aspects of the collections (DP068698 Licensor NCAP/aerial.rcahms.gov.uk)

2.5.2 Funding

This is not the place to dwell on the history of TARA, beyond identifying that the lack of established funding has forced it to sustain itself on its income. This has created a basic tension between servicing income-generating users, such as unexploded ordnance experts, and more research-based users, such as archaeologists, whose interest in the collection may be more speculative and certainly is less well funded. With the move of the collection to RCAHMS, this basic tension remains, as the collection did not come with any funding. With limited or no finding aids for much of TARA, opening up access thus presents a significant challenge. For RCAHMS, the key to developing the use of the collection will be to build partnerships across the globe to further research into content and to explore funding opportunities. Enabling access to the whole of TARA will take decades and may ultimately never be complete, but, with collaborative projects, there is the opportunity to frame research questions that can inform future directions.

Beyond the fundamental issue of funding, there remain other practical considerations that bear directly on the effective exploitation of the collection as a data mine for archaeologists, amongst whom there is a patchy recognition of their importance. These issues include the nature of finding aids, where they exist, the scale of photography and the appropriateness of photographs to the task in hand, such as time of year and conditions of, for example, vegetation cover.

2.6 Developing Best Practice in the Use of Historic Aerial Photographs

The potential problems of inappropriate or ill-informed uses of aerial photographs have been commented on above. Here, aspects of work practice that may contribute to this area are discussed, including the understanding of scale and the general issue of matching source material to intended purpose.

2.6.1 *Scale*

In the experience of the authors, the importance of scale of photographs is poorly appreciated by many archaeologists, and this is a major factor in the inappropriate application of aerial imagery, with a high rate of misidentification of features. Routinely users of aerial photographs of Scotland in the public search room at RCAHMS consult contact prints at scales varying from 1:5,000 to 1:24,000 and larger. At 1:5,000, a feature 5 m across will measure 1 mm across, or about the size of a blunt pencil point. At 1:10,000, the same feature will measure 0.5 mm across, and at smaller scales (e.g. 1:25,000, 1:100,000), the feature will be vanishingly small (Lillesand and Kiefer 2000: 136–41).

This may sound like a statement of the obvious, but it is reiterated here because 'sites' are regularly identified by professional archaeologists from aerial photographs where interpretation completely fails to take account of scale. By way of example, in eastern Scotland, a desk-based assessment of aerial photographs identified a long-cist cemetery of potentially mid-first millennium AD date. Such features, comprising clusters of 'maggot-shaped' cropmarks, some 2 m in length and about 1 m across formed over the graves of extended inhumations, have been identified from aerial photographs (e.g. Cowley 2009), but in this case, the identification was made from 1:24,000 scale contact prints, and the potential grave pits identified, when scaled correctly, were found to measure about 20 m across, and thus, whatever else they might be, they were clearly not the remains of a cemetery.

A related issue is that, while in most instances hand magnifiers may be used to examine prints, only a small proportion uses stereoscopic-viewers through which significantly more information can be captured and very rarely are enlargements or reprints requested (Fig. 2.7). In the case of reprints or enlargements, there is obviously a time/cost consideration, but in the case of stereoscopic viewers, there appears to be a basic lack of education about how to use aerial photographs effectively (e.g. see Palmer and Cowley 2010).

Scale, or resolution, is also at the heart of a sometimes-competitive relationship between satellite imagery and traditional aerial photographs. For instance, the positing of 'satellite archaeology' has led to the overstatement of the importance of satellite imagery over aerial photographs (Parcak 2009; Wiseman and El Baz 2007), when they should be seen as complementary and best suited for particular purposes.

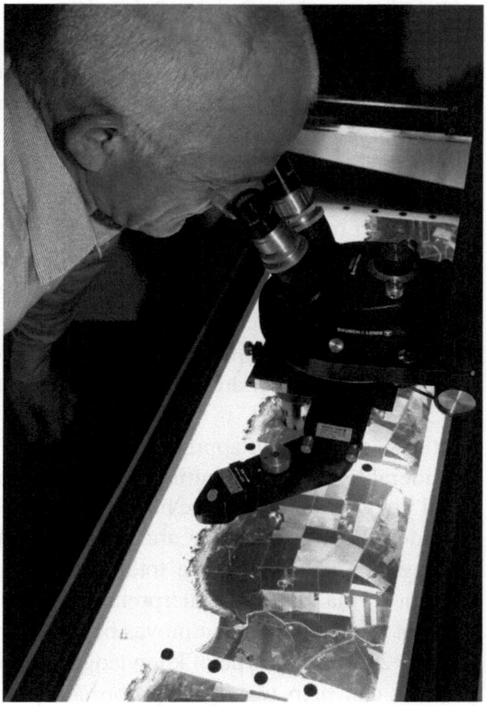

Fig. 2.7 The use of stereo-viewers is sadly not routine but carries with it an enormous advantage in interpreting aerial photographs (DP068705 Licensor NCAP/aerial.rcahms.gov.uk)

However, it is worth reiterating that historic aerial photographs could not be matched by satellite imagery for resolving detail until relatively recently (Fowler 2010) and will thus always be a unique resource for the period up to the mid-1960s (at the earliest).

2.6.2 Matching Source Material to Purpose

An understanding of the relationships between the nature of expected archaeological features (e.g. buried structures, earthworks) and the way in which they may register on aerial photographs is a further key consideration in the effective exploitation of a collection such as TARA. As with scale, the importance of these issues appears not to be as widely appreciated as one might hope (Palmer and Cowley 2010). Aerial images, including traditional photographs and satellite images, are often examined without full assessment of their appropriateness to the task in hand. Such basic considerations may include the following: photographs taken when vegetation is well developed and the sun is high in the sky will not reveal subtle earthworks the potential for crop/vegetation marks to reveal buried features is dependent on appropriate seasonal weather and cropping regimes; and sites such as artefact scatters will not be visible unless accompanied by gross contrasts with surrounding soil matrices.

In all cases, knowledge of the character of potential archaeological sites and the way in which they may register is vital to targeting appropriate photographs. If earthworks are expected, for example, winter conditions – when vegetation is low and lighting will tend to be oblique – are preferable to height of summer. The necessity for targeted approaches to making best use of large collections of aerial photographs has been illustrated in the English Heritage National Mapping Programme (Winton and Horne 2010: 11–15), where photographs with the highest potential information return are retrieved and examined, at the expense of those deemed to carry a lower potential for the purpose in hand (e.g. earthworks, cropmarks or military remains). At a European scale, mapping of the changing pattern of cropmark formation across the continent and using this as an index to identifying sorties with the highest potential information return provide a similar approach to targeting sorties that have a higher archaeological potential than others (Cowley and Stichelbaut 2012).

It is vital that such approaches should not be prescriptive, as that would run the considerable risk of becoming a self-fulfilling process, where the potential of the aerial imagery to expand the boundaries of knowledge becomes limited. To this can be added the importance of archaeologists' understanding modern and recent land use regimes and taking the totality of the information on the photographs into account in making their interpretations – an approach that requires an integrated study of the landscape that moves beyond narrow period specialisation (Palmer and Cowley 2010). This type of knowledge, combined with considerations such as scale (above), can help to target appropriate photographs and, hopefully, ensure time spent examining them produces a worthwhile return.

2.6.3 Traditions in Power

The 'competitive' relationship between aerial photographs and satellite images has been commented on above, but is only one example of methodological 'silos' that can become unhelpful if 'fitness for purpose' and integrated approaches are not foregrounded. Generally, it can be seen that varying research traditions in archaeology have a direct impact on the willingness of archaeological practitioners to explore different data sources. In the UK, for example, one can identify an early and sustained emphasis on observer-directed aerial survey and oblique photography but limited engagement with block-coverage vertical photographs (Palmer 2005). The exception to this, beyond small-scale developer-led mapping (e.g. Palmer 2010), is English Heritage's National Mapping Programme (Bewley 2003; Hegarty and Newsome 2007; Horne 2009, 2011; Winton and Horne 2010). This programme has made sustained and productive use of historic aerial photographs since the early 1990s to map large parts of England, though the most well-known product of this work was a pilot project in the Yorkshire Wolds (Stoertz 1997) based largely on oblique photographs with mapping completed during the winter of 1988–1989. In Scotland and Wales, the use of historic aerial photographs remains somewhat piecemeal and project based, while both countries maintain active programmes of observer-directed aerial survey. Such observer-directed approaches have been highly productive (e.g. British Academy 2001), but the limited engagement of parts of the archaeological establishment with collections of historic aerial photographs is illustrative of a tradition or agenda establishing, and then maintaining itself, in the face of a struggle for funding.

This emphasis on oblique aerial photographs collected through observer-directed survey is also evident in textbooks, which invariably make use of oblique views (Wilson 2000). This in part reflects an understandable desire to emphasise the importance of an initially 'new', and then later under-valued, technique in the face of limited funding. Indeed, the coining of the term 'aerial archaeology' in part at least reflects the need to create identity and relevance, though it can also serve to encourage the development of a ghetto mentality. A more recent parallel to this process is the defining of 'satellite archaeology' as an approach (e.g. Parcak 2009). To some extent, while such labels and approaches are inevitable, they can become unhelpful, identifying fringe interests or silos, or as established traditions embedded in power become a dominant and exclusive agenda (e.g. see Rączkowski 2005, 2011). The lack of appreciation of the value of historic collections of aerial photographs is certainly in part a victim of a tendency in academic and professional archaeology to undervalue this source material, but, with an increased number of related publications, that is gradually changing. Beyond this approach to hearts and minds, the issues of access and the appropriate and educated use of aerial photographs remain.

2.7 Using TARA: Some Thoughts on Developing Access

Even the limited examples of what TARA imagery can yield illustrated above demonstrate its value. However, there remain the general problems of how to facilitate access. The importance of partnerships in developing the academic applications of TARA cannot be stressed enough. For the most accessible parts of TARA – essentially those with the most comprehensive finding aids – access can be improved in the short and medium term, especially to those areas that have already been digitised. Elsewhere, in areas where there are limited or no finding aids, there is considerable potential for symbiotic relationships between TARA and practicing archaeologists. 'Local' experience and knowledge from particular areas of the world are invaluable to the identification and basic cataloguing of the less accessible parts of the collection. An aspiration, should funding be found, would be to create a 'baseline' coverage for all the areas covered by the collections. Indeed, for the major cities of Scotland, this has already been undertaken, providing time-lapsed views of changing urban landscapes through the twentieth century. The exploration of metadata, such as dates of photography for example, will be invaluable in facilitating targeted, efficient access based on understanding the appropriateness of the photographs to the task in hand.

The opening up of access to historic aerial photograph collections is also embedded as one of the key objectives of a 5-year, multi-partner European Union funded project – *ArchaeoLandscapes Europe*. This indicates that, in parts of Europe at least, the primary importance of historic aerial photographs is now recognised, though from a global perspective there clearly remains some progress to be made.

2.8 Conclusion

There is little sense in pretending that TARA, or indeed other archives such as NARA, is a 'magic bullet' solution to the problems of archaeologists across the globe. However, for many parts of the world, TARA's holdings are the earliest detailed record of its landscapes, with the added value of predating the often wholesale changes to landscapes of the twentieth century. Such photographs are uniquely rich tools for discovering sites and for observing material changes in the condition of known monuments. For the archaeologist, especially those that work in a landscape context, such records are a major component of the toolbox that can be brought to bear in understanding the past – demonstrated to best effect when data sources are integrated in the context of changing landscapes and aerial photographs (both historic and recent) are one of a complementary suite of approaches. And while photographs from the mid-twentieth century will remain a valuable view into the past, archives like TARA are not fossils and continue to grow with deposits of more recent imagery, all of which has a value to the landscape archaeologist.

Acknowledgments Our thanks to Otto Braasch, Andreas Buchholz, Pete Horne, Rog Palmer, Robin Standring and Jack Stevenson for their assistance in producing this chapter.

Bibliography

Babington Smith, C. (1957). *Evidence in camera: The story of photographic intelligence in world war II*. London: Chatto and Windus.

Bewley, R. H. (2003). Aerial survey for archaeology. *The Photogrammetric Record, 18*(104), 273–292.

Bewley, R. H., & Rączkowski, W. (Eds.). (2002). *Aerial archaeology developing future practice* (Nato Science Series). Amsterdam: IOS Press.

Braasch, O. (2002). Goodbye cold war! Goodbye bureaucracy? Opening the skies to aerial archaeology in Europe. In R. H. Bewley & W. Rączkowski (Eds.), *Aerial archaeology developing future practice* (Nato Science Series, pp. 19–22). Amsterdam: IOS Press.

Bradford, J. (1957). *Ancient landscapes. Studies in field archaeology*. London: J. Bell & Sons.

British Academy. (2001). *Aerial survey for archaeology. Report of a British Academy working party 1999*. Compiled by Robert Bewley. London: British Academy.

Brugioni, D. A. (1989). The serendipity effect of aerial reconnaissance. *Interdisciplinary Science Reviews, 14*, 16–28.

Ceraudo, G., & Shepherd, E. J. (2010). Italian aerial photographic archives: Holdings and case studies. In D. Cowley, R. Standring, & M. Abicht (Eds.), *Landscapes through the lens: Aerial photographs and the historic environment* (pp. 237–246). Oxford: Oxbow.

Cowley, D. (2009). Early Christian cemeteries in south-west Scotland. In J. Murray (Ed.), *St Ninian and the earliest Christianity in Scotland, papers from the conference held by the Friends of the Whithorn Trust in Whithorn on September 15, 2007* (BAR, British Series 483, pp. 43–56). Oxford: Archaeopress.

Cowley, D., & Ferguson, L. (2010). Historic aerial photographs for archaeology and heritage management. In M. Forte, S. Campana, & C. Liuzza (Eds.), *Space, time, place. Third international conference on remote sensing in archaeology* (BAR, International Series 2118, pp. 97–104). Oxford: Archaeopress.

Cowley, D. C., & Stichelbaut, B. (2012). Historic aerial photographic archives for European archaeology: Applications, potential and issues. *European Journal of Archaeology*, 15(2).

Cowley, D., Standring, R., & Abicht, M. (Eds.). (2010a). *Landscapes through the lens: Aerial photographs and the historic environment*. Oxford: Oxbow.

Cowley, D., Standring, R., & Abicht, M. (2010b). Landscapes through the lens: An introduction. In D. Cowley, R. Standring, & M. Abicht (Eds.), *Landscapes through the lens: Aerial photographs and the historic environment* (pp. 1–6). Oxford: Oxbow.

Ferguson, L. (2011). Aerial archives for archaeological heritage management: The aerial reconnaissance archives – A shared European resource. In D. C. Cowley (Ed.), *Remote sensing for archaeological heritage management* (pp. 205–212). Budapest: Archaeolingua.

Fowler, M. J. F. (2004). Archaeology through the keyhole: The serendipity effect of aerial reconnaissance revisited. *Interdisciplinary Science Reviews, 29*(2), 118–134.

Fowler, M. J. F. (2010). Satellite imagery and archaeology. In D. Cowley, R. Standring, & M. Abicht (Eds.), *Landscapes through the lens: Aerial photographs and the historic environment* (pp. 99–110). Oxford: Oxbow.

Going, C. J. (2002). A neglected asset. German aerial photography of the second world war period. In R. H. Bewley & W. Rączkowski (Eds.), *Aerial archaeology developing future practice* (Nato Science Series, pp. 23–30). Amsterdam: IOS Press.

Hegarty, C., & Newsome, S. (2007). *Suffolk's defended shore: Coastal fortifications from the air*. Swindon: English Heritage.

Horne, P. D. (2009). *A strategy for the National Mapping Programme*. Swindon: English Heritage. http://www.english-heritage.org.uk/content/imported-docs/a-e/astrategyforthenationalmappingprogramme2009.pdf. Accessed March 2010.

Horne, P. (2011). The English Heritage National Mapping Programme. In D. C. Cowley (Ed.), *Remote sensing for archaeological heritage management* (pp. 143–151). Budapest: Archaeolingua.

Jones, G. D. B. (1987). *Apulia 1: Neolithic settlement in the Tavolieri*. The Society of Antiquaries. London: Thames & Hudson.

Lillesand, T. M., & Kiefer, R. W. (2000). *Remote sensing and image interpretation* (4th ed.). New York: Wiley.

Oltean, I. A., & Hanson, W. S. (2007). Reconstructing the archaeological landscape of Southern Dobrogea: Integrating imagery. In M. Ehlers & U. Michel (Eds.), *Remote sensing for environmental monitoring, GIS applications, and geology VII* (Proceedings of SPIE Vol. 6749, no. 674906). Bellingham: Society of Photo-Optical Instrumentation Engineers.

Palmer, R. (2005). If they used their own photographs they wouldn't take them like that. In K. Brophy & D. Cowley (Eds.), *From the air: Understanding aerial archaeology* (pp. 91–116). Stroud: Tempus.

Palmer, R. (2010). Uses of vertical photographs for archaeological studies in parts of lowland England. In D. Cowley, R. Standring, & M. Abicht (Eds.), *Landscapes through the lens: Aerial photographs and the historic environment* (pp. 43–54). Oxford: Oxbow.

Palmer, R., & Cowley, D. (2010). Interpreting aerial images: Developing best practice. In M. Forte, S. Campana, & C. Liuzza (Eds.), *Space, time, place. Third international conference on remote sensing in archaeology* (BAR, International Series 2118, pp. 129–135). Oxford: Archaeopress.

Parcak, S. H. (2009). *Satellite remote sensing for archaeology*. Abingdon/New York: Routledge.

Rączkowski, W. (2004, September). Dusty treasure: Thoughts on a visit to The Aerial Reconnaissance Archives at Keele University (UK). *AARGnews (The newsletter of the Aerial Archaeology Research Group) 29*, 9–11.

Rączkowski, W. (2005). Tradition in power: Vicious circle(s) of aerial survey in Poland. In K. Brophy & D. Cowley (Eds.), *From the air: Understanding aerial archaeology* (pp. 151–167). Stroud: Tempus.

Rączkowski, W. (2011). Integrating survey data – The Polish AZP and beyond. In D. C. Cowley (Ed.), *Remote sensing for archaeological heritage management* (pp. 154–160). Budapest: Archaeolingua.

Stoertz, C. (1997). *Ancient landscapes of the Yorkshire Wolds*. Swindon: English Heritage.

Walton, A. (1975). *After the battle number 7: Keele air photo archive*. London: Battle of Britain Prints International.

Wilson, D. R. (2000). *Air photo interpretation for archaeologists*. Stroud: Tempus.

Winton, H., & Horne, P. (2010). National archives for national survey programmes: NMP and the English Heritage aerial photograph collection. In D. Cowley, R. Standring, & M. Abicht (Eds.), *Landscapes through the lens: Aerial photographs and the historic environment* (pp. 7–18). Oxford: Oxbow.

Wiseman, J., & El Baz, F. (Eds.). (2007). *Remote sensing in archaeology*. New York: Springer.

Chapter 3
Blitzing the Bunkers: Finding Aids – Past, Present and Future

Peter McKeague and Rebecca H. Jones

Abstract Aerial photographic collections developed from the need for military intelligence, for cartographic mapping, for commercial gain or for specific targeted research. Over time, the value of historical aerial photography has been appreciated far beyond its original purpose, as it provides an irreplaceable record of the ever-changing landscapes and townscapes that surround us. Key to the reuse of these resources is access to the finding aids that index the individual photographs. It is argued that the potential of the information on traditional ledgers and sortie traces can be, and should be, unlocked through digitisation, to provide spatial indexes that may be accessed through remote Geographic Information Systems as part of integrated information resources delivered through spatial data infrastructures. The role of new and disruptive technologies is also considered to demonstrate the potential for accessing historical mosaicked imagery in browsers such as Google Earth.

3.1 Introduction

Interest in early aerial photography rightly focuses on individual iconic images such as Giacomo Boni's photographs in the forum in Rome, taken from a balloon in 1899 (Ceraudo 2005: 74), or Lieutenant P. H. Sharpe's study of Stonehenge in 1906 (Capper 1907; Wilson 2000: 16–17): the latter evidently as part of an aerial reconnaissance exercise on Salisbury Plain. With the need for military intelligence during the First World War as a catalyst, there were rapid improvements in equipment and techniques, as well as the scale of aerial reconnaissance undertaken. Given both the extent of the theatre of war and the quantity of photographs taken, finding aids were essential in the organisation and retrieval of the photographs. Many of the photographs from this

P. McKeague (✉) • R.H. Jones
Survey and Recording, Royal Commission on the Ancient and Historical Monuments of Scotland, 16 Bernard Terrace, Edinburgh EH8 9NX, UK
e-mail: peter.mckeague@rcahms.gov.uk; Rebecca.Jones@rcahms.gov.uk

W.S. Hanson and I.A. Oltean (eds.), *Archaeology from Historical Aerial and Satellite Archives*, DOI 10.1007/978-1-4614-4505-0_3,

period have been lost, but where they survive, such as the 'Box Collection' held by the Imperial War Museum at Duxford and London or those of the Belgian Air Force, the Aviation Militaire Belge, held in the Belgian Royal Army Museum (Koninklijk Legermuseum Brussel – Musée Royale de l'Armée) and the Belgian military archives in Brussels (Service Général de Renseignements et Sécurité Section Archives) (Stichelbaut 2006; Stichelbaut et al. 2010; Chap. 5 by Stichelbaut et al. this volume), they are amongst the earliest formal collections of aerial photography.

At the end of the war, the skills learnt in military conflict were soon put to commercial use. Francis Wills, who saw active service as an Observer with the Royal Naval Air Service, and Claude Grahame-White established Aerofilms in 1919,[1] which became the most successful commercial aerial photographic company in Britain.

By the Second World War, the role of aerial reconnaissance in military intelligence had expanded considerably, unintentionally leading to the creation of significant photographic archives of the landscape of Europe and beyond; for instance, the National Collection of Aerial Photography held at the Royal Commission on the Ancient and Historical Monuments of Scotland (http://aerial.rcahms.gov.uk. Accessed 24 May 2011) houses millions of Second World War and Cold War aerial reconnaissance photographs of locations throughout the world declassified by the UK Ministry of Defence (Chap. 2 by Cowley et al., this volume).

After the Second World War, the Royal Air Force (RAF) in Britain commenced a national mapping programme of aerial photography, with the mantle continued by the Ordnance Survey. The Ordnance Survey appreciated the value of vertical air photographs as a cartographic surveying tool, and today, photogrammetric mapping from aerial survey, supplemented by ground-based revision, ensures the currency of their map products. Archaeologists, too, were quick to appreciate the wider application of aerial photographs as a technique for discovering and mapping new sites. Prospective aerial reconnaissance for archaeological sites was pioneered by a handful of individuals such as O. G. S. Crawford, the first archaeological officer of the Ordnance Survey, between the two world wars. However, analysis took off in earnest in 1945, led by the Cambridge University Committee for Aerial Photography (CUCAP) and various fliers around the country. In Scotland, Kenneth Steer, who had been trained as an aerial photographic intelligence officer in the Second World War, returned to his post at RCAHMS where he applied his skills to discover large numbers of subtly preserved settlements in the uplands of the Scottish Cheviots and elsewhere (Steer 1947; RCAHMS 1956). The Royal Commissions for England, Scotland and Wales started their own aerial survey programmes in 1967, 1976 and 1986, respectively (British Academy 2001); these photographs form a key part of the national collections held by these organisations. In Scotland, the photographs are indexed as part of the associated collections by site record in Canmore (an inventory of Scotland's historic environment) (http://canmore.rcahms.gov.uk/. Accessed 24 May 2011).

[1] The early part of this archive was acquired by English Heritage and the Royal Commissions on the Ancient and Historical Monuments of Scotland and Wales (RCAHMS and RCAHMW) in 2007–2008 (http://www.britainfromabove.org.uk/. Accessed 16 July 2012), with the later years accessible through Blue Sky. www.oldaerialphotos.com. Accessed 4 May 2011.

In the later twentieth century, bespoke aerial surveys were commissioned by both the public and private sectors for a wide range of uses (e.g. trunk road upgrading). From 1987 to 1989, the Scottish Office commissioned an 'All Scotland Survey', the last comprehensive coverage of Scotland using analogue film. The number of collections of air photography held in archives across the United Kingdom is detailed in a directory published in 1999 (NAPLIB 1999).

Recently, digital technologies have revolutionised the way that users take aerial photography, with almost all coverage now taken using digital formats. Many commercial companies have been quick to develop businesses providing seamless orthoimagery of landscapes for use within Geographic Information Systems (GIS). In turn, the availability of easily accessible photographic imagery on the Internet (for instance, through Google Earth or Microsoft's Bing) has increased both user awareness and user knowledge of aerial photography, as well as raising expectations regarding access to information.

Thus, aerial photographic collections have evolved for a variety of reasons, primarily military but latterly for cartographic purposes. However, reuse extends far beyond the original purpose. They are essential for the identification of archaeological sites and understanding the evolution of the historic landscape, but equally for monitoring the natural environment or the physical landscape. Easily sourcing collections and identifying relevant photography is the essential first step in this research. This chapter does not, and cannot, attempt to deliver a comprehensive review of finding aids, but instead looks at how information about aerial photographic collections might be delivered for the benefit of the user in the future.

3.2 Traditional Finding Aids

Large collections of air photographs need finding aids to enable the user to gain access to imagery for specific places and themes. These can vary from index cards or simple ledgers listing targets (e.g. much of the Aerofilms collection), to annotated maps (e.g. the RAF collection). Oblique air photographs taken specifically for archaeological purposes have been indexed in a variety of ways, ranging from simple lists of targets by site or place name to the geographic centre point of the photograph.

In Britain, the Imperial War Museum 'Box Collection' comprises over 80,000 glass plate negatives from the First World War, and organisation of the associated finding aids has been recently discussed by Stichelbaut et al. (2010) who demonstrate that the value in opaque historical archives can be unlocked and used in a contemporary manner. The index cards are organised by year from 1915 until the end of the war. Each index card (Fig. 3.1) records the series (the unit or wing that undertook the reconnaissance) and lists up to five images, recording the box and index number of each negative, its size and the date of the photography. The photography is referenced to a contemporary series of trench maps, based on the pre-war Belgian national survey, produced by the geographical section of the War Office General

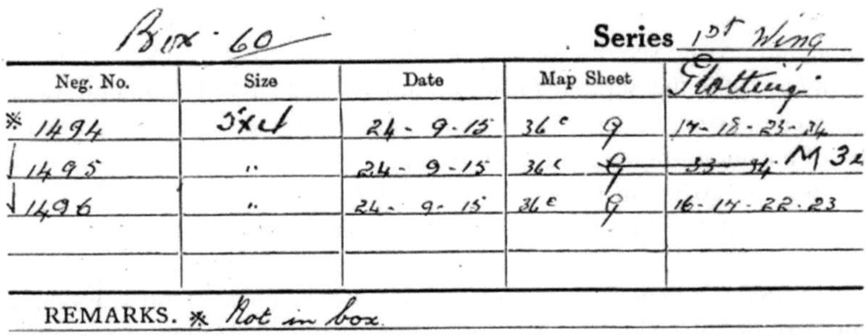
Box 60 Series 1st Wing

Neg. No.	Size	Date	Map Sheet		Plotting
* 1494	5x4	24-9-15	36c	9	17-18-23-24
1495	"	24-9-15	36c	9	33-34 M 32
1496	"	24-9-15	36c	9	16-17-22-23

REMARKS. * Not in box.

Fig. 3.1 An example of an Imperial War Museum 'Box Collection' index card from the 1915 drawer of vertical aerial photographs (Stichelbaut 2009)

Staff. Although appearing archaic compared to modern coordinate systems, effective accurate referencing was achieved through the subdivision of each map sheet into 24-lettered 6,000 yard squares, further subdivided into 30- or 36-numbered 1,000 yard squares each, split into quadrants and further subdivided to enable plotting up to within 5 yards (Anderson 2008). However, whilst the physical character and organisation of the original finding aids may be of interest to archivists, Stichelbaut demonstrates that the information preserved on the card indexes may be organised for use with modern information systems.

The finding aids for the national aerial survey, undertaken by the RAF across Britain from the mid-1940s to late 1950s, simply comprise either the sortie line and centre points of photographs, the outlines of the approximate extent of the coverage or the footprint of individual photographs in a sortie. Representation of stereoscopic coverage in this manner results in the photographs plotted as overlapping rectangles which may sometimes be virtually impossible for all but the specialist to decipher (Fig. 3.2). In the late 1990s, the flight lines and centre points for the majority of this collection was digitised by RCAHMS to aid internal retrieval of the associated searches within a GIS.

The Aerofilms Collection includes 1.26 million negatives and more than 2,000 photograph albums documenting the changing townscapes and landscapes of Britain. As a commercial photography company, its library was organised thematically as much as geographically. The Aerofilms Book of Aerial Photographs (Aerofilms 1965) provides a sampler, organised thematically, for the range of material in its collection. There are sections on archaeology, architecture, communications, sport and recreation, as well as topical and national events, weather features and miscellaneous subjects. Through its publications, Aerofilms showcased its collections by area or by subject. By 2005, its book on football grounds had reached its 13th edition (Aerofilms 2005).

Finding aids are traditionally held with the collections and may be consulted on application to the curatorial staff. More often than not, however, it is the staff who conduct the searches through the collections; this can be quite labour intensive and often incurs a charge to the enquirer. If the user of the archive has travelled some distance to view the collection, they are unlikely to want to spend valuable time

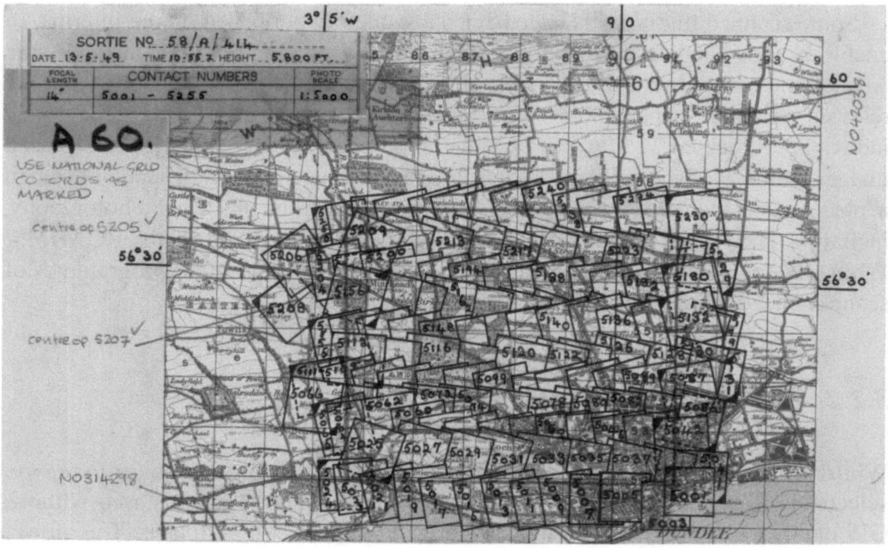

Fig. 3.2 The flight diagram, for RAF sortie 58/A/414 (13 May 1949) over the City of Dundee, illustrates the potential difficulties presented to the reader from both overlapping runs and individual photographic frames (Licensor NCAP/aerial.rcahms.gov.uk)

searching through the finding aids if this could be undertaken in advance of the visit. In a review of air photograph indexes undertaken by English Heritage, it was noted that 40% of enquiries to the National Monuments Record Centre (NMRC) in Swindon were for cover searches to identify the availability of air photographic holdings for a study area. This totalled over 3,500 enquiries in 2004–2005 (Shalders 2007: 425). Although the figures are not directly comparable, Airphotofinder (www.airphotofinder.com. Accessed 4 May 2011), the RCAHMS online index to its aerial photographic holdings for Scotland, received 11,694 visits during 2005 whilst curatorial staff dealt with 1,600 enquires. In 2008, the website attracted 40,169 visits leading to 10,467 searches (26%) with minimal promotion.

3.2.1 Going Digital: Raster Images

Latterly, efforts have been made to digitise and georeference finding aids and photographs from several collections, initially to aid curatorial staff dealing with public enquiries. The process of digitally photographing, or scanning, analogue pictures such as finding aids or the photography itself creates a digital record of the original material, saving wear and tear on the original archive. For historic sorties, scanning the finding aids has the bonus of capturing a cartographic view of the target area more closely contemporary with the date of the photography, important in areas where either the urban or rural landscape has radically changed.

Some scanned finding aids have been subsequently published on the Internet, to enable a wider audience to conduct their own cover searches. For instance, the Australian Geoscience website enables the user to select an area of the map and see relevant scanned paper plots (http://www.ga.gov.au/apps/aerial-flight-diagrams/index.php. Accessed 24 May 2011). Other approaches include making the digitised images themselves accessible, perhaps plotted against a map background or through a place name indicator, as with the National Collection of Aerial Photography, including The Air Reconnaissance Archives (TARA), now held by the RCAHMS. Whilst the increased visibility of imagery tantalises the user, major programmes of scanning need to be undertaken to realise the potential.

3.2.2 Going Digital: Vector Data

Whilst scanned or raster images of original finding aids enable the viewer to browse selected images individually on the Internet or overlay other digital datasets within a GIS environment, they are no more than static pictures or wallpaper. The viewer cannot interact with or select from the scanned image other than identify values stored within the pixels that define the image. In contrast, vector data offers a more powerful interactive environment, incorporating attribute data stored in an associated table that can be searched and selected by the user – enabling more powerful online catalogues, for example, flight lines, centre points and buffers – and can offer the opportunity to select an area and produce a digital report listing individual photographs (Moloney 1997a, b). For example, the CUCAP catalogue, held by the Unit for Landscape Modelling in Cambridge, has been online since 2001. Using a web-based GIS, the user is able to conduct their own cover search against over one million vertical and oblique photographs taken since 1945 and retrieve listings of both vertical and oblique photographs, indexed by target location (http://venus.uflm.cam.ac.uk/website/cucap/viewer.htm. Accessed 4 May 2011). In a sign of the times, however, the web-based GIS application has recently been replaced by a simpler 'Point and Click' map browser (http://www.geog.cam.ac.uk/cucap/. Accessed 16 July 2012).

In Scotland, Airphotofinder (Fig. 3.3) (www.airphotofinder.com) was designed to provide a similar portal for the Scottish vertical collection, held at RCAHMS, together with some of the RAF coverage for Wales (McKeague 2005). In addition, oblique coverage by the Luftwaffe and RAF in Scotland is also available,[2] referenced on the location of principal targets. Such detailed cataloguing immediately unlocks the potential of the archive imagery by making the interpreted images more readily accessible (immediately now that the photography is created digitally). By linking the imagery to sites, researchers can browse the collections using a variety of methods, ranging from searches via a map, place name or administrative area to searches by archaeological keywords, for example, using the thesaurus of monument types.

---

[2] The targeted oblique photographs conducted through RCAHMS' own programmes of aerial survey are indexed by site through the Canmore database (http://canmore.rcahms.gov.uk), online since 1998; digital photographs can also be viewed online.

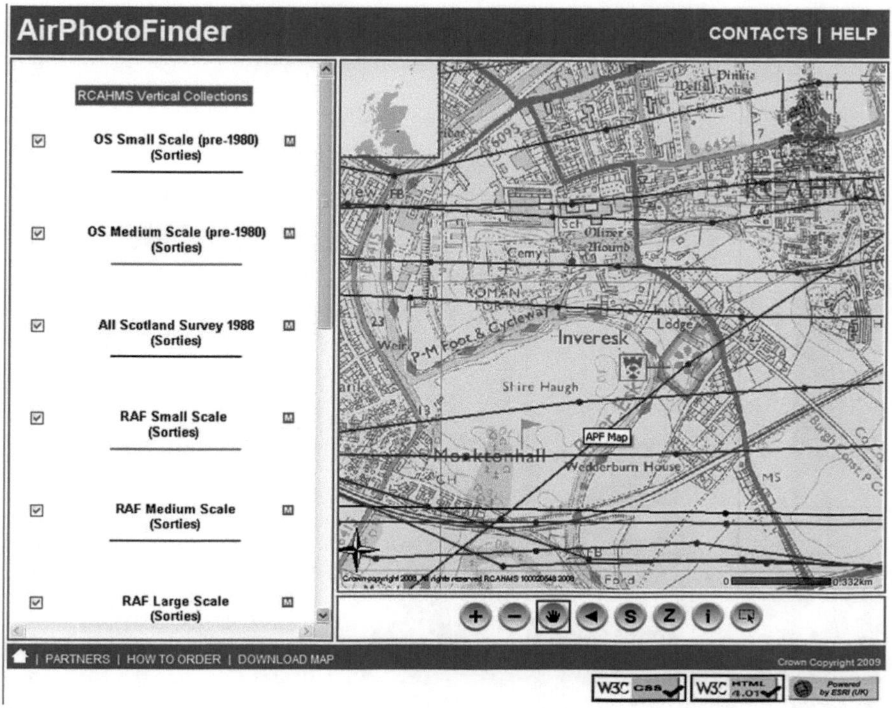

Fig. 3.3 Airphotofinder provides the user with the opportunity to view and identify coverage of flight lines and photographic centre points for the Vertical Aerial Photographic Collection held at RCAHMS. This example illustrates the coverage over Inveresk, Musselburgh

Whereas raster images allow the user to view information about a single sortie in turn, both the CUCAP online catalogue and Airphotofinder enable the user rapidly to access and search information from all available digitised photographic targets within a geospatial environment. Thus, a user can quickly retrieve information for a selected area across *any* relevant sorties, adding a temporal dimension to searching. For instance, in Airphotofinder, the user can quickly assess available coverage for a particular area from the RAF sorties of the 1940s and 1950s, Ordnance Survey flying from the 1960s to 1980 or from the 1987 to 1989 All Scotland Survey. More sophisticated searches of the data could be developed to filter photography by month or season.

3.3 To the Future…

Other datasets which relate to the collections of aerial imagery and their archaeological interpretation could also be added to browsers in the future. This could include the flight plots from targeted oblique aerial survey (gathered using global positioning systems) and the interpreted archaeological features (transcriptions) (Fig. 3.4). From

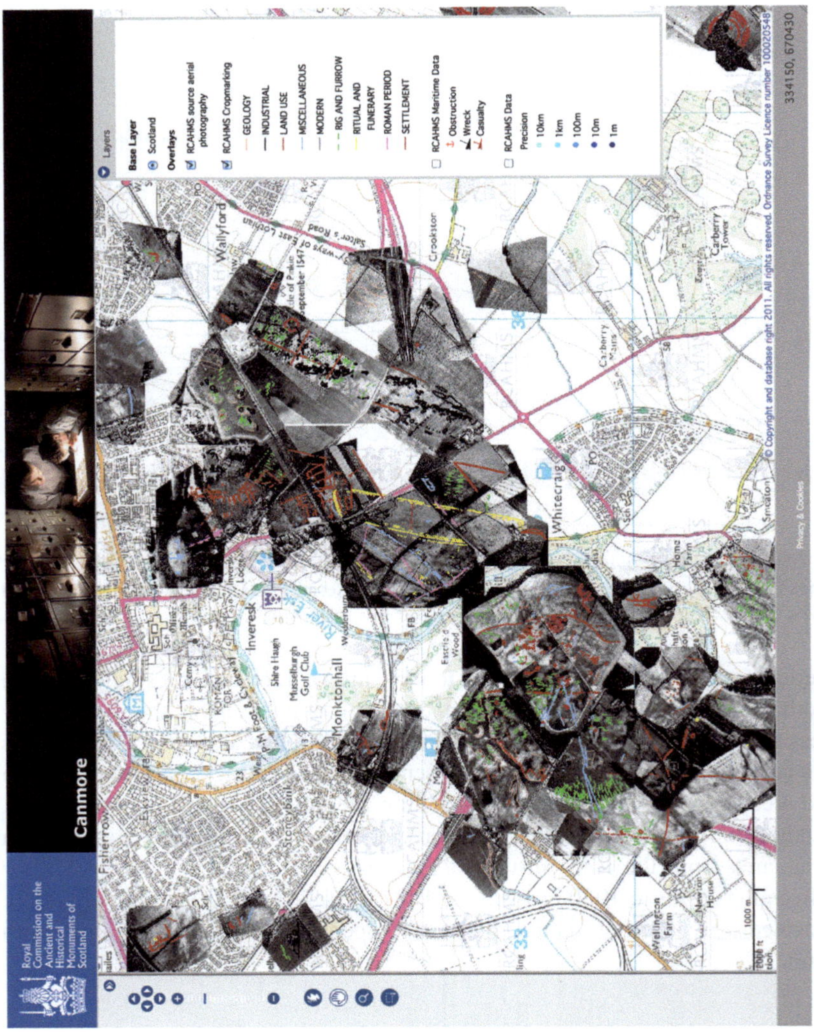

Fig. 3.4 Screenshot showing the integration of oblique aerial photographic imagery and mapping information of the archaeological landscape at Inveresk, Musselburgh, provided as a Web Map Service, within the RCAHMS Canmore portal

serving the information through web-based browsers, the next step is to enable the user to retrieve the data through their own systems. This can be achieved through the provision of Web Map and Web Feature Services (WMS, WFS),[3] which can be consumed by external applications, including GIS. For instance, scanned raster plots detailing historic aerial imagery could be georeferenced and served as WMS. Many of these include contemporary background mapping which can provide valuable additional information. However, the benefits to the remote user are limited. WMS provides information as a raster image – essentially a wallpaper that the user can view their data against but not otherwise interact with. Furthermore, the user may already have access to historic background mapping, obviating the desire to see this on the original plots. In contrast, vector data (e.g. digital flight lines, centre points and/or extents, archaeological transcriptions) provided as WFS would enable a remote user to interact with the archives and perform sophisticated queries and cover searches.

These ideas may sound radical, but they draw inspiration from the European INSPIRE Directive,[4] which has been transposed to national law by individual European governments and adopted by both the Westminster and Scottish governments through Statutory Instruments in 2009.[5] INSPIRE addresses the need to ensure that the spatial data infrastructures of the member states are compatible and usable in a community and trans-boundary context. INSPIRE requires that a common approach is adopted with regard to discovering information through metadata, viewing and downloading services, whereby users can see and use data and, potentially, develop new services. In the context of this chapter, there is a requirement that public authorities across Europe provide such services, as appropriate, for newly collected and extensively restructured orthoimagery from December 2014 (http://inspire.jrc.ec.europa.eu/index.cfm/pageid/44. Accessed 10 June 2011). The status of historic datasets, including scanned, geo-rectified imagery and digital finding aids, remains unclear and dependent on national and regional implementations of spatial data infrastructures. The indices 'published' through Airphotofinder (www.airphotofinder.com) already partially meet these requirements in providing detailed metadata (Appendix 1), describing and providing the ability to view both the flight paths and photographic centre points (Fig. 3.3).

In the wider context of the management of geographical information, for example, the finding aids discussed in this chapter, there is a long-standing recognition of the need to share spatial information beyond organisational use (Chorley 1987; Scottish Executive 2005). Bruce Gittings, in a recent paper reflecting on 40 years of geographic information in Scotland (2009), discusses the need to break down the

[3] A Web service is a software system designed to share information over a network or the Internet using a standardised set of shared data fields. Web Map Services deliver map data over the web by converting the host data into a picture that can be loaded into the user's GI system. The user can view the data and overlay it with their own data. Web Feature Services provide the actual geographic information with which users can interact with points and polygons as if they were hosted on their own system.

[4] Infrastructure for spatial information in Europe (http://inspire.jrc.ec.europa.eu/).

[5] http://www.opsi.gov.uk/si/si2009/uksi_20093157_en_1 and http://www.opsi.gov.uk/legislation/scotland/ssi2009/ssi_20090440_en_1

barriers to sharing information between individual 'data silos', that is, the individual organisations that gather, use, curate and manage the rights of information in their charge. Whilst acknowledging the development of websites to showcase work, Gittings observes that they are invariably accompanied by a level of organisational branding, which necessarily promotes the proprietorial interests of organisations. In effect, such developments can accentuate existing barriers to sharing the data by turning the 'data silos' into fortresses of data that may be breached individually, but which fail to allow cross-searching (Gittings 2009: 88). However, a glimmer of hope is in sight, pioneered by the ScotlandsPlaces development between RCAHMS and the National Archives of Scotland (www.scotlandsplaces.gov.uk. Accessed 24 May 2011, Beamer and Gillick 2010: 225–239), a web application which presents data from different sources together in a meaningful way for the user. Yet for this ideology truly to succeed, we need to move collectively beyond the bunker mentality of individual organisations gathering, creating and protecting *their* information, to opening up data to wider audiences. As Gittings observes, 'geography provides the key to penetrating the data fortresses and providing the integrated information resource, which could benefit Scotland's knowledge economy' (2009: 88).

Aerial photographic survey information discussed in this chapter and archaeological data in general are prime candidates for dissemination through WMS and WFS for integration with other data. It is already possible to add the site locations from the RCAHMS Canmore database as a WMS in remote GIS applications, and it is the intention to deliver the aerial photographic transcription information initially as a WMS, with interactive WFS to follow once the technology is robust enough. Thus, a student in Glasgow or Germany will have the opportunity to view, and interact with, live up-to-date information and research from the RCAHMS databases and GIS within their own computer workstations in their own offices. Similarly, remote access to the flight line and photograph indices from vertical aerial photographic collections, as WFS, would provide the tools to assess the quantity of data; adding thumbnails of images, where available, could address issues of quality and provide further information to the user.

The benefits are greater access to information. Currently, this relates to data that has already been collected digitally (e.g. vector flight lines and photo points), but in the future could be modified to encourage the creation of vector indices. The creation of good metadata is key to opening up remote access to such information. INSPIRE and modern web developments require us to fully document the structure and expected content of our datasets. Geographic information metadata standards use high-level terminology to help the user find out what broad types of data are available for a particular geographic extent, whilst further details document the content. An example of the completed metadata documenting the extent of cover for the All Scotland Survey flight lines is shown in Appendix 1. Much of the information gathered conforms to the UK Gemini I data standard current when the metadata was first compiled and published on the now decommissioned GIGateway website.[6]

[6] www.GIGateway.org.uk was decommissioned on 31 March 2011. Replacement arrangements for access to metadata are currently being developed through UK Location.

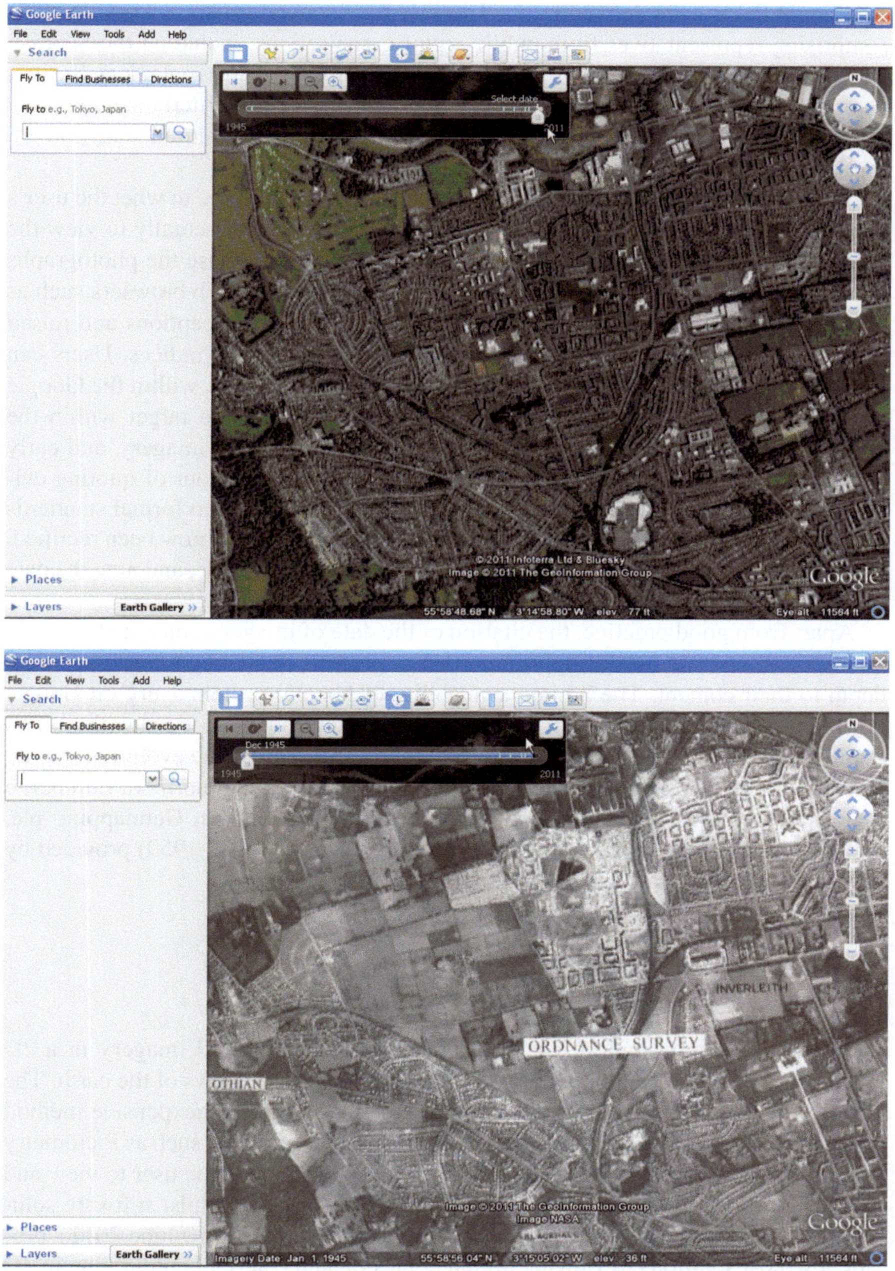

Fig. 3.5 (**a**) Contemporary and (**b**) historical imagery available through Google Earth ((**a**) © 2011 Google Earth; © 2011 Infoterra Ltd. and Bluesky; © 2011 The Geoinformation Group; (**b**) © 2011 Google Earth; © 2011 The Geoinformation Group; © NASA)

Additional fields were defined by the Imagery for Scotland group, a working party comprising representatives from public sector organisations, including the Scottish Executive, the Ordnance Survey, Scottish Enterprise, British Geological Survey, Scottish Natural Heritage and RCAHMS, to document the key criteria including the date of photography, time of flying, photographic scale, flight altitude and camera focal length.

Yet, metadata standards and indices to material held only serve to whet the user's appetite, providing a taster of what may be available. In order actually to view the images, the user must either visit the relevant archive or purchase the photographs remotely. Disruptive innovation, through the development of web browsers such as Google Earth and Microsoft's Bing, has radically changed perceptions and raised user expectations of viewing imagery as opposed to searching indices. Users can share the location of discoveries simply by creating place marks within the Google Earth application. Whilst ideal for taking users directly to the target within the browser, place marks neither record the date nor source of the imagery, and early iterations of the applications initially lowered the academic rigour of quoting evidence through a lack of available metadata. Although there are no formal standards for citations, the ability to at least cite the date of the imagery has now been rectified. Users are able both to reference location, in latitude and longitude, and note the date stamp and imagery source 'published' on the browser.

Apart from good practice, the citation of the date of imagery consulted over the Internet will become increasingly important as projects like Google Earth add time depth to their content. The limited availability of historic orthoimagery on Google Earth 5 demonstrates the potential for integrating data from collections within mass popular browsers, enabling the user to appreciate and analyse the evolution of both urban and rural landscapes. For instance, the urban expansion of northern Edinburgh is documented through comparison of 2011 orthoimagery from Getmapping plc, with historic photo-mosaicked Ordnance Survey imagery (1944–1950) provided by The Geoinformation Group (Fig. 3.5a, b).

3.4 The Third Dimension

In the modern digital world, users generally access mosaicked imagery in a 2D environment, yet aerial photography is ideal for creating 3D views of the earth. The creation of anaglyphs from stereoscopic imagery provides an inexpensive method of creating 3D views of the subject matter. Commercial ventures, such as Pictometry (http://www.pictometry.com/. Accessed 24 May 2011), enable the user to view and navigate around iconic landmarks and cityscapes in faux 3D, whilst software solutions, such as Purview (http://www.mypurview.com/. Accessed 10 June 2011), provide low-cost solutions to viewing digital copies of stereo aerial photography in 3D within GIS applications. Yet these developments are only the prologue to a true 3D experience as 3D capabilities of televisions and computer screens are developed.

But before we can move into such utopia for our historic aerial collections, resources must be found to create adequate indexing systems that meet the requirements of modern users.

Acknowledgments The authors would like to thank Birger Stichelbaut for permission to reproduce the sample card index from the Box Collection. We are also grateful to Jack Stevenson and Dave Cowley for comments on earlier drafts of this chapter, but the views expressed in this chapter are those of the authors.

Appendix 1. Metadata used to describe an index map to the 1988 All Scotland Survey flight lines

All Scotland Survey 1988 lines	
Title	All Scotland Survey 1988
Alternative title	None
Originator	National Monuments Record of Scotland, Royal Commission on the Ancient and Historical Monuments of Scotland
Abstract	Centre of photograph from All Scotland Survey
Data capture period	Known
Status of start date of capture	Known
Start date of capture	19880101.0
Status of end of capture	Known
End date of capture	19890101.0
Frequency of update	Never
Presentation type	Map
Access constraint	Internal (RCAHMS and HS) use
Use constraint	All RCAHMS data are Crown Copyright, Licenced copyright may be sought from the Curator, National Monuments Record of Scotland
Keywords	Photography
Geographic extent	Known
Spatial referencing by coordinates	Known
System of spatial referencing	British National Grid
Bounding rectangle	Known
West bounding coordinate	0.0
East bounding coordinate	500.0
North bounding coordinate	1300.0
South bounding coordinate	500.0
Spatial referencing by geographic identifiers	Known
National extent	Scotland
Administrative area extent	Scotland
Postcode area extent	Scotland
Spatial reference system	British National Grid
Level of spatial detail	Null
Supply media 1	Magnetic
Supply media 2	Optical
Supply media 3	Online
Data format 1	Genamap vector map
Data format 2	ArcView shapefile
Data format 3	DXF
Data format 4	MapInfo

(continued)

Appendix (continued)

All Scotland Survey 1988 lines	
Additional information	#http://www.rcahms.gov.uk#
Dataset association	Unknown
Supplier	National Monuments Record of Scotland, Royal Commission on the Ancient and Historical Monuments of Scotland
Contact name or title	The Curator, National Monuments Record of Scotland, Royal Commission on the Ancient and Historical Monuments of Scotland
Full postal address of supplier	John Sinclair House, 16 Bernard Terrace, Edinburgh
Postcode of supplier	EH8 9NX
Telephone of supplier	0131-662-1456
Fax of supplier	0131-662-1477
Email of supplier	info@rcahms.gov.uk
Web of supplier	#http://www.rcahms.gov.uk#
Date of metadata update	20000701.0
Originating source	RCAHMS
Dataset name	All Scotland Survey Points (1988–9)
Date of photography	1988 (with some 1989)
Time of flying	Variable
Photo scale	1:24,000
Flight altitude	12,000 ft
Camera focal length	151.84, 151.91, 152.24, 152.31 or 152.57 mm
Spatial resolution	Unknown
Format of original imagery	Black and white, some colour

All Scotland Survey 1988 points	
Title	All Scotland Survey 1988
Alternative title	None
Originator	National Monuments Record of Scotland, Royal Commission on the Ancient and Historical Monuments of Scotland
Abstract	Centre of photograph from All Scotland Survey
Data capture record	Known
Status of start date of capture	Known
Start date of capture	19880101.0
Status of end of capture	Known
End date of capture	19890101.0
Frequency of update	Never
Presentation type	Map
Access constraint	Internal (RCAHMS and HS) use
Use constraint	All RCAHMS data are Crown Copyright, Licenced copyright may be sought from the Curator, National Monuments Record of Scotland

(continued)

Appendix (continued)

All Scotland Survey 1988 points	
Keywords	Photography
Geographic extent	Known
Spatial referencing by coordinates	Known
System of spatial referencing	British National Grid
Bounding rectangle	Known
West bounding coordinate	0.0
East bounding coordinate	500.0
North bounding coordinate	1300.0
South bounding coordinate	500.0
Spatial referencing by geographic identifiers	Known
National extent	Scotland
Administrative extent	Scotland
Postcode area extent	Scotland
Spatial reference system	British National Grid
Level of spatial detail	Null
Supply media 1	Magnetic
Supply media 2	Optical
Supply media 3	Online
Data format 1	Genamap vector map
Data format 2	ArcView shapefile
Data format 3	DXF
Data format 4	MapInfo
Additional information	#http://www.rcahms.gov.uk#
Dataset association	Unknown
Supplier	National Monuments Record of Scotland, Royal Commission on the Ancient and Historical Monuments of Scotland
Contact name of title	The Curator, National Monuments Record of Scotland, Royal Commission on the Ancient and Historical Monuments of Scotland
Full postal address of supplier	John Sinclair House, 16 Bernard Terrace, Edinburgh
Postcode of supplier	EH8 9NX
Telephone of supplier	0131-662-1456
Fax of supplier	0131-662-1477
Email of supplier	info@rcahms.gov.uk
Web of supplier	#http://www.rcahms.gov.uk#
Date of metadata update	20000701.0
Originating Source	RCAHMS
Dataset name	All Scotland Survey Points (1988–9)
Date of photography	1988 (with some 1989)
Time of flying	Variable
Photo scale	1:24,000
Flight altitude	12,000 ft
Camera focal length	151.84, 151.91, 152.24, 152.31 or 152.57 mm
Spatial resolution	Unknown
Format of original imagery	Black and white, some colour

Bibliography

Anderson, H. (2008). How to read a trench map. http://www.westernfrontassociation.com/great-war-mapping/mapping-the-front-great-war-maps-dvd/great-war-map-lists/855-read-a-trench-map.html. Accessed 24 May 2011.

Aerofilms. (1965). *The Aerofilms book of aerial photographs*. London: Aerofilms Ltd.

Aerofilms. (2005). *Aerofilms guide. Football grounds* (13th ed.). Hersham: Ian Allen.

Beamer, A. & Gillick, M. (2010). Scotlandsplaces: Accessing remote Digital Heritage Datasets using Web Services. In M. Ioannides, D. Fellner, A. Georgopoulos & D.G. Hadjimitsis (Eds). Lecture notes in Computer Science, 2010, Volume 6436/2010, (225–239). DOI: 10.1007/978-3-642-16873-4_17

British Academy. (2001). *Aerial survey for archaeology. Report of a British Academy working party 1999*. Compiled by Robert Bewley. London: British Academy.

Capper, J. C. (1907). Photographs of Stonehenge as seen from a war balloon. *Archaeologia, 60*, 571.

Ceraudo, G. (2005). 105 Years of Archaeological Aerial Photography in Italy (1899-20040. in J.Bourgeois and M.Meganck (Eds.), Aerial Photography and Archaeology 2003. A century of Information. (pp.73–86) Archaeological Reports Ghent University 4.

Chorley, R. R. E. (1987). *Handling geographic information. Great Britain. Committee of enquiry into the handling of geographic information*. London: HMSO.

Gittings, B. (2009). Reflections on forty years of geographic information in Scotland: Standardisation, integration and representation. *Scottish Geographical Journal, 125*(1), 78–94.

McKeague, P. (2005). Indexing vertical aerial photograph collections: An introduction to www.airphotofinder.com. *AARG News (The newsletter of the Aerial Archaeology Research Group), 30*, 28–31.

Moloney, R. (1997a). Flying too close to the sun? Air photography and GIS. *AARG News (The newsletter of the Aerial Archaeology Research Group), 15*, 13–14.

Moloney, R. (1997b). The air photographs collection. In *Monuments on record: RCAHMS annual review 1996–7* (pp. 42–45). Edinburgh: RCAHMS.

NAPLIB. (1999). *Directory of aerial photographic collections in the United Kingdom* (2nd ed.). Dereham: NAPLIB.

RCAHMS. (1956). *An inventory of the ancient and historical monuments of Roxburghshire with the fourteenth report of the Commission*. Edinburgh: HMSO.

Scollar, I., & Palmer, R. (2008). Using Google Earth imagery. *AARG News (The newsletter of the Aerial Archaeology Research Group), 37*, 15–21.

Scottish Executive. (2005). 'One Scotland – One Geography' a Geographic Information Strategy for Scotland. http://www.scotland.gov.uk/Publications/2005/08/31114408/44098. Accessed 4 May 2011.

Shalders, H. (2007). Approaches to web access to aerial photos held at the NMRC. In J. T. Clark & E. M. Hagemeister (Eds.), *Digital discovery. Exploring new frontiers in human heritage. CAA 2006 computer applications and quantative methods in archaeology.* Proceedings of the 34th conference, Fargo, USA, April 2006 (pp. 423–428). Budapest: Archaeolingua.

Steer, K. (1947). Archaeology and the national air-photograph survey. *Antiquity, 21*(81), 50–53.

Stichelbaut, B. (2006). The application of First World War aerial photography to archaeology: The Belgian images. *Antiquity, 80*(307), 161–172.

Stichelbaut, B. (2009). *World War One aerial photography: An archaeological perspective*. Unpublished Ph.D. dissertation, Ghent University.

Stichelbaut, B., Gheyle, W., & Bourgeois, J. (2010). Great War aerial photographs: The Imperial War Museum's Box Collection. In D. C. Cowley, R. A. Standring, & M. J. Abicht (Eds.), *Landscapes through the lens. Aerial photographs and historic environment* (pp. 225–236). Oxford: Oxbow.

Wilson, D. R. (2000). *Air photo interpretation for archaeologists.* Stroud, Tempus Publishing Ltd.

Chapter 4
Declassified Intelligence Satellite Photographs

Martin J.F. Fowler

Abstract The declassification at the end of the last century of over 900,000 photographs acquired by the CORONA, ARGON, LANYARD, GAMBIT and HEXAGON US photo reconnaissance satellite programmes between 1960 and 1980 has resulted in an archive of declassified intelligence satellite photographs (DISP) that is both global in scale and easily accessible. As a source of low-cost, relatively high-resolution satellite imagery, the DISP archive is being used extensively by archaeologists to investigate landscapes in the arid regions of Asia Minor and the Middle East, as well as in more temperate regions. In this chapter, the nature and archaeological uses of the various DISP products are described, and representative examples are provided in order to permit the reader to appreciate their archaeological potential.

4.1 Introduction

On 22 February 1995, President William J. Clinton signed Executive Order 12951 directing the declassification of photographs acquired by the first generation of US film-return photoreconnaissance satellites. Over 860,000 photographs were subsequently transferred to the National Archives and Records Administration (NARA) and copies made available to the public for purchase at relatively modest cost through the United States Geological Survey (USGS). Five years later, a further 48,000 photoreconnaissance satellite photographs were approved for declassification and were subsequently handed over to the NARA and USGS in 2002. The resulting archive of declassified intelligence satellite photographs (DISP) is global in scale and comprises photographs acquired by five satellite programmes that were operational between 1960 and 1980.

M.J.F. Fowler (✉)
Les Rocquettes, Orchard Road, South Wonston, Winchester SO21 3EX, UK
e-mail: satarchuk@btinternet.com

W.S. Hanson and I.A. Oltean (eds.), *Archaeology from Historical Aerial and Satellite Archives*, DOI 10.1007/978-1-4614-4505-0_4,

Originally acquired for intelligence, mapping and geodesy purposes (McDonald 1997; Peebles 1997), the archaeological potential of the DISP archive was quickly appreciated after their declassification (e.g. Kennedy 1998). As a source of low-cost, relatively high-resolution overhead photographs, they are now being used by archaeologists to investigate archaeological features in the arid regions of Asia Minor and the Middle East as well as in more temperate regions such as Southern England. In this chapter, the availability, nature and archaeological uses of the various types of photographs present in the DISP archive are described, and representative photographs provided in order to permit the reader to appreciate their potential.

4.2 Availability

The DISP archive can be searched online using the USGS *EarthExplorer* website (http://earthexplorer.usgs.gov/). The metadata associated with each photograph includes details of the mission and camera system used to acquire the photograph, the date of acquisition and an estimation of the coordinates of the four corners of the ground footprints. The latter can be displayed on a Google Maps backdrop to confirm frame coverage, and low-resolution browse images can be viewed to assess the degree of cloud cover. Whilst the ground coverage footprints have been estimated using a rigorous modelling approach (Selander 1997), the corner coordinates have an error potential of up to 20 km from their actual ground positions (USGS 2006), and therefore, it is prudent to consider purchasing adjacent frames in order to guarantee coverage of any particular location.

The photographs are provided in the form of high-resolution digital scans that can be purchased through the website at a modest cost (currently US $30 per frame), with payment by means of credit card. Images are delivered over the Internet and are subsequently included in a growing data archive that can be downloaded free of charge. The digital scans are provided in the form of uncompressed 8-bit greyscale, or 32-bit colour, TIF files and are available at 3 levels of resolution: 21 μm (1,200 dpi), 14 μm (1,800 dpi) and 7 μm (3,600 dpi). For archaeological applications, the latter is the resolution of choice for the majority of the products, although it is less than the 4 μm (6,300 dpi) recommended for lossless digitisation when compared with the original hard copy photographs (Leachtenauer et al. 1997).

4.3 DISP Products and Collateral Information

The photographs in the DISP archive were acquired by systems associated with programmes code-named CORONA, ARGON, LANYARD, GAMBIT and HEXAGON. The latter has not been released officially, but is understood to be correct (Richelson 2002: 129). As well as having different programme names, the products of the various satellite systems were also given KH (Keyhole) designators with, for example, the product from the CORONA J-3 camera being designated as KH-4B. Different names were also given to the cameras carried by the satellites as well as the number series for the particular missions. A concordance of these various naming conventions is given in Table 4.1 together with details of the various camera systems, frame formats and numbers of photographs in the archive.

In addition to the DISP products, over 2,000 documents relating to the CORONA, ARGON and LANYARD programmes were declassified by the National Reconnaissance Office (NRO) in 1997. Although the majority relate to programmatic issues, for most missions, photographic evaluation reports (e.g. NRO 1972) were prepared that describe the overall quality of the photographs that were acquired and provide details of the specific camera systems that were carried. PDF copies of the documents can be downloaded from the NRO FOIA Special Collections webpage (http://www.nro.gov/foia/declass/collections.html).

Whilst the metadata associated with the photographs in the USGS archive gives only the date that the photographs were acquired, computer printouts of frame ephemeris data detailing times of acquisition of the individual CORONA photographs are available from NARA. Alternatively, the acquisition times can estimated with an accuracy of the order of ±20 seconds by using readily available historical orbital ephemeris data for CORONA missions in order to recreate the orbit of the satellite in both space and time (Fowler 2006, 2011). Finally, the CIA Records Search Tool (CREST) located at NARA includes a large number of declassified photographic interpretation reports that were prepared by the National Photographic Interpretation Center (NPIC) from the exploitation of the satellite products. They illustrate vividly the enormous wealth of intelligence that was acquired by these systems throughout the Cold War and can be searched online through the CIA 25-year Programme Archive search page (http://www.foia.cia.gov/search_archive.asp).

4.3.1 CORONA

CORONA was the US Intelligence Community's first successful photoreconnaissance satellite system and provided a broad area search capability over the denied territories of the former Soviet Union, Communist China and other countries between 1960 and 1972 (McDonald 1997). The 826,553 photographs that were acquired by the CORONA programme represent over 95% of the DISP archive. Carrying a single panoramic camera (see Madden 1999; Dashora et al. 2007 for technical descriptions of the various CORONA camera models), the first successful mission flew on 18 August 1960 and returned to Earth some 1432 exposed frames covering more than 4,000,000 km^2 of Soviet territory. This single mission provided more overhead coverage of the Soviet Union than the manned U-2 aircraft had acquired previously in all of its 24 flights.

The spatial resolution of the KH-1 film product from the original C camera was of the order of 12 m, but within 2 years, the resolution had improved to about 4 m

Table 4.1 US declassified intelligence satellite photographs

Parameter	CORONA						ARGON	LANYARD	GAMBIT	HEXAGON
Camera model	C	C′	C‴	M	J-1	J-3	ARGON	LANYARD	Unknown	Unknown
Product designator	KH-1	KH-2	KH-3	KH-4	KH-4A	KH-4B	KH-5	KH-6	KH-7	KH-9
Period of operation	1960	1960–1961	1961–1962	1962–1963	1963–1969	1967–1972	1962–1964	1963	1963–1967	1973–1980
Mission Series			9000		1000	1100	9000A	8000	4000	1200
Camera type		Mono panoramic			Stereo panoramic		Frame	Panoramic	Strip	Frame
Focal length (mm)				609.6			76.2	1,676.4	Unknown	Unknown
Film type	1188	J 22-7600	9029	4404		3404/3414	Unknown	Unknown	Unknown	Unknown
Wratten filter	W21	W21	W21	W21	W25 (Fwd)/W21 (Aft)		Unknown	Unknown	Unknown	Unknown
Film format (mm)				55 × 757			114 × 114	114 × 635	228 × >440	228 × 463
Film resolution (l/mm)			50–100		120	160	30	160	nk	nk
Ground resolution (m)	12	8	4–8	3–8	2.7–8	1.8	140	1.8	1.2	6–9
Frame ground coverage (km^2)			2,520–24,360		3,927	2,632	230,400	768	1,000–14,000	4,273,480
Number of successful missions	1	3	5	21	49	16	6	1	30	12
Total number of frames	1,432	7,246	9,918	101,743	517,688	188,526	38,578	910	19,000	29,000

for the KH-3 photographs acquired by the C''' (C Triple Prime) camera. The next model of the satellite brought the added advantage of stereoscopic photography taken by the M (Mural) camera system and was subsequently followed by the workhorses of the CORONA programme in the form of the Janus J-1 and J-3 cameras. These satellites had dual film recovery systems that eventually extended mission life up to 18 days and produced the KH-4A and KH-4B products with nominal spatial resolutions of the order of 2.7 and 1.8 m, respectively. Each frame from the J-3 cameras covered approximately 3,500 km^2 of the Earth's surface, and on a typical mission, each pair of cameras returned up to 12,000 photographic frames. In total, 95 successful CORONA missions were flown between October 1960 and May 1972 delivering an estimated 800 million km^2 of cloud-free photography of the Earth's surface.

The panoramic cameras used throughout the CORONA programme were based around a 24-in. (609.6-mm) focal length lens (Smith 1997; Dashora et al. 2007). The first CORONA missions, utilising the C and C′ cameras, employed a four-element Tessar lens design with a relative aperture of f/5.0, whereas the C''' and subsequent camera systems employed a five-element Petzval lens design with a relative aperture of f/3.5. Whilst the physical characteristics of the Petzval lenses remained virtually unchanged to the end of the CORONA programme, improvements in materials, fabrication and test techniques, and an overall tightening of lens tolerance, improved significantly the performance of the lenses with the result that the final Type IV Petzval lens used by the J-3 camera system was achieving film resolutions of the order of 160 lines/mm compared with 50–100 lines/mm for the earlier C''' camera.

The field of view of all the CORONA cameras was approximately 6° along the track of the satellite and 70° cross track. Stereo photographs were acquired by the Mural and Janus systems by using a pair of panoramic cameras, one of which looked approximately 15° forward and the other 15° aft of the direction of satellite motion. The ground footprints that were covered by the cameras depended on the altitude of the satellite, and in the case of the KH-4B product, this equated to an area of some 16 by 217 km for an acquisition altitude of 157 km.

Inherent in all of the CORONA photographs is the distortion that comes from the use of a panoramic camera carried by a moving platform (for a technical description, see Goossens et al. 2006; Casana and Cothren 2008). This distortion takes the form of the displacement of ground points from their expected perspective as a result of a number of factors including the cylindrical shape of the negative film surface, the scanning action of the lens and the movement of the satellite in its orbit during the brief time taken to acquire each photograph. The result is that the ground footprints, rather than being approximately rectangular in nature as suggested by the four corner locations given on *EarthExplorer*, have a characteristic 'bow tie' shape and the shapes of features on the ground can be distorted significantly depending on their location on the film frame. Whilst rigorous models to address the distortion are not yet available in commercial photogrammetry packages, because of the long focal length of the CORONA panoramic cameras, a small sub-image (approximately 6,000 by 6,000 pixels when scanned at 7 μm) may be treated as a frame camera (Altmaier and Kany 2002) and oriented to the ground using ground control points derived from differential GPS surveys (e.g. Goossens et al. 2006) or orthorectified satellite images (e.g. Casana and Cothren 2008).

The film carried by the CORONA missions was predominantly a 70-mm-wide polyester-based panchromatic, black-and-white negative aerial film having extended red sensitivity and medium speed and was used in conjunction with red (Wratten 25) and orange (Wratten 21) haze reduction filters. Such a combination gave the film a peak sensitivity in the region of 680–700 nm of the electromagnetic spectrum, which is on the boundary of the 'red edge' of healthy vegetation, thus making it sensitive to the detection of crop-marked features through differences in the infrared reflectance of healthy and stressed plants (Fowler and Fowler 2005). Details of the actual film/filter combinations used in a particular mission can be found in their respective photographic evaluation reports. In addition to black-and-white film, a small amount of colour film was used experimentally on a number of KH-4B missions, usually at the end of the primary mission film loads of their aft-looking cameras (NRO 1969). Missions 1105, 1106 and 1108 acquired conventional colour photographs on passes over the USA, the former Soviet Union, China and Korea. Likewise, a small amount of false colour infrared film was carried on Mission 1104, and photographs were acquired over the USA, the Soviet Union, China, Vietnam, Israel and Jordan. Whilst coverage plots for these photographs can be found in a declassified summary report (NRO 1969), copies of the photographs are unfortunately not available for purchase from the USGS website and have to be obtained from the Special Media Archives Service at NARA (for details see http://www.archives.gov/research/order/still-pictures.html).

4.3.2 ARGON

ARGON was a frame camera system with a focal length of 3 in. (76.2 mm) that was intended to provide mapping and geodesic data on the Soviet Bloc. Six successful missions were flown between 1961 and 1964 and returned 38,578 frames. Whilst the area of the Earth's surface covered by each KH-5 frame was high, of the order of 550 by 550 km, the spatial resolution was low being of the order of 140 m, and consequently, the KH-5 product has limited utility for archaeological purposes. However, it has been used effectively in studies of environmental change, including monitoring changes in the Antarctic ice sheet (Bindschadler and Vornberger 1998), where high spatial resolution is not required.

4.3.3 LANYARD

LANYARD was a short-lived, high-resolution surveillance satellite programme that was based on leftover hardware from the previously cancelled SAMOS E-5 programme (Day 1997). Carrying a 66-in. (1676.4-mm) focal length panoramic camera designed to achieve a ground resolution of 1.5 m and the ability to look ±30° either side of the ground track, only one of the three KH-6 satellites that were launched in 1963 successfully returned film; this mission was itself only partially successful as the camera failed after 32 hours in orbit. The LANYARD programme was terminated after this mission as a result of the success of the KH-7 GAMBIT system that was also under development in 1963.

The 908 frames that were returned by LANYARD Mission 8003, each having a ground footprint of the order of 114 by 635 km, covered targets in the former Soviet Union and China and demonstrated in a few instances that the system had the potential to achieve a ground resolution of the order of 1.5 m (NRO 1963). Whilst the coverage of the frame footprints can be searched using *EarthExplorer*, preview images are not available. Since cloud cover during the mission varied between 25 and 87.5%, with an average of 53.7%, there is a better than even chance that any location covered by a LANYARD frame will be obscured by cloud. Nevertheless, some indication of cloud-free areas can be gained from the Mission Coverage Index (NPIC 1963) which lists some 107 intelligence targets that were successfully photographed in Latvia, Lithuania, Belarus, Ukraine, Turkmenistan, China and Russia.

4.3.4 GAMBIT

The KH-7 GAMBIT was the first US high-resolution surveillance or 'spotting' satellite and complemented the area search capability of CORONA with photographs that had initially a best spatial resolution of 1.2 m, but by 1966 had increased to 0.6 m (Richelson 2003). Whilst little technical information has been released about the GAMBIT camera (NRO 2006), it would appear to have comprised a strip camera that exposed a continuous photograph of the terrain below the satellite by passing the film over a stationary slit in the focal plane of the lens at a speed synchronised with the velocity of the ground image across the focal plane (Richelson 2003). As the satellite moved forward, a long continuous photograph would have been 'painted' onto the film covering an area some 20 km wide and normally 50 km long, but occasionally extending to over 700 km in length. The camera appears to have been capable of acquiring both vertical and oblique photographs either side of the ground track, as well as stereo photographs, and used predominantly 228-mm-wide black-and-white film, although a small number of colour photographs was also acquired in 1966 of parts of the former Soviet Union and China on Mission 4030. In contrast to colour photographs that were acquired by CORONA, these photographs can be purchased from the USGS, although the scanning resolution is limited to 1,200 dpi in order to limit the size of the TIF files to a practical size.

Some 19,000 frames were returned by the 34 successful KH-7 CORONA missions that were flown between 1963 and 1967 and cover mainly the former Soviet Union and China, although parts of other territories were also covered, albeit to far lesser extent. GAMBIT missions continued after 1967 using the improved KH-8 system and produced photographs with a spatial resolution in the region of 15 cm (Richelson 2003); however, the products from these missions have yet to be released.

4.3.5 *HEXAGON Mapping Camera*

The KH-9 Mapping Camera System was a frame camera that was flown on 12 missions between 1973 and 1980 (NRO 2006) in addition to the two high-resolution cameras that were carried by the system (Richelson 2002: 158–159), the products of which have yet to be declassified. The photographs acquired by the camera were devoted to mapping, charting and geodesy and covered ground footprints of the order of 130 by 260 km with a spatial resolution of initially 9 m that improved to 6 m on later missions (Surazakov and Aizen 2010). The coverage of the 29,000 frames that were acquired is worldwide with most areas being covered by up to three different frames that were acquired at different dates. However, those frames that would have covered Israel and its neighbouring states do not appear to have been released into the public domain[1].

4.4 Comparison of DISP Products

In order to provide a comparison of the quality of the major DISP products, extracts from photographs acquired by the KH-9 HEXAGON, KH-4B CORONA and KH-7 GAMBIT satellites covering the location of the former Assyrian city of Nineveh are shown in Fig. 4.1. Situated on the east bank of the Tigris at its junction with the Rover Khosr and now overlain in part by the suburbs of the city of Mosul, ancient Nineveh's ruins cover an area of over 700 ha and include the ancient tells of Tell Kuyunjik and Tell Nabi Yunus, together with a 12-km rampart with 15 gates and an external moat and canal (Fig. 4.1a).

Considering the DISP products in order of increasing spatial resolution, the characteristic outline of Nineveh's ramparts can be readily seen on the KH-9 HEXAGON mapping camera photograph acquired on 9 May 1974 (Fig. 4.1b), and the location of Tell Kuyunjik can be broadly recognised. However, smaller features are more difficult to discern on this 10-m resolution image, and at this level of enlargement, the film is somewhat 'grainy'. By way of contrast, the KH-4B CORONA photograph acquired on 4 June 1970 (Fig. 4.1c) has a spatial resolution of the order of 2–3 m and is capable of discerning smaller detail including buildings and surface irregularities such as those of Tell Kuyunjik, which includes the site of Sennacherib's Southwest Palace that was built at the end of the eighth century BC. Elements of the former moat and canal to the east of the river Khosr can be discerned through differences in vegetation cover. Finally, the KH-7 GAMBIT photograph that was acquired on 20 August 1966 has the highest spatial resolution of the three products, estimated to be of the order of 1 m, and shows considerable fine detail of topographical and surface

[1] Since preparing this chapter, technical details of the GAMBIT and HEXAGON camera systems have been declassified by the NRO and confirm the descriptions provided above. Details can be found at http://www.nro.gov/foia/declass/GAMBHEX.html.

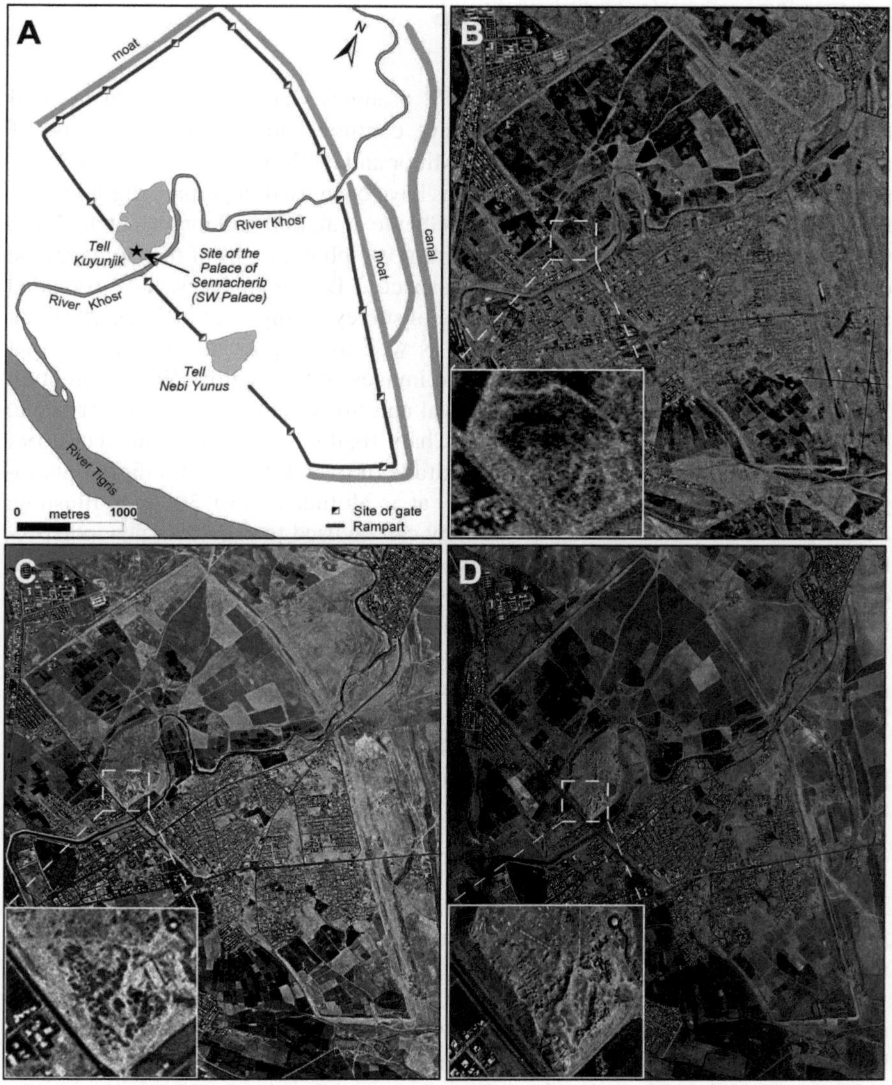

Fig. 4.1 Comparison of DISP products. (**a**) Outline plan of the ancient city of Nineveh; (**b**) KH-9 HEXAGON mapping camera photograph from Mission 1208-5; (**c**) KH-4B CORONA photograph from Mission 1110-2; (**d**) KH-7 GAMBIT photograph from Mission 4031. Insert enlargements cover the approximate location of Sennacherib's Southwest palace on Tell Kuyunjik (Data available from the US Geological Survey, EROS Data Center, Sioux Falls, SD, USA)

features, small subdivisions within fields, individual buildings and vehicles on the modern roads (Fig. 4.1d). As such, the GAMBIT product is broadly comparable with a conventional medium-scale vertical aerial photograph, but it covers a considerably larger area on a single frame (of the order of some 20 by 50 km).

4.5 Archaeological Uses of DISP

Since the archaeological uses of the DISP products were last reviewed in 2004 (Fowler 2004a), CORONA photographs have continued to be used in at least 21 archaeological studies, primarily in Asia Minor and the Middle East (Table 4.2). Of particular note, stereo CORONA photographs have been used to prepare digital elevation models for use in landscape studies (Gheyle et al. 2004; Goossens et al. 2006; Casana and Cothren 2008), and studies of CORONA photographs of the Homs region of Syria, together with more recent multispectral Ikonos imagery, have provided important advice for the design and costing of surveys using satellite imagery (Beck et al. 2007 and Chap. 15, this volume), as well as insights into the factors that influence the visibility of sites on satellite images in this region (Wilkinson et al. 2006). In a study of a prehistoric, Medieval and Soviet era landscape in Armenia, Faustmann and Palmer (2005 and Palmer Chap. 16, this volume) combined CORONA satellite photographs acquired from an altitude of over 150 km with digital photographs taken from a two-person paramotor at an altitude of 250–300 m to illustrate how the extremes of overhead photography can be used to provide a new perspective on a previously uninvestigated landscape. Increasingly, orthorectified historical CORONA photographs are being combined with more recent high-resolution imagery to map ancient landscapes to a high degree of precision (e.g. Bitelli and Girelli 2009; Castrianni et al. 2010), and in a *tour de force*, CORONA photographs have been used to map over 6,000 km of premodern trackways in the vicinity of one of the largest Bronze Age sites in northern Mesopotamia at Tell Hamoukar (Ur 2010).

As an illustration of the archaeological potential of CORONA photographs in the Middle East, Fig. 4.2 shows an extract from a KH-4B photograph that was acquired by Mission 1115–2 on 29 September 1971 and which covers the environs of the Roman legionary fortress at El-Lejjun in Jordan. Whilst the quality of the photograph is by no means as good as conventional aerial photographs (cf. Fig. 10.5a,b) in Kennedy and Bewley 2004), the outline of the walls of the fortress, together with many of the U-shaped interval towers, can be discerned. Within the fortress, the ruins of the *principia* and several rows of barrack blocks can be discerned, all of which have been robbed extensively for stone in the past. To the west of the fortress, two lines of buildings represent the remains of early-twentieth-century Late Ottoman barracks, and approximately 1 km to the northwest lies the Roman fort of Khirbet el-Fityan. Additional remains from Rome's desert frontier in this region can be readily identified on this and other frames acquired by the mission (Fowler 2004b).

In addition to recording archaeological features in relief or through changes in soil tone as a result of the presence of former human habitation, in more temperate regions such as Southern England, CORONA photographs acquired in late summer have been shown to be capable of recording plough-levelled features in the form of cropmarks (Fowler and Fowler 2005). To further illustrate the archaeological potential of CORONA photographs in temperate regions, Fig. 4.3 shows an extract from a KH-4B photograph that was acquired by Mission 1104-2 on 17 August 1968 and covering an area in the vicinity of Rowbury Copse, some 3 km to the northeast of

Table 4.2 Recent archaeological uses of DISP

Location	Mission(s)	Archaeological use	Reference(s)
Syria	Not specified	KH-4B photographs were used to investigate the topography and cultural landscape of the early Islamic city of al-Raqqa in northern Syria	Challis et al. (2004), Challis (2007)
Russian Federation	1042	KH-4A photographs were used as part of an archaeological survey of Bronze Age, Iron Age and Turkish monuments in the Altai Mountains of South Siberia	Gheyle et al. (2004), Goossens et al. (2006)
Armenia	1115	KH-4B photographs were used together with digital photographs taken at low altitude from a paramotor were used to investigate a prehistoric, Medieval and Soviet-era landscape in Armenia	Faustmann and Palmer (2005)
Iraq	1102, 1108, 1117, 4017, 4031	KH-4B and KH-7 photographs were used to describe canals and other archaeological features related to water management and settlement in the landscape in the vicinity of the Assyrian city of Nineveh	Wilkinson et al. (2005), Ur (2005)
England	1104	Ring ditches and a possible oval enclosure were detected as crops marks on a KH-4B photograph acquired in August 1968	Fowler and Fowler (2005)
Iraq	Not specified	CORONA satellite photographs were used to supplement aerial photography to produce a photogrammetric plan of the ancient city of Samarraon the east bank of the Tigris	Northedge (2005)
Turkey	1112	Used KH-4B photographs together with archaeological survey data to investigate aspects the settlement history of the Amuq Valley in the northern Levant including the archaeological landscape of Late Roman Antioch	Casana (2004, 2007)
Iran	1103, 1110	Mapped using KH-4B photographs pastoral and irrigation landscapes in north-western Iran that have been virtually obliterated since the photographs were acquired	Alizadeh and Ur (2007)
Iran	1052	KH-4A photographs were used to map the natural landscape of the Bushehr peninsular of Iran and to highlight changes to the landscape caused by modern water extraction and agriculture and to map individual monuments within the landscape	Challis (2007)
Syria	1108, 1110	Archaeological sites captured on KH-4B were accurately located using a recent Ikonos satellite image of the Homs Region of western Syria and provides advice in the design and costing of surveys using satellite imagery	Wilkinson et al. (2006), Beck et al. (2007)
Egypt	1049	KH-4A photographs that revealed surface features that have been obscured by recent agricultural developments, together with field investigations of Holocene sedimentary deposits, were used to create a palaeogeographic map that places Late Bronze Age archaeological sites in Northwest Sinai in their environmental context	Moshier and El-Kalani (2008)

(continued)

Table 4.2 (continued)

Location	Mission(s)	Archaeological use	Reference(s)
Russia	4028	A KH-7 photograph was used to describe the material culture of the Cold War installations around Moscow that were associated with the former Soviet Union's first generation of surface to air missile (SAM) system	Fowler (2008)
Turkey and Syria	1105	The stereo capabilities of KH-4B photographs were utilised to produce high-resolution digital elevation data and maps of archaeological landscapes in the Levant	Casana and Cothren (2008)
Turkey	Not specified	Orthorectified KH-4A imagery was combined with Landsat Thematic Mapper, QuickBird and ASTER imagery to investigate the archaeological site at Tilmen Höyük (south-eastern Turkey)	Bitelli and Girelli (2009)
Syria	1046, 1105	The Tell Tuqan Survey Project used multi-temporal KH-4A and KH-4 photographs together with QuickBird imagery to study the tell and the surrounding area to inform the reconstruction of the ancient topography of the region	Castrianni et al. (2010)
Italy and Turkey	9022, 1043, 1049, 1103, 1107, 1109, 1111, 4036, 1210, 1216	KH-3, KH-4A, KH-4B, KH-7 and KH-9 photographs were used to illustrate the importance of historical aerial and satellite photographs support archaeological and geo-archaeological research in Italy and turkey	Scardozzi (2010)
Iran	1035, 1045	The evolution of the Lower Khuzestan plain in SW Iran was investigated using KH-4A photographs, Landsat imagery together with archaeological, geological and historica; datasets	Walstra et al. (2010)
Sudan	Not specified	KH-4B and QuickBird imagery to support archaeological research and Cultural Resource Management in the Sudanese Middle Nile	Edwards (2010)
Latvia	4031, 1110, 1210	KH-7, KH-4B and KH-9 photographs were used to describe the development of the Hen House ballistic missile early warning radars at Skrunda in Latvia between 1966 and 1975	Fowler (2010)
Syria	1021, 1102, 1105, 1108, 1117	Over 6,000 km of premodern trackways in the vicinity of Tell Hamoukar, one of the largest Bronze Age sites in northern Mesopotamia, were identified and mapped from KH-4A and KH-4B photographs	Ur (2010)
Israel	1111, 1115	A comparison of two KH-4B photographs of the Roman siege-works at the fortress of Masada shows the impact of the solar lighting conditions at different times of the day on the appearance of the landscape and upstanding archaeological features	Fowler (2011)

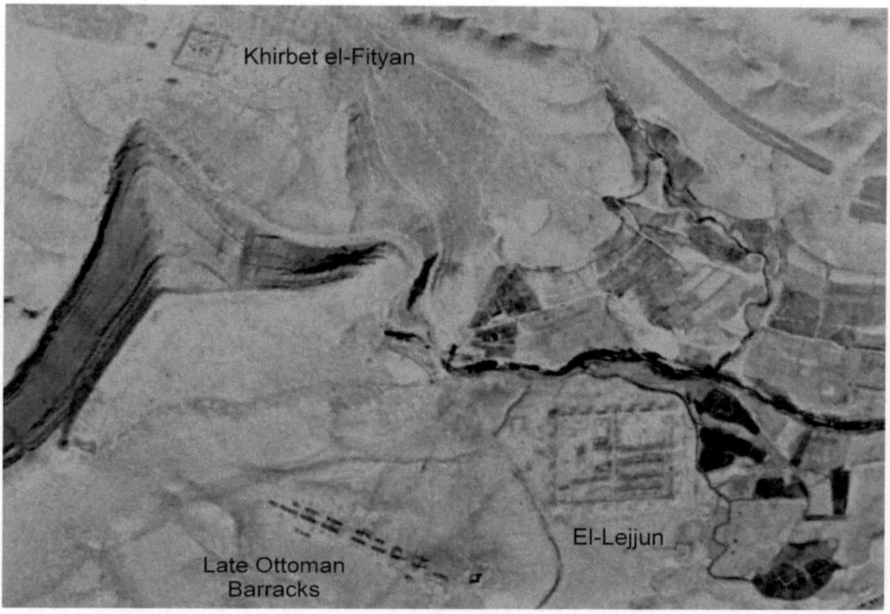

Fig. 4.2 The Roman legionary fortress at El-Lejjun, Jordan. Extract from KH-4B CORONA photograph acquired by Mission 1115-2 on 29 September 1971 (Data available from the US Geological Survey, EROS Data Center, Sioux Falls, SD, USA)

Fig. 4.3 Plough-levelled lynchets of the prehistoric fields in the vicinity of Rowbury Copse, Hampshire. Extract from KH-4B photograph acquired by Mission 1104-2 on 17 August 1968 (Data available from the US Geological Survey, EROS Data Center, Sioux Falls, SD, USA)

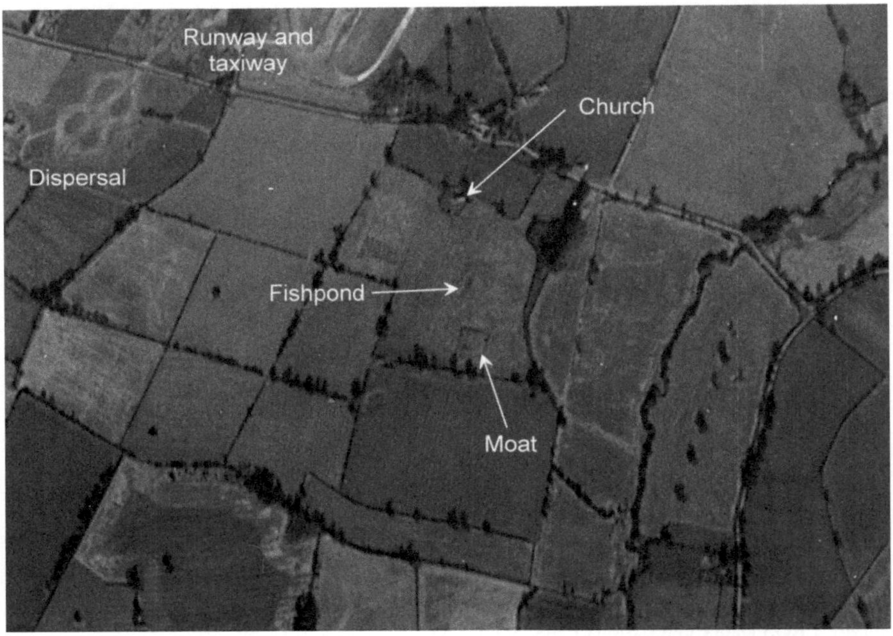

Fig. 4.4 Medieval moat and fish pond near the deserted medieval village at Stratton Magna, Leicestershire. Extract from KH-7 GAMBIT photograph acquired by Mission 4011 on 24 September 1964 (Data available from the US Geological Survey, EROS Data Center, Sioux Falls, SD, USA)

the well-known Iron Age hill fort at Danebury, Hampshire. Visible on the extract are many of the ploughed-out remains of the lynchets of the prehistoric field system that was mapped by Palmer (1984) using conventional aerial photographs.

In contrast to CORONA, little use has been made to date of the other DISP products either because of their low spatial resolution (ARGON and HEXAGON Mapping Camera) or their limited ground coverage (LANYARD and GAMBIT). Nevertheless, high spatial resolution GAMBIT photographs were found to be useful in a study of canals and other archaeological features associated with water management and settlement in the landscape in the vicinity of Nineveh (Ur 2005).

Figures 4.4 and 4.5 show two extracts from a KH-7 GAMBIT photograph that was acquired by Mission 4011 on 24 September 1964 of the area between Leicester and Rugby in England. Figure 4.4 covers the deserted medieval village of Stratton Magna, some 7 km to the east of Leicester. Whilst the location of the restored Church of St. Giles can be identified on the somewhat grainy photograph, with the exception of a rectangular moated site (70×60 m) and a fish pond (35×12 m) that formed part of a manorial complex to the south of the village, none of the hollow ways, house platforms and enclosures that comprise the remains of the deserted village can be discerned with any degree of certainty, although possible features are hinted at in the fields near the church, and medieval ridge and furrow open fields can be found in the surrounding fields. The inability to see some of the features on the

Fig. 4.5 Medieval ridge and furrow open field cultivation near Great Glen, Leicestershire. Extract from KH-7 GAMBIT photograph acquired by Mission 4011 on 24 September 1964 (Data available from the US Geological Survey, EROS Data Center, Sioux Falls, SD, USA)

photograph is more likely to be because of the unsuitability of the lighting rather than the resolution of the film (cf. Wilson 2000: 40–41). To the northwest of the church, a spectacle-type dispersal area and part of a taxiway and runway associated with the former Royal Air Force Station Leicester East (now Leicester Airport) can be identified.

Further evidence of ridge and furrow can be seen some 3 km to the south in the fields that surround the shrunken medieval village at Great Glen (Fig. 4.5) as well as at numerous other locations on the GAMBIT photograph. At a cost of $30 (US), the photograph represents a highly cost effective source of historical aerial coverage with which to map the distribution of surviving ridge and furrow in order to supplement existing studies in the English Midlands (e.g. Hall and Palmer 2000).

Since GAMBIT photographs were acquired specifically to gain high-resolution surveillance images of military and other installations in primarily the former Soviet Union and China, they are well suited to support Cold War archaeological studies, much as aerial photographs acquired some 50 years earlier are now being used to study aspects of the archaeology of the First World War (Stichelbaut 2006 and Chap. 5, this volume). The utility of GAMBIT photographs to studies of Cold War material culture has been demonstrated recently in a case study of the installations associated with the first generation surface to air missile (SAM) system that was built in the late 1950s to protect Moscow from aerial attack (Fowler 2008) and the development

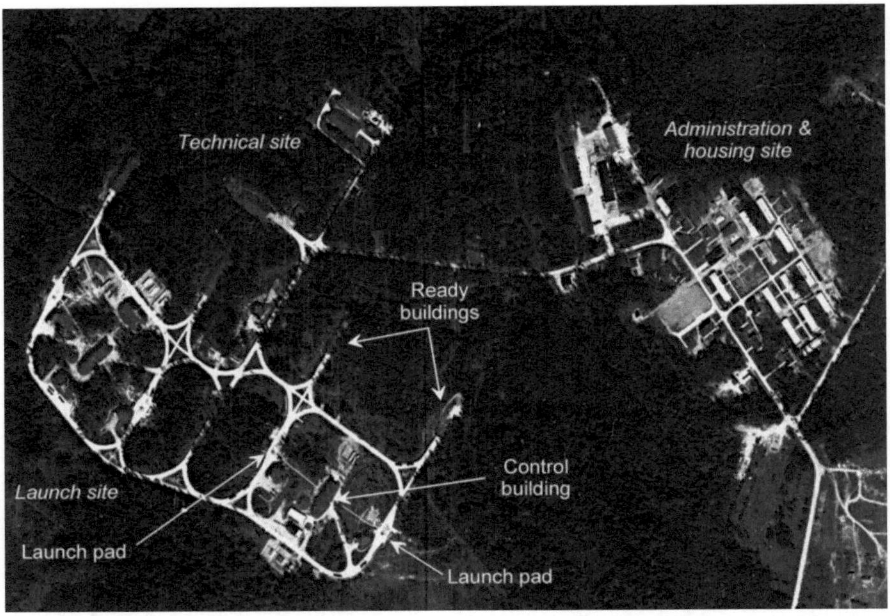

Fig. 4.6 The Soviet SS-14 IRBM launch site at Nigrande, Latvia. Extract from KH-7 GAMBIT photograph acquired by Mission 4032 on 20 August 1966 (Data available from the US Geological Survey, EROS Data Center, Sioux Falls, SD, USA)

of a ballistic missile early warning radar facility in Latvia (Fowler 2010). A further example of this application is shown in Fig. 4.6 which comprises an extract from a GAMBIT photograph that was acquired by Mission 4031 on 20 August 1966 of the Intermediate Range Ballistic Missile (IRBM) launch site at Nigrande in Latvia. Guided by a declassified 1964 Photographic Interpretation Report of the site from the CREST archive (NPIC 1964), the four elongated launch pads for the R-14/SS-5 SKEAN missiles can be readily identified together with four 'ready buildings' in which the missiles were stored and the two control bunkers between each pair of launch pads. Some 300 m to the northeast of the launch site, the technical section of the site support facility comprising three large buildings and several smaller buildings can be identified. The administration and housing section can be identified some 1,000 m to the east of the launch site. As many of the military installations that form the twentieth century's 'defence heritage' become lost, contemporary GAMBIT and CORONA photographs represent a unique source for use in the study of the material culture of this period.

Finally, with a spatial resolution comparable to that of the SPOT satellite panchromatic sensor which has been shown to be capable of detecting archaeological features (Shennan and Donoghue 1992; Fowler 1993), the KH-9 HEXAGON mapping camera has the potential to detect archaeological features such as pre-Roman rectilinear field systems on the island of Hvar, Croatia (Fowler 2004a), and the ramparts of the city of Nineveh, as shown in Fig. 4.1. Whilst its low spatial resolution constrains it to the detection of relatively large archaeological features, when rectified and georeferenced using orthorectified Landsat 7 imagery (available online from the NASA Global Land Cover Facility http://glcf.umiacs.umd.edu/data/landsat/), it can be used to provide a 10-m resolution base map for use in a Geographical Information System and as an alternative to SPOT panchromatic imagery to rectify CORONA photographs without recourse to survey on the ground (Casana and Cothren 2008).

4.6 Looking Ahead

The DISP archive is now well established as a source of low-cost, relatively high-resolution overhead photographs for use in archaeological studies in primarily Asia Minor and the Middle East. The finding that both CORONA and GAMBIT photographs can record plough-levelled archaeological features in more temperate regions opens up the potential to use the DISP archive to supplement conventional aerial photography in areas such as Eastern Europe where civilian aerial photography has been prohibited until only recently (Braasch 2002; See Chap. 18 by Oltean and Hanson, this volume). Importantly, the DISP archive represents an invaluable source of overhead coverage that predates the growth in urban expansion that took place in the latter part of the twentieth century and has the potential to contain a record of archaeological features that have since been destroyed. Given the great utility of the DISP products, it is to be hoped that additional obsolete photoreconnaissance satellite photographs, including those acquired by the KH-8 GAMBIT and KH-9 HEXAGON systems, will soon be declassified and added to the existing DISP archive for use by archaeologists and researchers in other disciplines.

Acknowledgments This chapter is dedicated to the memory of the late Ernest Fowler, 1920–1998. Special thanks are due to Rog Palmer for his critical comments, to Jeff Hartley of NARA for providing copies of CREST records and to Linda Hathaway of the Information Access and Release Team at the NRO for providing copies of original records relating to the CORONA, ARGON and LANYARD programmes.

Since preparing this chapter, technical details of the GAMBIT and HEXAGON camera systems have been declassified by the NRO and confirm the descriptions provided above. Details can be found at http://www.nro.gov/foia/declass/GAMBHEX.html.

Bibliography

Alizadeh, K., & Ur, J. A. (2007). Formation and destruction of pastoral and irrigation landscapes on the Mughan Steppe, north-western Iran. *Antiquity, 81*, 148–160.

Altmaier, A., & Kany, C. (2002). Digital surface model generation from CORONA satellite images. *ISPRS Journal of Photogrammetry and Remote Sensing, 56*, 221–235.

Beck, A., Philip, G., Abdulkarim, M., & Donoghue, D. (2007). Evaluation of corona and Ikonos high resolution satellite imagery for archaeological prospection in western Syria. *Antiquity, 81*, 161–175.

Bindschadler, R., & Vornberger, P. (1998). Changes in the West Antarctic ice sheet since 1963 from declassified satellite photography. *Science, 279*, 689–692.

Bitelli, G., & Girelli, V. A. (2009). Metrical use of declassified imagery for an area of archaeological interest in Turkey. *Journal of Cultural Heritage, 10S*, e35–e40.

Braasch, O. (2002). Goodbye cold war! Goodbye bureaucracy? Opening the skies to aerial photography in Europe. In R. H. Bewley & W. Raczkowski (Eds.), *Aerial archaeology – Developing future practice* (Nato science series, pp. 19–22). Amsterdam: Ios Press.

Casana, J. (2004). The archaeological landscape of Late Roman Antioch. In I. Sandwell & J. Huskinson (Eds.), *Culture and society in later Roman Antioch: Papers from a colloquium London, 15th December 2001* (pp. 102–125). Oxford: Oxbow Books.

Casana, J. (2007). Structural transformations in settlement systems of the Northern Levant. *American Journal of Archaeology, 112*, 195–221.

Casana, J., & Cothren, J. (2008). Stereo analysis, DEM extraction and orthorectification of CORONA satellite imagery: Archaeological applications from the near East. *Antiquity, 82*, 732–749.

Castrianni, L., Di Giacomo, G., Ditaranto, I., & Scardozzi, G. (2010). High resolution satellite ortho-images for archaeological research: Different methods and experiences in the near and middle East. *Advances in Geosciences, 24*, 97–110.

Challis, K. (2007). Archaeology's cold war windfall – The corona programme and lost landscapes of the near East. *Journal of the British Interplanetary Society, 60*, 21–27.

Challis, K., Priestnall, G., Gardner, A., Henderson, J., & O'Hara, S. (2004). corona remotely-sensed imagery in dryland archaeology: The Islamic city of al-Raqqa, Syria. *Journal of Field Archaeology, 29*(2002–04), 139–153.

Dashora, A., Lohani, B., & Malik, J. N. (2007). A repository of earth resource information – CORONA satellite programme. *Current Science, 92*, 926–932.

Day, D. A. (1997). A failed phoenix: The KH-6 LANYARD reconnaissance satellite. *Spaceflight, 19*, 170–174.

Edwards, D. N. (2010). Exploring rural landscapes in Sudanese Nubia. In D. Cowley, R. A. Standring, & M. J. Abicht (Eds.), *Landscapes through the lens. Aerial photographs and historic environment* (pp. 167–178). Oxford: Oxbow.

Faustmann, A., & Palmer, R. (2005). Wings over Armenia: Use of a paramotor for archaeological aerial survey. *Antiquity, 79*, 402–410.

Fowler, M. J. F. (1993). Stonehenge from space. *Spaceflight, 35*, 130–132.

Fowler, M. J. F. (2004a). Archaeology through the keyhole: The serendipity effect of aerial reconnaissance revisited. *Interdisciplinary Science Reviews, 29*, 118–134.

Fowler, M. J. F. (2004b). Declassified CORONA KH-4B satellite photography of remains from Rome's desert frontier. *International Journal of Remote Sensing, 24*, 3549–3554.

Fowler, M. J. F. (2006). Modelling the acquisition times of CORONA KH-4B satellite photographs. *AARG News, 33*, 34–40.

Fowler, M. J. F. (2008). The application of declassified KH-7 GAMBIT satellite photographs to studies of Cold War material culture: A case study from the former Soviet Union. *Antiquity, 82*, 714–731.

Fowler, M. J. F. (2010). The Skrunda hen houses: A case study in cold war satellite archaeology. In D. Cowley, R. A. Standring, & M. J. Abicht (Eds.), *Landscapes through the lens. Aerial photographs and historic environment* (pp. 287–293). Oxford: Oxbow.

Fowler, M. J. F. (2011). Modelling the acquisition times of CORONA satellite photographs: Accuracy and application. *International Journal of Remote Sensing, 32*(23), 8865–8879.

Fowler, M. J. F., & Fowler, Y. M. (2005). Detection of archaeological crop marks on declassified CORONA KH-4B intelligence satellite photography of southern England. *Archaeological Prospection, 12*, 257–264.

Gheyle, W. R., Trommelmans, R., Bourgeois, J., Gossens, R., Bourgeois, I., De Wulf, A., & Willems, T. (2004). Evaluating CORONA: A case study in the Altai Republic (South Siberia). *Antiquity, 78*, 391–403.

Goossens, R., De Wulf, A., Bourgeois, J., Gheyle, W., & Willems, T. (2006). Satellite imagery and archaeology: The example of CORONA in the Altai mountains. *Journal of Archaeological Science, 33*, 745–755.

Hall, D., & Palmer, R. (2000). Ridge and furrow survival and preservation. *Antiquity, 74*, 29–30.
Kennedy, D. (1998). Declassified satellite photographs and archaeology in the Middle East: Case studies from Turkey. *Antiquity, 72*, 553–561.
Kennedy, D., & Bewley, R. (2004). *Ancient Jordan from the air*. London: The Council for British Research in the Levant.
Leachtenauer, J. C., Daniel, K., & Vogl, T. P. (1997). Digitising corona imagery: Quality vs. cost. In R. A. McDonald (Ed.), *corona between the sun and the earth: The first NRO reconnaissance eye in space* (pp. 189–203). Bethesda: American Society for Photogrammetry.
Madden, F. (1999). The CORONA camera system: Itek's contribution to world security. *Journal of the British Interplanetary Society, 52*, 379–396.
McDonald, R. A. (Ed.). (1997). *corona between the sun and the earth: The first NRO reconnaissance eye in space*. Bethesda: American Society for Photogrammetry and Remote Sensing.
Moshier, S. O., & El-Kalani, A. (2008). Late bronze age paleogeography along the ancient ways of Horus in Northwest Sinai, Egypt. *Geoarchaeology, 23*, 450–473.
Northedge, A. (2005). Remarks on Samarra and the archaeology of large cities. *Antiquity, 79*, 119–129.
NPIC. (1963). *Mission coverage index mission 8003, 31 July–1 August 1963*. CREST record CIA-RDP78B04560A000800010068-6 approved for release on 21 August 2001. College Park: National Archives and Records Administration.
NPIC. (1964). *Nigrande IRBM complex, USSR*. CREST record CIA-RDP78B04560A002100010042-9 approved for release on 3 March 2003. College Park: National Archives and Records Administration.
NRO. (1963). *System photographic evaluation report mission 8003*. Record 4/A/0033 of the collection of CORONA, ARGON and LANYARD records declassified on 26 November 1997. Chantilly: National Reconnaissance Office.
NRO. (1969). *Summary report – Special purpose photographic techniques for overhead reconnaissance*. Record 5/D/0032 of the collection of CORONA, ARGON and LANYARD records declassified on 26 November 1997. Chantilly: National Reconnaissance Office.
NRO. (1972). *Photographic evaluation report mission 1115*. Record 6/A/0029 of the collection of CORONA, ARGON and LANYARD records declassified on 26 November 1997. Chantilly: National Reconnaissance Office.
NRO. (2006). *Review and redaction guide for automatic declassification of 25-year-old information (Version 1.0)*. Chantilly: National Reconnaissance Office. http://www.nro.gov/foia/docs/foia-rrg.pdf. Accessed 15 May 2009.
Palmer, R. (1984). *Danebury: An aerial photographic interpretation of its environs*. London: Royal Commission on Historical Monuments (England).
Peebles, C. (1997). *The* corona *project: America's first spy satellites*. Annapolis: Naval Institute Press.
Richelson, J. T. (2002). *The Wizards of Langley: Inside the CIA's Directorate of Science and Technology*. Oxford: Westview Press.
Richelson, J. T. (2003). A 'rifle' in space. *Air Force Magazine, 86*, 72–75.
Scardozzi, G. (2010). The contribution of historical aerial and satellite photos to archaeological and geo-archaeological research: Case studies in Italy and Turkey. *Advances in Geosciences, 24*, 111–123.
Selander, J. M. (1997). Image coverage models for declassified corona, Argon and Lanyard satellite photography – a technical explanation. In R. A. McDonald (Ed.), *corona between the sun and the earth: The first NRO reconnaissance eye in space* (pp. 177–188). Bethesda: American Society for Photogrammetry.
Shennan, I., & Donoghue, D. N. M. (1992). Remote sensing in archaeological research. *Proceedings of the British Academy, 77*, 223–232.
Smith, F. D. (1997). The design and engineering of corona's optics. In R. A. McDonald (Ed.), *corona between the sun and the earth: The first NRO reconnaissance eye in space* (pp. 111–120). Bethesda: American Society for Photogrammetry.

Stichelbaut, B. (2006). The application of First World War aerial photography to archaeology: The Belgian images. *Antiquity, 80*, 161–172.

Surazakov, A., & Aizen, V. (2010). Positional accuracy evaluation of declassified Hexagon KH-9 mapping camera imagery. *Photogrammetric Engineering and Remote Sensing, 76*, 603–608.

Ur, J. (2005). Sennacherib's Northern Assyrian canals: New insights from satellite imagery and aerial photography. *Iraq, 67*, 317–345.

Ur, J. A. (2010). *Urbanism and cultural landscapes in Northeastern Syria: The Tell Hamoukar Survey, 1999–2001*. Chicago: The Oriental Institute of the University of Chicago.

USGS. (2006). *Declassified satellite imagery – 1 (1996)*. http://eros.usgs.gov/#/Find_Data/Products_and_Data_Available/Declassified_Satellite_Imagery_-_1. Accessed 19 June 2011.

Walstra, J., Verkindren, P., & Heyvaert, M. A. (Eds.). (2010). *Reconstructing landscape evolution in the Lower Khuzestan plain (SW Iran): Integrating imagery, historical and sedimentary archives*. Oxford: Oxbow Books.

Wilkinson, T. J., Wilkinson, E. B., Ur, J., & Altaweel, M. (2005). Landscape and settlement in the Neo-Assyrian Empire. *Bulletin of the American Schools of Oriental Research, 340*, 23–55.

Wilkinson, K. N., Beck, A. R., & Philip, G. (2006). Satellite imagery as a resource in the prospection for archaeological sites in central Syria. *Geoarchaeology, 21*, 735–750.

Wilson, D. R. (2000). *Air photo interpretation for archaeologists*. Stroud: Tempus.

Part III
Historical Aerial and Satellite Photographs in Archaeological Research

Chapter 5
First World War Aerial Photography and Medieval Landscapes: Moated Sites in Flanders

Birger Stichelbaut, Wim De Clercq, Davy Herremans, and Jean Bourgeois

Abstract During the First World War, millions of aerial photographs were taken by all fighting countries. Aerial photographs were taken extensively across the different theatres of war documenting a conflict landscape from which the relicts often remain visible as scars on the landscape. The aerial photographs which were taken during the conflict also provide an unparalleled record of the landscape at the beginning of the previous century. This chapter explores the potential of these historical aerial photographs to reveal previously unknown archaeological sites, highlighting some of the interpretational issues involved. A case study focuses on the landscapes and moated sites of the medieval period in Flanders (Belgium).

5.1 Introduction

The potential of military aviation for intelligence purposes was considered for the first time during the First World War. Shortly after the Battle of the Marne at the end of 1914, the concept of a moving war came to a standstill in the trenches. The traditional eyes of the army – the cavalry and espionage – failed to provide the necessary information and thus created an opening for aerial photography (Carlier 1921). Aerial photography recorded far more information than the aerial observers could see, providing photographs which were indisputable documents able to confirm the accounts of the aviators. As progress was made in the technical aspects of aerial photography,

B. Stichelbaut (✉) • W. De Clercq • D. Herremans • J. Bourgeois
Department of Archaeology, University of Ghent, Sint-Pietersnieuwstraat 35, B-9000 Ghent, Belgium
e-mail: birger.stichelbaut@ugent.be; W.DeClercq@UGent.be; Davy.Herremans@UGent.be; Jean.Bourgeois@UGent.be

W.S. Hanson and I.A. Oltean (eds.), *Archaeology from Historical Aerial and Satellite Archives*, DOI 10.1007/978-1-4614-4505-0_5,

so the art of reading aerial photographs also advanced (Wrigley 1932: 53). These extraordinary sources document a European landscape of horror that stretches from the North Sea in Belgium to the French-Swiss border, from the Black Sea in the south to the Baltic in the north, comprising parts of Italy, the Balkans and even more distant areas of the world. Aerial photographs were taken all over these different theatres of war, documenting a cultural landscape from which the relicts often remain visible as scars on the landscape, frequently concealed from the untrained eye.

Many archaeological papers often acknowledge the importance of the First World War in the development of aerial photography, yet without going into detail. The same relationship can be noticed between the historiography of the First World War in the air and the attention granted to the role of aerial photography (Streckfuss 2009). In addition to this, the discipline of historical aerial photography has rarely made use of aerial photographs dating from these early years. This contrasts sharply with the frequent use of Second World War aerial photographs in aerial archaeological practice (see for instance Going 2002; Raczkowski 2004; Hegarty and Newsome 2007).

Archival research focused on gaining insight into the dispersal and distribution of surviving collections of First World War aerial photographs revealed that large collections still remain accessible and are spread across Europe, the USA, and even Australia. A GIS plot of most of the collections has been carried out (Stichelbaut and Bourgeois 2009). This was not just to give an idea of the dispersal of these sources through Europe, but where possible, the intention was also to indicate blind spots and hot zones, to enable a realistic assessment of future research areas. The outcome was astonishing because the quantities of preserved aerial photographs are enormous and much more consolidated than expected. The most important collections are in the Belgian Royal Army Museum (48,500 photographs), the Imperial War Museum (133,000), the Australian War Memorial (16,000), the Bavarian War Archive (300,000) and the US National Archives (16,000).

This detailed archival knowledge enables us to select specific aerial photographs of large areas in France and Belgium, as well as some other theatres of war. It allows us to explore the use of the First World War aerial photographs, not only as extraordinary illustrative documents for specific sites, but rather as a unique source covering an entire research area. Thousands of historical aerial photographs covering large areas along the Western Front can be selected which have the potential to reveal previously unknown archaeological sites. First of all, the photographs were taken in the period before the large landscape changes which took place in Europe after the Second World War. The extent of villages and cities is still rather small, and the field systems often go back to older periods. In addition, the photographs represent the landscape before the major impact of the conflict. Large areas of the Western Front were completely devastated and transformed into a poisonous lunar landscape. On the early aerial photographs (dating from 1915 until the start of the artillery preparation for the Third Battle of Ypres in June 1917), many sites are still undamaged, whereas they are no longer visible on photographs taken in late 1917 and 1918. Finally, the photographs are special and of value because for large areas of Europe, they are the first complete aerial coverage in history and have, therefore, an important documentary value.

5.2 Conflict Archaeology

The aerial photographs are one of the sources par excellence for the study of a new and rapidly developing scientific field of research: the archaeology of the First World War. This conflict left its mark as scars on the landscape. Some of the features, such as mine craters and bunkers, are still visible on the current landscape, although the majority are preserved beneath the surface. Professional and scientific archaeological research on the material remains of the First World War can add a whole new layer of information which is currently hidden in the historical records, such as unrecorded behaviour and activities, the treatment of the dead (Desfossés et al. 2003; see also Chap. 6 by Pollard and Barton, this volume), the detailed typology of trenches and other features, daily life at the front, the analysis of the multi-layered conflict landscape and much more. The archaeology of the First World War, and twentieth-century conflict archaeology in general, is still in its infancy but is certainly a rapidly evolving field of research. A multidisciplinary aerial photographic study using modern techniques can provide more adequate information on such issues than any other source. The long-distant views provided by aerial photographs enable a redirection of the focus of First World War archaeology from a site-directed approach to research on a landscape scale. The aerial archaeological research on the Western Front using contemporary aerial photographs provides an accurate insight into the density, distribution and diversity of war features (Stichelbaut 2006, 2011). This approach facilitates analysis of how the conflict landscape was organised and where certain types of features were situated, providing a detailed level of information which cannot be found in any other historical source. The focus of this chapter, however, is on the application of First World War aerial photographs to traditional archaeology in Flanders (Belgium).

5.3 Cropmarks, Soilmarks and Dampmarks

If the First World War aerial photographs are studied in detail, cropmarks and soilmarks relating to archaeological sites can be discovered hidden in the fields. But because of the high altitude at which the reconnaissance normally took place – between 2 and 5 km – the recognition of these archaeological remnants in Flanders is not easy. Also, the fact that the pictures are panchromatic and not necessarily taken at the ideal time of year for cropmarks and soilmarks does not help in their interpretation and use for traditional archaeological purposes.

On the other hand, the continuous artillery fire destroyed the field boundaries and drainage ditches of the fields. This fact, combined with their deliberate inundation from some rivers (such as the IJzer and the Handzame) for strategic purposes, provided ideal circumstances for the detection of dampmarks, which become apparent because of the different levels of dampness in places where ditches or moats were once situated. Literally, hundreds of medieval moated sites can be detected, many of which were destroyed during the war and are now no longer visible.

It is sometimes difficult to make a distinction between soilmarks and cropmarks on panchromatic aerial photographs. The small scale of some of the aerial photographs and the lack of colours often renders it impossible to identify bare ground, as, for example, in Fig. 5.1a where a large moated site is visible. The outer moat can still be seen as an oval field boundary, inside of which a smaller island of approximately 35 by 45 m is visible as a soilmark. In Fig. 5.1b, some (positive) cropmarks of linear features can be noted, which may relate to an old system of field boundaries of unknown date or origin. Unfortunately, no other photographs of this area are available, and modern prospection flights are impossible because of the nearby location of a major airport at Ostend. Thus, areas inaccessible to modern oblique aerial photographic reconnaissance can be studied with remotely sensed data using historical aerial photographs. The disadvantage, on the other hand, is that there is no guarantee that a single historical aerial photograph would have been taken at a favourable moment.

The use of historical aerial photographs also presents a number of interpretational issues. Although some of the visible cropmarks have the characteristics of archaeological sites, they actually have their origins in the First World War. For example, at first sight, horse-riding rings leave the same traces (circular cropmarks or soilmarks) on the aerial photographs as well-known Bronze Age burial sites in Flanders.[1] Many aerial photographs show traces in fields which are circular or rectangular (with rounded corners). These are always lighter in colour than the surrounding area and indicate intense movement. The shape and size of these features allows us to interpret them as paddocks or horse-riding rings, which may indicate the presence of artillery troops with horse-drawn carts (de Bissy 1916: 2), rather than as Bronze Age mounds. Figure 5.1c is one such example of a dubious case. Its appearance could be interpreted as a cropmark of a possible circular archaeological feature. One argument in favour of this interpretation is the date of the exposure: 22 July 1917. In Flanders, this is traditionally a good period for the detection of archaeological sites as cropmarks on aerial photographs. But we have some reservations about its interpretation as a Bronze Age barrow because it has not been recorded on any other historical aerial photograph or modern oblique photograph. We consider it more likely to be a paddock which remained visible after a period of disuse.

Figure 5.1d shows the two different manifestations of a paddock and possible archaeological cropmark on the same aerial photograph. This photograph was also taken in the summer (July 1918). The outline of the white circle, which is clearly visible, is too prominent to be of archaeological origin and presumably is related to horse-riding activities. The second circle (see lower right inset) is less visible and could be a circular archaeological site. However, it could as well be a disused paddock. More research is necessary to disclose the true nature of this site, but this

[1] The phenomenon of circular features on the oblique aerial photography has already been studied at the Department of Archaeology of Ghent University. One of the main conclusions is that many of these are Bronze Age burial sites (see, for instance, Ampe et al. 1996; Bourgeois et al. 2002).

Fig. 5.1 Examples of soilmarks and cropmarks recorded on First World War aerial photographs (Source: Belgian Royal Army Museum): (**a**) soilmark, (**b**) cropmark recorded near Ostend, (**c**) Bronze Age circle or disused paddock, (**d**) First World War German paddock and probable Bronze Age site

example highlights one of the dangers of using historical aerial photographs. A thorough knowledge of what military features look like on the aerial photographs contributes to the interpretation of visible features.

5.4 Earthworks and Other Extant Sites

Many aerial photographs along the Western Front were taken in circumstances which were ideal for the detection of standing earthworks. Various inundations and the destruction of the drainage systems in the fields by intensive artillery fire created favourable wet conditions which resulted in many sites being recorded unintentionally on the aerial photographs. Even shallow depressions, which are invisible from the air in dry periods, can become visible when they are filled with water. This is particularly important for the study of landscapes and moated sites of the medieval period (see case study below).[2] These sites, which were detected because of wet conditions and the presence of water, are also sometimes referred to as 'watermarks'. The enigmatic circular feature in the centre of Fig. 5.2 is a splendid example of how historical photographs can contribute to the detection of these sites. The site, possibly dating from the medieval or post-medieval period, was unknown until the discovery of this picture. Additional research on the site is no longer possible as an industrial complex has been built on it without any prior archaeological investigation.

Medieval and post-medieval city walls and fortifications can be studied in many aerial photographs. Their detection is important in instances where they had already been destroyed before any archaeological investigation had taken place. The age of the First World War aerial photographs makes them a valuable record of the landscape at the start of the twentieth century. Fortifications and city walls can be recorded in often better conditions than can be done by using modern documents. In a number of cases, the sites have even totally disappeared as a result of city enlargements and modern agriculture.

A first site is clearly visible at Nieuwpoort, close to the North Sea. The city was founded in 1163 by the Count of Flanders as a strategically situated new town. It played an important role as strategic military stronghold during the Spanish-Dutch and French-Spanish Wars and was repeatedly fortified between the mid-sixteenth and early eighteenth century by Spanish, French and Austrian military engineers, resulting in an impressive complex of bastions, ravelins and hornworks surrounding the town. During the Dutch period (1815–1830), a new line of fortifications (the Wellington barrier) was built to protect the city against an imminent French invasion (de Vos et al. 2002: 164–165). The traces of a large bastion are visible on Fig. 5.3a, which illustrates the advantage of using historical aerial

[2] The use and potential of historical aerial photography (both First and Second World War) at high water level for the study of medieval settlement history has already been explored by F. Verhaeghe and J. Bourgeois in the 1970s and 1980s (e.g. Verhaeghe 1978, 1986; Bourgeois 2003).

Fig. 5.2 High-quality aerial photograph (24 March 1917) revealing a circular 'watermark' site (Source: Belgian Royal Army Museum)

photography because it enables the detection, mapping and documentation of sites which have now been destroyed or made inaccessible. There is now an industrial estate on top of this fortification, making any further research impossible. This is just one of many fortifications which are visible on the aerial photographs in the area being studied.

A second example is the fortification called Fort Nieuwendamme (3 km east of Nieuwpoort), also dating back to the Spanish-Dutch Wars and still clearly visible in the present landscape (Fig. 5.3b). However, looking at the historical aerial photograph, the organisation of the outer defences of the small fortress can be understood much more easily. When comparing these photographs with a modern oblique aerial photograph of the same site, the advantage of using the old aerial photographs becomes self-evident. On the 1916 (French) aerial photograph, the small Spanish fortress dating back to 1581 is clearly visible. A detailed outline of the fieldwork, constructed according to the so-called 'Old-Dutch system' of fortifications, is still present. The bastions, breastworks on the curtain and even traces of a ravelin can be seen. If, on the other hand, the photograph is compared with a modern aerial photograph taken in 2004, it is evident that the preservation of this fortification has changed dramatically. The moat has been backfilled and the ravelin levelled out. If the modern oblique aerial photograph is studied carefully, however, we can still identify some features, such as a soilmark in the upper left corner. The historical aerial photographs also explain why concrete bunkers are found on the ramparts, as

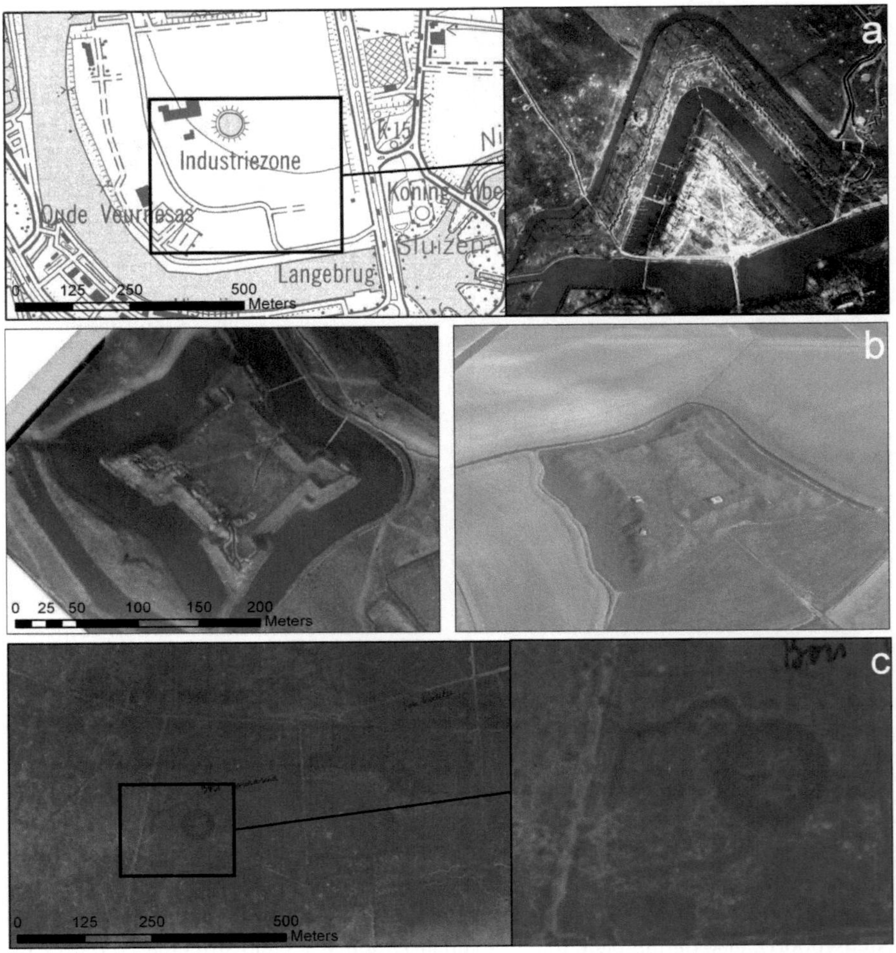

Fig. 5.3 (**a**) Aerial photograph of a large bastion at Nieuwpoort (September 1916) (Source: Belgian Royal Army Museum), (**b**) Fort Nieuwendamme comparing April 1916 and August 2005, (**c**) Hoge Mote at Merkem (late 1918) (Source: Belgian Royal Army Museum)

this fortification was incorporated into the German network of fieldworks during the First World War because of its elevation on the flat polder landscape.

Another category of earthworks which are easily visible are the medieval motte castles, such as the Hoge Mote situated at Merkem. This site was in use until the end of the thirteenth century/beginning of the fourteenth century (De Meulemeester and Termote 1981: 126). Some of the moats around this site are clearly visible. These features on the aerial photograph, however, represent only a part of the archaeological monument. The bailey was originally half as large and was also delimited by a moat, as can clearly be observed on the First World War aerial photographs (Fig. 5.3c).

Moated sites are more common on these photographs (Fig. 5.4).[3] Literally, hundreds are known. Some can be observed only on the aerial photographs, while others are still visible in the present landscape or in LIDAR height data. Such sites are quite a common feature in the medieval landscape. They have already been studied thoroughly in some parts of Flanders, as in the Veurne area (Verhaeghe 1978) or the Comines-Warneton area (Bourgeois 2003). They are rural sites that comprise at least one settlement area (or island) and a moat. These rural settlements are part of the medieval colonisation of large areas in Flanders and are generally dated between the late twelfth/early thirteenth century and the fifteenth century. They are quite often the seat of a medieval fief or even sub-fief, having seigniorial rights, but their sheer number in these areas tends to show that they were not solely limited to the higher levels of society.

However, it is necessary to remain critical when studying moated sites. For instance, one moat visible on the historical remote-sensing data need not imply that no additional moats will be revealed during archaeological excavations. This was clearly illustrated at the Jonkershof, excavated in 1981. On First World War aerial photographs, just one large moat can be seen. The excavation plan, on the other hand, shows a more complex twofold situation (Termote 1981: 133). In general, the early date of the photographs can certainly give us major advantages, but we have to be sure not to rely too much on this source. The aerial photographs enable previously unknown sites to be detected or sites to be observed in a better-preserved state than is currently possible. However, it is evident that they cannot provide us with a level of information which is comparable to excavation data. In order to try to understand more readily the potential of First World War aerial photographs for the medieval sites, especially moated sites, a case study based on recent PhD research (Stichelbaut 2009) is provided below.

5.5 Case Study: Medieval Moated Sites Along the Former Belgian-German Front

A large study area (443.5 km²) was selected in order to examine the archaeological application of First World War aerial photographs. This area comprises parts of 13 amalgamated municipalities (Diksmuide, Gistel, Hooglede, Houthulst, Ichtegem, Koekelare, Kortemark, Langemark-Poelkapelle, Lo-Reninge, Middelkerke, Nieuwpoort, Oostende and Staden) between the North Sea coast at Nieuwpoort and Houthulst in the south. The study area is situated on the Belgian coastal plain and consists of several distinct (agricultural) areas: dunes, polders and sands. The area studied is covered by 4,270 georectified aerial photographs taken between April 1915 and October 1918, all of which were obtained from the Belgian Royal Army Museum.

[3] For general information about the phenomenon of moated sites, see Aberg (1978).

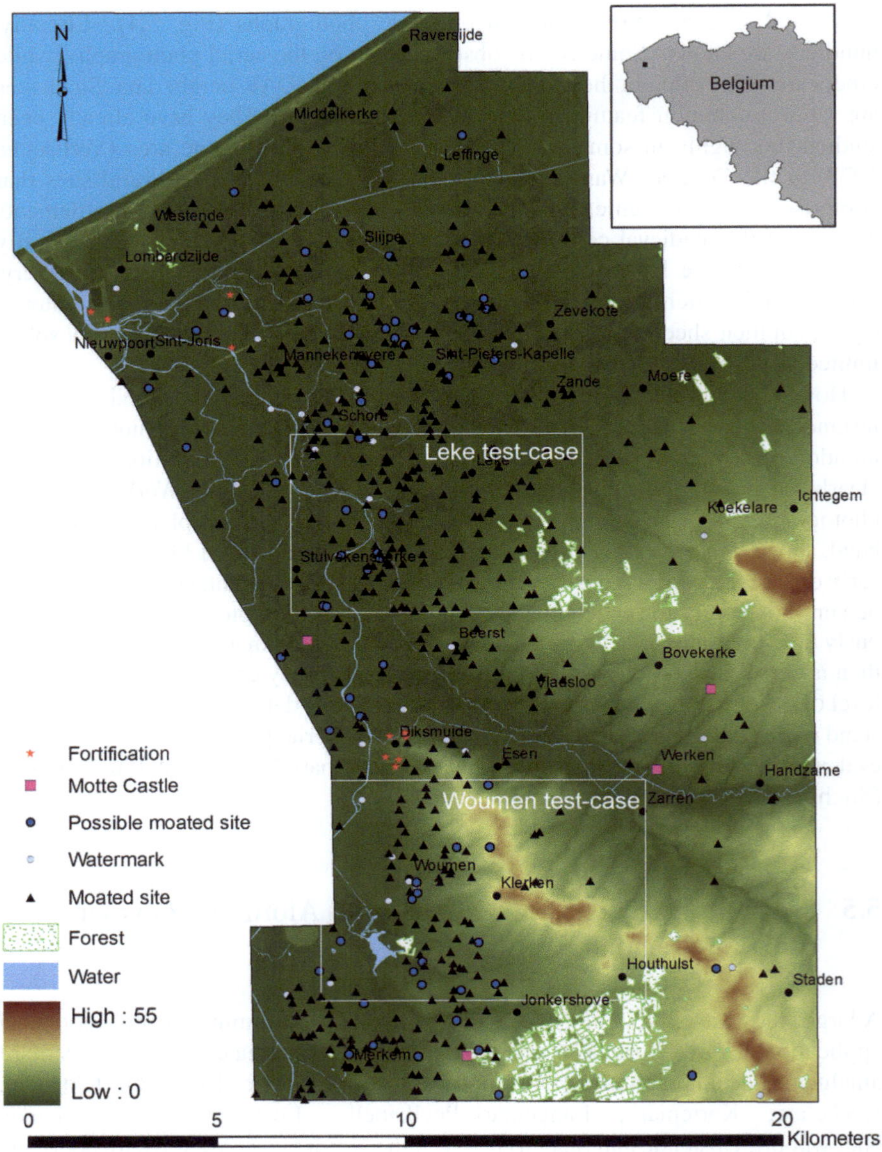

Fig. 5.4 Traditional archaeological sites mapped in the studied area along the former Belgian-German front (Source: authors and Vlaamse Landmaatschappij and OC GIS-Vlaanderen 2004)

5.5.1 Documented Sites

The examination of the First World War aerial photographs resulted in an inventory of 489 features identified as moated sites (Fig. 5.4). In addition, some 76 features were listed as possible moated sites, but because their interpretation is less certain, they are not considered further here.

The area of the case study is close to the focus of research undertaken by Frans Verhaeghe in the 1970s and 1980s (Verhaeghe 1981, 1986). Historically, the research area comprises parts of the castellany of Brugge (Brugse Vrije) and Veurne-Ambacht separated by the IJzer and a small part of the castellany of Ieper in the southern part of the research area. The relative empty zone in the south-eastern part of the area is related to the presence of Houthulst Forest, a remnant of the much larger *Vrijbos* which is mentioned as early as 1096 (de Meyer and Demeyere 2006: 119). Only eight archaeological sites (two prehistoric and six early and late medieval sites) were recorded near the forest during an inventory of archaeological sites in Houthulst. The medieval sites are probably related to the reclamation of the forest.

All sites detected have been checked on the Cabinet map of the Austrian Netherlands produced by Count de Ferraris (dated 1777) as an aid to their interpretation (Lemoine-Isabeau 1978). However, many of the sites were not visible on this historical map, a situation which is also observed in a case study in East-Flanders (De Mulder 2001, 2005). This is probably because the sites were not recognised as such, or had no military importance, since the primary use of the map was for military purposes (Beyaert et al. 2006: 11–12).

5.5.2 *Detailed Study: Leke and Woumen*

To further test the potential of First World War aerial photographs for detecting previously unknown moated sites, two smaller study areas have been researched in greater detail, concentrating on the areas around Leke and Woumen (Fig. 5.4).

The first, in the area north of Diksmuide, comprises the south-western sheet of the Gistel map sheet on the Cabinet map of the Austrian Netherlands. All sites detected on the aerial photographs were checked against the Cabinet map, the *Carte Topographique de la Belgique en couleurs* of 1873 and the modern Belgian 1:10,000 topographic maps, with quite interesting results. Some 122 moated sites, identified on the basis of their aerial photographic signature, were located (Fig. 5.5). Only 53 sites are still visible in the landscape as depicted on the modern Belgian topographic map. The Cabinet map of the Austrian Netherlands depicts 42 sites, and the topographic map of 1873 shows 58. Twenty-one sites are visible on just one of the cartographic sources consulted, 30 on two and 24 sites are visible on all three maps. The aerial photographs enable in this case a more precise localisation of the sites and are an important photographic record of their condition at the start of the last century. On the other hand, 47 sites (38%) detected on the aerial photographs are not visible on any of the maps (Fig. 5.5). It might be suggested that too many doubtful sites are mapped from the photographs. Yet, close examination of the aerial photographs shows the sites have the clear characteristics of moated sites and could easily fit into the typology developed by Frans Verhaeghe (1981). Figure 5.6 shows 12 such examples, all of which have clear characteristics of moated sites, but were not recorded as such on the cartographic data.

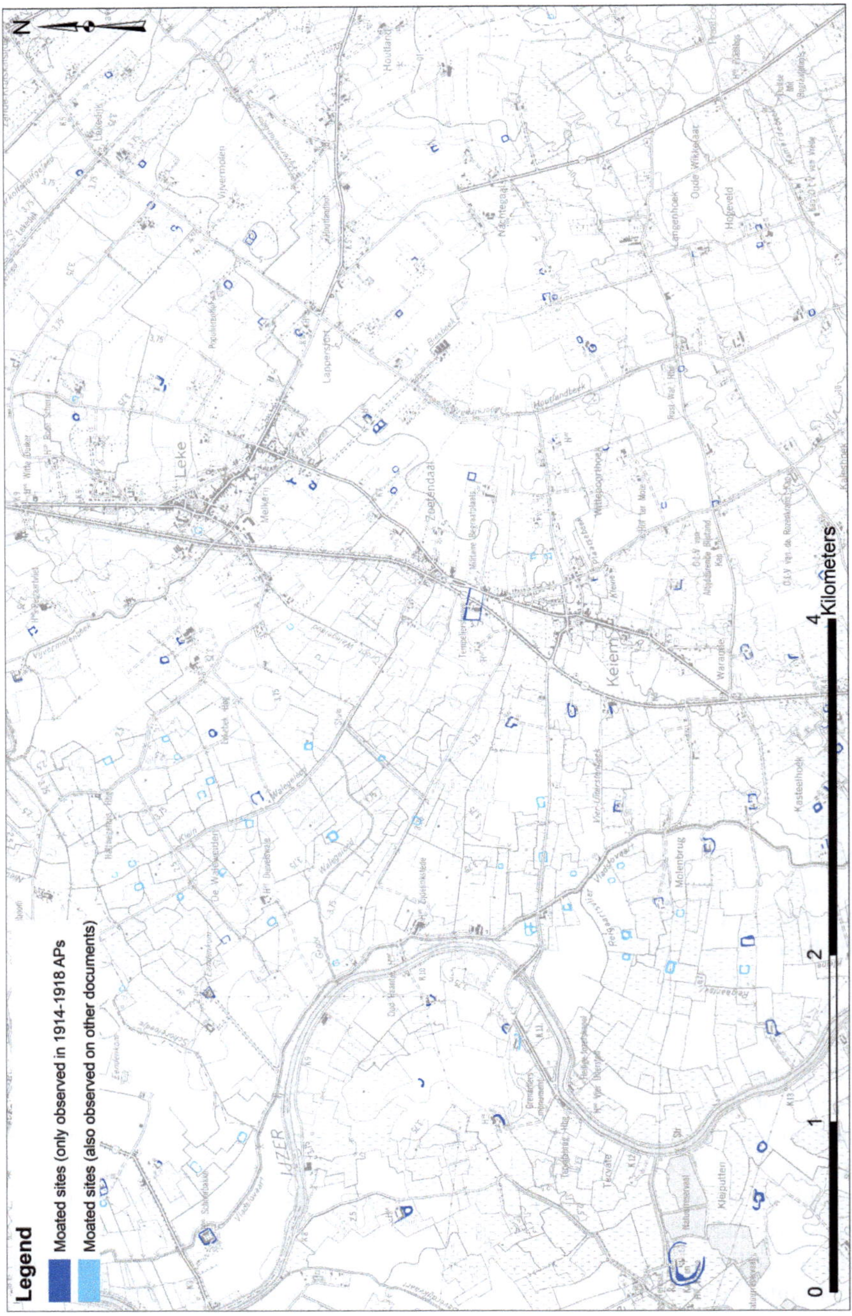

Fig. 5.5 Moated sites mapped from the aerial photographs

The same approach was undertaken for a second case study near Woumen (south of Diksmuide). All detected sites were checked against the same cartographic sources. In this test case, 83 sites were recorded as a moated sites, of which 12% are visible only on the aerial photographs. Though this number is significantly lower than in the Leke test case, there are also fewer sites observed in general.

These two small test cases illustrate the potential of these historical aerial photographs for the detection of previously unknown archaeological sites. Our hypothesis is that many of these sites are more easily detected on the First World War aerial photographs as a result of the special circumstances at that time. The strategic inundation of the landscape and the artillery fire destroying drainage systems created the wet conditions ideal for the detection of the sites as watermarks (earthworks). A further benefit is the age of the photographs. Because they are almost a century old, they have also the potential to show a number of sites which have since been destroyed (Fig. 5.6). This case study highlights only the potential for the detection of new sites; a more in-depth study could perhaps reveal other mechanisms. Additional fieldwork (for instance, augering, field walking or geophysical survey techniques) should be used to confirm if all the recorded examples are correctly identified as moated sites, as this was not possible within the framework of this piece of research.

5.5.3 Other Archaeological Sites Within the Case Study Area

In addition to rural sites, those with a military function have also been recorded. Near Nieuwpoort and Diksmuide, the remnants of ramparts and bastions of different periods are documented on the aerial photographs. In addition, Fort Nieuwendamme (east of Lombardsijde and Nieuwpoort) and a small fortification just south of it have been mapped using aerial photographs. This latter site is identified as a fieldwork and on several historical maps is indicated as a fortress.[4]

Four sites date to the High Middle Ages and are interpreted as motte castles: the Hoge Andjoen Motte in Werken (De Meulemeester and Vanthournhout 1986); the Hoge Mote in Merkem (De Meulemeester and Vanthournhout 1986); the motte castle at Oud-Stuivekenskerke (Dewilde et al. 2003); and the Vrouwenhillewal site near Bovekerke. Based solely on aerial photographic data, it is difficult to discern between some moated sites and motte castles because both types can have a twofold structure and are (partially) surrounded by a moat.

Finally, 24 sites are listed as watermarks, presumably identifiable as archaeological sites of various types. The aerial photographic interpretation enables us to detect and map these features, but cannot reveal further details without additional fieldwork or archival research, so they are merely mentioned here.

[4] Cabinet map of the Austrian Netherlands by Count de Ferraris (1777); Map of Chanlaire-Capitaine (end eighteenth century).

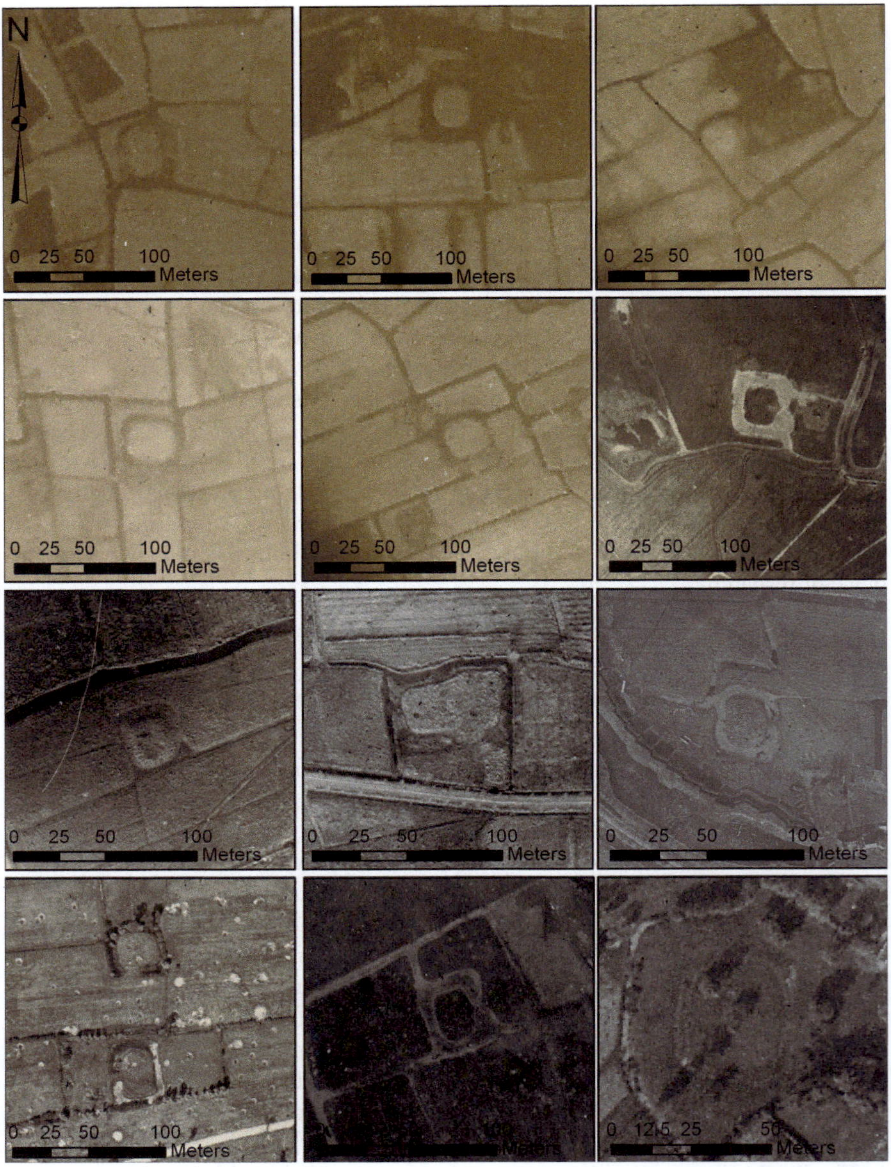

Fig. 5.6 Examples of sites which were detected only on the First World War aerial photographs (Source: Belgian Royal Army Museum)

The detection of cropmarks on the historical aerial photographs is problematic. Partly, this is due to the small scale of some of the aerial photographs; but the soil conditions in the area (silt loam and polders) are far from ideal for the generation of cropmarks. An additional problem is that the aerial photographs were not taken for archaeological purposes, and the probability of detecting archaeological sites in this way depends much 'on whether or not the imagery was acquired at a propitious time of day and year' (Scollar et al. 1990: 26). Therefore, we must be very careful when it comes to interpreting crop- and soilmarks. As regards the coastal area in Flanders, these aerial photographs will not be the most suitable source for studying pre- or early historic sites and landscapes, as the heavy soils (polders with clay and the silt loam region) create difficult cropmark conditions. This phenomenon has already been documented with oblique aerial prospection of Bronze Age circular sites in Flanders (Bourgeois and Cherretté 2005: 50).

A superficial comparison with an inventory of archaeological sites in Koekelare, based on the collection of oblique aerial photographs of the Department of Archaeology (UGent) by Marc Meganck, shows that there are plenty of archaeological sites in the area. These oblique aerial photographs were specifically taken for archaeological purposes. At least 42 Bronze Age burial mounds and many other archaeological traces are recorded within the boundaries of the community of Koekelare (see also Bourgeois et al. 1995), but it was not possible to detect these sites on the First World War aerial photographs. On the other hand, some of the medieval moated sites were not recorded on the oblique aerial coverage. In these cases, the historical aerial photographs provide an additional source of information. It is also important to note that almost no traces of the conflict landscape of the First World War are documented on oblique coverage, with the exception of some bunkers. This points to the complementary nature of the two different sources. Nevertheless, we strongly believe that there is a potential for this kind of research in other areas of the front where soil conditions are more favourable for the detection of cropmarks.

5.6 Conclusion

Aerial photographs dating to the First World War are available in large numbers for vast amounts of the Western Front in Belgium and France and in more limited numbers for other theatres of war. Based on the knowledge of these archives, it is possible to select a large number of historical aerial photographs. These are a major contemporary source for the study of the material remains and landscape of the First World War. Yet, the photographs are much more than this, for they are also a valuable record of the high- and post-medieval landscape along the former front line. The pictures were taken in large numbers at different times of the year and cover huge areas. They are a unique and privileged source for the detection of earthworks and watermarks because of the special circumstances that were created by the ongoing war. The case study of Leke and Woumen indicated that the pictures have huge potential for the detection of a large number of earthworks and watermarks. On the other hand, they suffer from the disadvantage – like other historical aerial photographs – that they were not taken for archaeological purposes. This is especially the case for recording cropmarks. In addition to the special circumstances in which they were taken, the pictures have the further advantage that they represent the landscape at the beginning of the twentieth century, so that archaeological sites which have been destroyed during the major landscape changes since the Second World War might still be detected or recorded in better conditions. These pictures can therefore be regarded as a new and largely forgotten source of information for archaeologists.

Bibliography

Aberg, F. A. (Ed.). (1978). *Medieval moated sites* (CBA-Research Reports 17). London: CBA.

Ampe, C., Bourgeois, J., Crombé, P., Fockedey, L., Langohr, R., Meganck, M., Semey, J., Van Strydonck, M., & Verlaeckt, K. (1996). The circular view. Aerial photography and the discovery of Bronze Age funerary monuments in East- and West-Flanders (Belgium). *Germania, 74*(1), 45–94.

Beyaert, M., Antrop, M., De Maeyer, P., Vandermotten, C., Billen, C., Decroly, J.-M., Neuray, C., Ongena, T., Queriat, S., Van den Steen, I., & Wayens, B. (2006). *België in Kaart: de Evolutie van het Landschap in Drie eeuwen Cartografie*. Tielt: Lannoo.

Bourgeois, J. (2003). Les sites fossoyés médiévaux de la région de Comines-Warneton (Province de Hainaut, Belgique). *Revue du Nord, 353*, 141–159.

Bourgeois, J., & Cherretté, B. (2005). L'Age du Bronze et le Premier Age du Fer dans les Flandres occidentale et orientale (Belgique): un état de la question. In J. Bourgeois & M. Talon (Eds.), *L'Age du Bronze du Nord de la France dans son contexte européen* (pp. 43–81). Paris: Éditions du Comité des Travaux Historiques et Scientifiques.

Bourgeois, J., Meganck, M., & Semey, J. (1995). *Cirkels in het land II: Een Inventaris van cirkelvormige structuren in de provincies Oost- en West-Vlaanderen*. Gent: Archeologische Inventaris Vlaanderen.

Bourgeois, J., Roovers, I., Meganck, M., Semey, J., Pelegrin, R., & Lodewijckx, M. (2002). Flemish aerial archaeology in the last 20 years: Past and future perspectives. In R. Bewley & W. Raczkowski (Eds.), *Aerial archaeology. Developing future practice* (NATO science series, Vol. 337, pp. 76–83). Amsterdam: Ios Press.

Carlier, A. (1921). *La photographie aérienne pendant la guerre*. Paris: Librairie Delagrave.

de Bissy, J. (1916). *Illustrations to Accompany Captain de Bissy's notes regarding the interpretation of aeroplane photographs*. London: Harrison & Sons.

De Meulemeester, J., & Termote, J. (1981). De Hoge Mote te Merkem. *Archaeologia Belgica, 247*, 125–129.

De Meulemeester, J., & Vanthournhout, C. (1986). De Hoge Andjoen-motte te Werken (W-Vl). *Archaeologia Mediaevalis, 9*, 3–15.

de Meyer, M., & Demeyere, F. (2006). De inventarisatie van de gemeente Houthulst (prov. West-Vlaanderen). In E. Meylemans (Ed.), *CAI-1: De Opbouw van een archeologisch beleidsinstrument* (pp. 43–74). Brussels: VIOE.

De Mulder, G. (2001). Relicten uit het Verleden. Een eerste verkennend onderzoek naar sites met walgracht te Zulte, Olsene en Machelen. *Bijdragen tot de Geschiedenis en de Folklore van Machelen, Olsene en Zulte, 20*, 74–85.

De Mulder, G. (2005). Sites met walgracht te Machelen. *Bijdragen tot de Geschiedenis en de Folklore van Machelen, Olsene en Zulte, 20*, 74–85.

de Vos, L., Gils, R., & Verbist, R. (2002). Inleiding versterkingen in de Nieuwste Tijd. In L. de Vos (Ed.), *Burchten en Forten en Andere Versterkingen in Vlaanderen* (pp. 157–170). Leuven: Davidsfonds.

Desfossés, Y., Alain, J., & Gilles, P. (2003). Arras "Actiparc", les oubliés du "Point du Jour". *Sucellus, 54*, 84–100.

Dewilde, M., Vanhoutte, S., & Wyffels, F. (2003). De Motte van Oud-Stuivekenskerke. *Archaeologia Mediaevalis, 26*, 36–37.

Going, C. (2002). A neglected asset. German aerial photography of the Second World War period. In R. Bewley & W. Raczkowski (Eds.), *Aerial archaeology. Developing future practice* (NATO science series, Vol. 337, pp. 23–30). Amsterdam: Ios Press.

Hegarty, C., & Newsome, S. (2007). *Suffolk's defended shore*. Suffolk: English Heritage.

Lemoine-Isabeau, C. (1978). *La cartographie au XVIIIe siècle et l'œuvre du comte de Ferraris (1726–1814)*. Brussels: Crédit communal de Belgique.

Raczkowski, W. (2004). Dusty treasure: Thoughts on a visit to The Aerial Reconnaissance Archives at Keele University (UK). *AARG News, 29*, 9–11.

Scollar, I., Tabbagh, A., Hesse, A., & Herzog, I. (1990). *Archaeological prospecting and remote sensing*. Cambridge: Cambridge University Press.

Stichelbaut, B. (2006). The application of First World War aerial photography to archaeology: The Belgian images. *Antiquity, 80*(307), 161–172.

Stichelbaut, B. (2009). *World War One aerial photography: An archaeological perspective*. Unpublished PhD dissertation, Ghent University.

Stichelbaut, B. (2011). The first thirty kilometres of the Western Front 1914–1918: An aerial archaeological approach with historical remote sensing data. *Archaeological Prospection, 18*(1), 57–66.

Stichelbaut, B., & Bourgeois, J. (2009). The overlooked aerial imagery of World War One: A unique source for conflict and landscape archaeology. *Photogrammetrie – Fernerkundung – Geoinformation, 3*, 231–240.

Streckfuss, J. (2009). Why the air war really mattered: A guide to understanding the significance of aviation in World War I. In B. Stichelbaut, J. Bourgeois, N. Saunders, & P. Chielens (Eds.), *Images of conflict: Military aerial photography and archaeology* (pp. 41–54). Newcastle-upon-Tyne: Cambridge Scholars Publishing.

Termote, J. (1981). Het Jonkershof te Jonkershove. *Archaeologia Belgica, 247*, 130–134.

Verhaeghe, F. (1978). De Laat-Middeleeuwse bewoning te Lampernisse en omgeving: het archeologisch onderzoek. *Westvlaamse Archaeologica, 5*, 30–64.

Verhaeghe, F. (1981). Medieval moated sites in coastal Flanders. In F. A. Aberg & A. Brown (Eds.), *Medieval moated sites in North-Western Europe* (BAR international series, Vol. 121, pp. 127–171). Oxford: Archaeopress.

Verhaeghe, F. (1986). Les sites fossoyés du Moyen Age en Basse et Moyenne Belgique: état de la question. In M. Bur (Ed.), *La Maison Forte au Moyen Age* (pp. 55–86). Paris: CNRS.

Vlaamse Landmaatschappij and OC GIS-Vlaanderen. (2004). *DHM-Vlaanderen (2001–2004): Digitaal Hoogtemodel Vlaanderen (Cd-rom)*. Brussels: Vlaamse Landmaatschappij and OC GIS-Vlaanderen.

Wrigley, H. (1932). *The battle below. Being the history of No. 3 Squadron A.F.C.* Sydney: Errol Know.

Chapter 6
The Use of First World War Aerial Photographs by Archaeologists: A Case Study from Fromelles, Northern France

Tony Pollard and Peter Barton

Abstract This chapter considers the use of aerial reconnaissance photographs taken on the Western Front during the First World War as a source of information for the growing number of archaeologists carrying out archaeological investigations on sites of conflict from this period. Hundreds of thousands of aerial photographs were taken by both sides during the war, a process which in itself provided the engine for the development of aerial warfare, because of the need to provide protection for military reconnaissance. One of the largest collections of Allied photographs resides in the Imperial War Museum in London which provided the examples used in this study.

These photographs relate to the suspected burial of Australian and British soldiers behind German lines following the Battle of Fromelles, in northern France, in July 1916. They show a series of pits adjacent to Pheasant Wood, to the north of the village of Fromelles, and in doing so provided a vital piece of evidence in the search for the graves in 2007 and 2008.

The chapter provides an introductory overview of the development of wartime aerial photography and, taking Fromelles as a case study, considers the role of aerial photographs, alongside other forms of evidence, in a programme of archaeological works which succeeded in the location and evaluation of mass grave pits which had lain unmarked since 1916. The result of this work was the recovery of 250 sets of human remains which have since been buried individually in a specially created Commonwealth War Graves Commission cemetery close to the original site.

T. Pollard (✉)
Centre for Battlefield Archaeology, University of Glasgow, Glasgow G12 8QQ, UK
e-mail: tony.pollard@glasgow.ac.uk

P. Barton
The Tunnellers Memorial Fund, 8 Egbert Road, Faversham, Kent ME13 8SJ, UK
e-mail: pb@parapet.demon.co.uk

W.S. Hanson and I.A. Oltean (eds.), *Archaeology from Historical Aerial and Satellite Archives*, DOI 10.1007/978-1-4614-4505-0_6,

> Photography has become a household word and a household want; is used alike by art and science, by love, business, and justice; is found in the most sumptuous salon, and in the dingiest attic – in the solitude of the Highland cottage, and in the glare of the London gin-palace, in the pocket of the detective, in the cell of the convict, in the folio of the painter and architect, among the papers and patterns of the millowner and manufacturer, and on the cold grey breast of the battlefield.
>
> (Eastlake 1857)

6.1 Introduction

The archaeological investigation of sites associated with the First World War, especially those on the Western Front of Belgium and France, has begun to make an important contribution to the burgeoning field of conflict archaeology. One reason perhaps that archaeology has come to play a major role in the study of the First World War is that veterans of the conflict which had such a profound impact on the political and social fabric of Europe and the world beyond are now an extinct species. Harry Patch, the last man alive to have fought in the trenches of the Western Front, and latterly known as the 'Last Tommy', died on 25 July 2009, aged 111. Making its own contribution to this field has been the work of the Centre for Battlefield Archaeology at the University of Glasgow. This chapter will discuss the Centre's investigation into mass graves at Pheasant Wood, in Northern France, with particular reference to the important role which archival aerial photographs played in that work.

6.2 The Evolution of First World War Battlefield Photography

Despite the capture of tens of millions of images, the First World War and its photographers feature surprisingly little in photographic history. The focus has settled on a relatively small band of world-renowned practitioners, such as Roger Fenton, Matthew Brady, Felice Beato, Margaret Bourke-White, Robert Capa and Bert Hardy, who together with others help shaped the iconography of warfare in the nineteenth and twentieth centuries. The First World War was overlooked. Yet it too had its highly distinguished practitioners, and their often entirely unattributed images are reproduced year on year as the flow of books on this world-changing conflict continues unabated. Today, nine decades later, the work of 'official' photographers immeasurably helps to mould our impressions of trench life. Indeed, without photographs, we would struggle to form any kind of mental image of those troubled times. Amongst the most anonymous of all the practitioners are those whose sole mission was to provide military intelligence and reconnaissance. Their photographs, often underused and underappreciated, are some of the most outstanding to emerge from the war.

When we look at 1914–1918 archive photographs, we are confident in believing what we see. Because the cameras of the early twentieth century are generally supposed not to 'lie', as they can be made to do so easily today, pictures do more than simply provide an instant and accessible communication with a difficult subject. This is quite often a fallacy, for both film and still photographs were frequently posed and/or, indeed, faked for propaganda purposes. Nevertheless, it is through them that we construe the ambience and character of the physical environment of the First World War – the battlegrounds.

Amalgamating all the photographic and document-based material from archives around the world relating to the First World War would produce an immeasurably vast collection. The interlocking archival jigsaw might appear to be as complete as one could possibly need to build an accurate picture. However, fundamental – and bluntly obvious – gaps in the puzzle exist. Apart from aerial photographs – an unearthly view – there were no representations of the full landscape of the various battlefields, ones to which we as latter-day observers might easily relate. Without these final pieces, our perceptions would always remain flawed and imperfect. Recent years have seen researchers, the second named author among them, excluding the personal to delve into areas that might once have been looked upon as academic, even peripheral: an appreciation of the landscape itself through the study of maps, aerial photographs and panoramic images, each critically integral to the activities of men on the ground, but in the eyes of many somehow slightly divorced. These images have become of great historical and archaeological importance.

Although the full potential of aerial reconnaissance was fully realised during the First World War, its origins go back further than aerial combat, with the use of observation balloons dating to the middle of the nineteenth century, when they were deployed by both sides in the American Civil War. The first photograph from an aeroplane was taken by Wilbur Wright in 1909, near Rome (Heiman 1969: 34). Credit for the first use of aircraft-based aerial photography in warfare goes to the Italians, who were also the first to drop a bomb from an aircraft, in what is now Libya in 1911 (Lundqvist 2002). Aerial reconnaissance gave the Italians prior warning of Turkish troop concentrations prior to the battles of Sciarra-Sciatt and Ain Zara, and Italian victories established beyond doubt the great potential of this new technique.

At the same time, the British were also making their first tentative steps; 1911 saw the first use of aircraft to engage in military training manoeuvres, following the establishment of an Air Battalion of the Royal Engineers that same year. The French and Germans were both quicker to build up air forces and to train their pilots in the use of aircraft in reconnaissance, taking aerial photographs and directing artillery fire. It was, however, the acquisition of overlapping vertical photographs by a British aviator, Sergeant A.V. Laws, in 1912 (taken from a Royal Navy airship) which marked the first real technological advance in aerial photography (McMaster University Libraries 2008). This technique allowed photographs to be overlapped, removing the problem of lens distortion at the edges and facilitating the creation of photo mosaics of extensive areas which provided a basis for accurate maps. It was also Laws who took an aerial photograph of a parade just as a sergeant major was chasing a dog away from the scene. Once the photograph was developed, Laws realised that he could see the footprints of the sergeant major left behind in the grass, an observation which strikes a chord with the present authors and their own interpretation of First World War aerial photographs (see below).

By September 1914, military photographers were already being asked for quality images of the landscape. Despite pre-war developments, systematic aerial photography was still not being deployed within the combat zone. Instead, commanders relied upon the word of observers and pilots, whilst military mapping on the embryonic Western Front was nascent and struggling. However, panoramic sketching and photography was an established medium for reconnaissance, and several photographic panoramas of the early battlefronts, such as the Aisne Valley, were produced (Barton 2005: 64). Such photographs were to become an integral part of the routine of trench warfare, allowing company commanders to reconnoitre the enemy terrain opposite their own trenches without risking death from a sniper's bullet as they poked their head above the parapet. Photographs from the air were also taken at this time, and indeed in the same place, the Aisne Valley, but not officially. Five images captured by Lt. G. Prettyman of 3 Squadron RFC on 15 September 1914 were intended only as a private record. Not long after, the onset of trench warfare was to provide the ideal environment within which aerial photography could prove its worth, and the first Air Photograph Section was formed in January 1915.

The first deliberate use of aerial photography in conjunction with wartime military manoeuvres was in preparation for the battle of Neuve Chapelle in March 1915 (Barton 2005: 212). At this time, the purpose was to assist in the production of accurate maps. Later, as the war expanded and deepened, but remained static in mutual siege, so the applications for such valuable intelligence grew, not least (in the same way as panoramas) because the picture and not the observer could be relied upon to tell the finer truth. By the end of 1915, both the quality of the images and the skills of the scalers and interpreters had already become remarkable. If the weather was favourable, photographs would be taken. It was reconnaissance, not dogfights, that occupied the vast majority of aircrews; indeed, it was only as a result of the desperate need for photographs that 'scout' aircraft – 'fighters' in Second World War parlance – became necessary and numerous. The primary task of the great aces of the war was to protect photo-reconnaissance aircraft from hostile molestation.

Aerial photography was undoubtedly the most valuable intelligence tool available to the Army during the war; indeed, it was indispensable in helping to precisely plot changing features on trench maps and fresh developments in defensive schemes. Aerials were particularly valuable when planning a raid or a more substantial attack that required the temporary capture of enemy support or reserve positions – positions that were easily distinguished from the air and could be plotted on maps for artillery (this being the major use to which aerial photographs were put).

But this intelligence gathering also allowed for the regular monitoring of the enemy. Noticeable concentrations of troops could provide early warning of enemy attack; fresh digging may provide evidence for the preparation of artillery positions and mines and so on. At times photographs were taken during battle, the resulting images showing troops advancing across no man's land and artillery shells bursting.

By the end of the war, millions of photographs had been taken, and today, these can be found in a variety of archives, including those in the Imperial War Museum London, which are the subject of this chapter. Many of these can be read in conjunction with the thousands of trench maps which were drawn up throughout the war, with the relevant map sheet sometimes appearing as an annotation in the corner of the photograph.

6.3 The Resource

The greatest and therefore most useful British collections of both aerial photographs and panoramas lie in the archives of the Imperial War Museum in London. Here, amongst a total of approximately 6,000,000 images, one may examine over 40,000 official First World War photographs but also the original negatives of almost 90,000 aerials in what is called the 'box collection'. In the same building, there are around 2,000 panoramas taken from the ground. It is from the box collection that aerial prints are ordered, the technicians thus going back to the original glass-plate negatives in order to achieve best quality. The Imperial War Museum also has a further vast collection of aerial prints, so many that the number has not yet been ascertained. These are housed at the Duxford facility and are yet to be catalogued. However, in the photograph archive in the annexe to the main Lambeth Road buildings, a large selection of useful prints (10× 8 in.) organised according to trench map sheet are available to view.

Aerial photographs exist in two basic categories: vertical photographs taken perpendicular to the landscape's surface and oblique images captured at a 'diagonal' (and varying angle) to the ground. As obliques are effectively a panorama taken from a height, certain features can shield others, so their utility from an archaeological viewpoint is lessened. In verticals, however, even minute alterations in earthworks can be discerned. Ninety percent of the Imperial War Museum collections are verticals.

Trench warfare prescribed that the time available for a relatively safe investigation of no man's land and enemy defences was restricted to the hours of darkness, a situation that clearly had its drawbacks in gathering data for maps and plans that demanded the finest detail: information such as grass length, depths of depressions and width of streams. The ubiquitous trench periscopes and loopholes, although widely used for observation, were less than satisfactory for photo-topographical work; nonetheless, the panoramas achieved are of astonishing quality. For the archaeologist, these panoramas offer perspective, placements, heights, depths, densities, dimensions, construction method and material, quantities employed, battlefield detritus and the incorporation of the structures in question into other man-made and natural features: something neither maps nor aerial photographs can do.

Place aerial photographs and panoramas together, and one has the most accurate multidimensional image of the landscape that it is possible to achieve. To the battlefield archaeologist, they are indispensable.

As the interest in the archaeology of the Western Front has grown within the evolving sub-discipline of conflict archaeology (Pollard 2008), so the potential of the available aerial photographic archives, including that held in the Imperial War Museum but also major collections in Germany, Belgium, Australia and the USA, has been recognised (e.g. Stichelbaut 2005 and Chap. 5 by Stichelbaut et al.). Along with trench maps and panoramic photographs, aerial photographs have the potential to add an incredible amount of interpretative detail to archaeological research.

6.4 Mass Graves at Pheasant Wood, Fromelles, France

In 2007 and 2008, the survey and evaluation of pits suspected to contain the bodies of British and Australian soldiers killed in the 1916 Battle of Fromelles was carried out by GUARD and the Centre for Battlefield Archaeology (Fig. 6.1). The grave pits were dug by German troops behind their front lines for the burial of enemy dead removed from the German trenches and from the no man's land in front of them and the ground behind. The project was funded by the Australian Army in response to repeated calls for an examination of the ground in the hope of locating the unmarked graves, largely on the part of a retired school teacher called Lambis Englezos. The project involved detailed historical research by the second named author, which included an examination of Bavarian military archives in Munich (Barton 2007).

The Battle of the Somme had been going badly for the British for almost 3 weeks when, 50 km to the north, the attacks at Fromelles were launched on the evening of 19 July 1916. The two confrontations, one vast, the other tiny by comparison, each epitomise the mass slaughter for which the First World War is remembered. Indeed, they are conjoined, for the Battle of Fromelles was an attempt by the allies to prevent German reinforcement of their Somme garrisons. The attack involved the British 61st Division and the Australian 5th Division, the latter having only recently arrived on the Western Front. The opposing lines were held by experienced troops, the 6 Bavarian Reserve Division; among their ranks was a young corporal by the name of Adolf Hitler.

Despite a preparatory bombardment by massed artillery, the German defences remained largely intact. Protection was effectively enhanced by a host of concealed concrete bunkers and underground dugouts (the presence of which does not seem to have been recognised, even with the aid of aerial reconnaissance). Although gains were made, they were meagre and disappointing. In places, men made it into the German front line and beyond, but Allied losses in no man's land were dreadful. The attacks took place at 6.00 pm in broad daylight. The Germans were ready, their machine guns and artillery mowing down the troops as they stepped into no man's land. Some reached the German trenches and began a fruitless search for their objective, the second enemy line. But it was not to be found, for it existed only as water-filled ditches and old abandoned trenches. A further attack was ordered for 9.00 pm; the order was then cancelled, but tragically, the news did not reach the Australian 15th Brigade which jumped off and again suffered shocking losses.

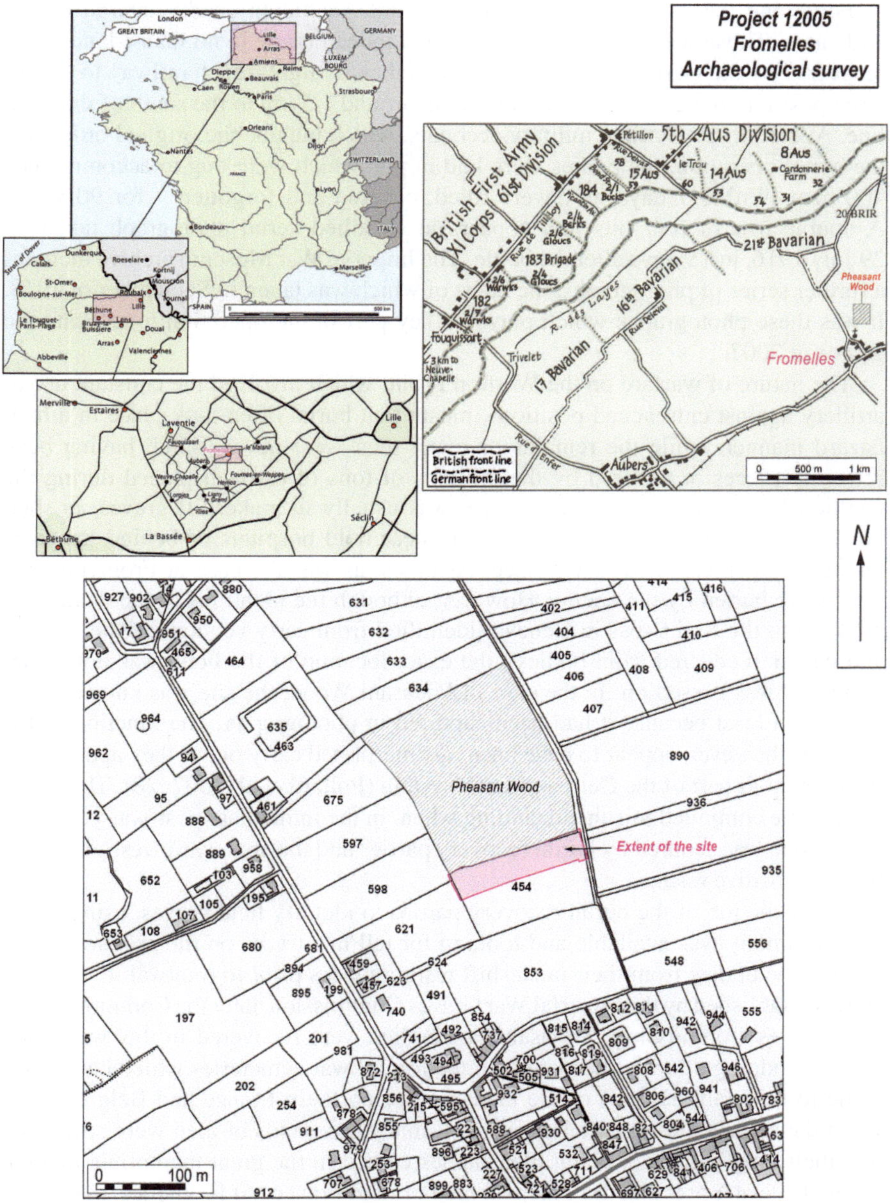

Fig. 6.1 Site location maps

In the early morning of 20 July, the Germans closed the noose, trapping all those Allied troops that had entered their lines. In the space of 15 hours, the Australians had lost around 5,500 killed, wounded, missing or captured and the British around 1,500. Total German casualties were less than 1,000.

Following the battle, there were British and Australian dead scattered over a wide area. Those within the enemy lines, and indeed many in no man's land, were recovered by the Germans. Some were then taken by light trench railway to a sheltered position to the rear of Pheasant Wood, around 1.5 km to the south of the front line. According to German military accounts, which include the original orders for the burial operation, the bodies were laid in pits which were dug to accommodate 400 men. Within 7 days they were filled, covered and forgotten – for 90 years. A double row of four pits first appears in an allied aerial photograph taken on 29 July 1916, just 9 days after the battle. The Imperial War Museum archive includes a further series of photographs, the latest of which was taken in September of 1918. It was these photographs which played a key part in the operation to relocate the graves in 2007.

The nature of warfare on the Western Front, which involved the constant use of artillery against entrenched positions, meant that burial often took place in a haphazard manner, while the remains of many men were never found, having been blown to pieces or covered by the millions of tons of earth displaced during the conflict. When burial did take place, it was usually in makeshift graves, in shell holes or abandoned trenches, in pits dug near field hospitals or behind the front where engagements had taken place. At times, as was the case at Fromelles, the dead were buried by the enemy. However, although the identities of the dead were reported to the Red Cross in Geneva (identified from early versions of the dog tag and papers recovered from bodies), the exact location of the body was not something that was passed on. In the case of Pheasant Wood, the site was known to the Allies, not least because it had been captured in photographs. The function of the pits does, however, appear to have been misinterpreted early on, as they appear on a trench map as part of the German trench system (Pollard et al. 2007: 28). There was to be some continued misunderstanding when, in the immediate post-war years, the battle zone was scoured by burial recovery parties and the area was investigated, but with no positive result.

It was the job of the burial recovery parties to identify field graves, using whatever testimony was available and looking for tell tale traces on the ground, and to recover the bodies from their makeshift resting places prior to removal to the cemeteries established by the Imperial War Graves Commission, later the Commonwealth War Graves Commission. Thousands of bodies were recovered in this way, each recovery adding a further headstone in the formal war cemeteries which have since come to be a familiar part of the landscape of northern France and Belgium. But when the searches were curtailed in 1921, many thousands of men were still missing, their only memorial being their names etched in the great memorials like the Menin Gate at Ypres and the Thiepval memorial in France. At Fromelles, the names of the missing were added to a memorial wall at the back of the cemetery at VC corner, which sits on the battlefield. This is no ordinary cemetery as it contains the bodies of 410 Allied soldiers (most thought to be Australian) buried in multiple, 10-man graves. The names of a further 1,200 men, all of them Australians, appear on the wall, and in addition to these, several hundred British soldiers killed in the battle have no known grave.

6.5 The Search for the Graves

Despite desperate pleas from relatives wanting to know where their husbands or sons had been buried, the matter was not given much thought after 1921, not until, that is, Lambis Englezos took an interest in the early 2000s. Referring to the aerial photographs, he lobbied the Australian government to instigate an investigation of the site, which finally came to fruition in 2007 when a team from the Centre for Battlefield Archaeology and GUARD was commissioned to carry out a non-invasive archaeological evaluation of the site. The main aim was to establish beyond reasonable doubt whether or not there were burials still lying beneath the ground in a farmer's field adjacent to Pheasant Wood.

An important element of this work was the historical research which sought to identify documentary evidence for the activities of German burial parties at Pheasant Wood. Much of this work took place in the Bavarian military archives in Munich, the original source of an important document first identified by Englezos. This was the original set of orders for the burials issued by Major von Braun of the 21st reserve Infantry Regiment and dated 21 July 1916. In this document, he states, 'The English bodies will be buried in mass graves immediately to the south of Pheasant Wood … h-Company is to excavate mass graves for approximately 400 bodies' (Pollard et al. 2008: 66). The term 'English' was used generically by the Germans to refer to both Australian and British troops, much in the same way that the present authors refer to the Bavarians as 'Germans'.

The first known aerial photograph of the pits resulting from von Braun's orders was taken on 29 July 1916 (Fig. 6.2). It shows eight long pits sitting in pairs and running parallel with the southern side of Pheasant Wood. Five of the pits have been backfilled, but the other three remain open and were to do so for the rest of the war – they can still be seen as such in the last photograph of the sequence taken on 16 September 1918. The pits sit immediately to the east of a light trench railway. Clearly visible in the photographs, this was used to carry supplies to the front line from the reserve areas. This hand-pushed line is mentioned in the orders, 'the dead are to be separated by nationality and suitably stacked close to the light railway', and there is a photograph which shows German soldiers standing around one of the open-sided wagons stacked full of corpses, which from the absence of boots appear to be Allied (the Germans envied the British and Australians for their superior footwear and often removed them from corpses for reuse).

First and foremost, the aerial photographs were invaluable in that they showed the location of the pits, which in 2007 corresponded to a strip of level ground sitting at the foot of the gentle slope which drops down from the northern edge of the village of Fromelles. The slope is regularly planted with a potato crop, but the area occupied by pits has been left fallow – it transpires that this was not due to respect for the presence of the graves, but because the level ground is largely composed of clay and so very wet and unsuitable for planting. Prior to the project, there was no folk memory of the graves in the village, which is totally understandable given that the entire population was moved away by the Germans during the war so that the

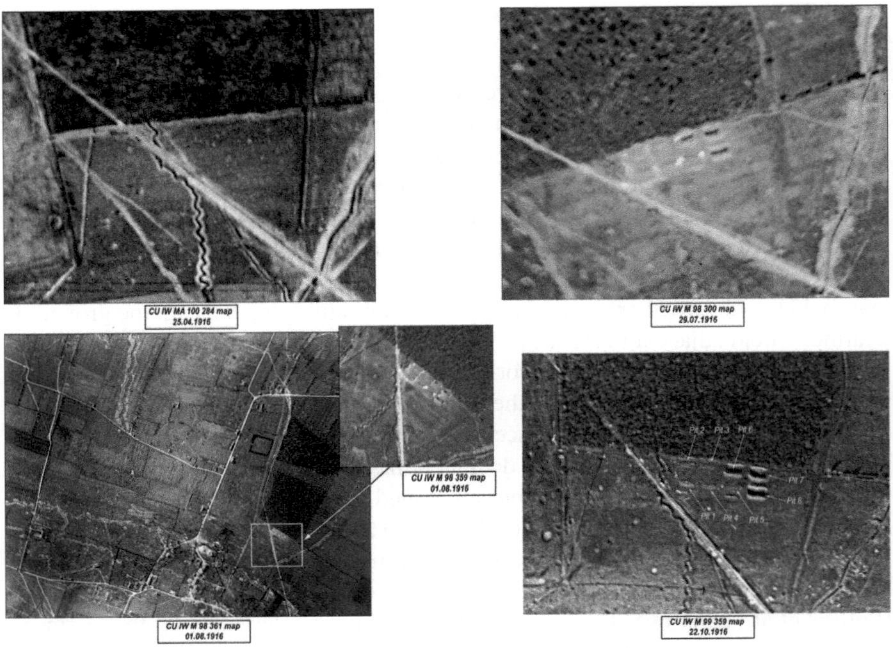

Fig. 6.2 The key aerial photographs showing grave pits (Courtesy of the Imperial War Museum)

village could be turned into a strong point on the high ground of the Aubers Ridge. By the time the locals returned after the cessation of hostilities, the backfilled pits would be all but invisible to the untrained eye and the open pits no more worthy of comment than any of the many other earthworks which over the coming years would themselves be backfilled as the land was slowly returned to agricultural use.

6.6 The Survey

With the search area clearly defined, the first phase of fieldwork aimed to locate the pits and ascertain, as far as would be possible without excavation, whether or not they were intact and had not been cleared of bodies after the war (it was possible that a recovery operation had taken place but the documentation relating to this mislaid or destroyed in the intervening years). A number of non-invasive techniques were utilised, including geophysics, GPR, topographic survey and metal detector survey. The heavy clay soils placed severe limitations on the effectiveness of the geophysics, but the topographic survey revealed a series of gentle undulations indicative of ground disturbance. But it was the metal detector survey which produced the most striking results, with a number of important artefacts recovered from the topsoil in the vicinity of the pits.

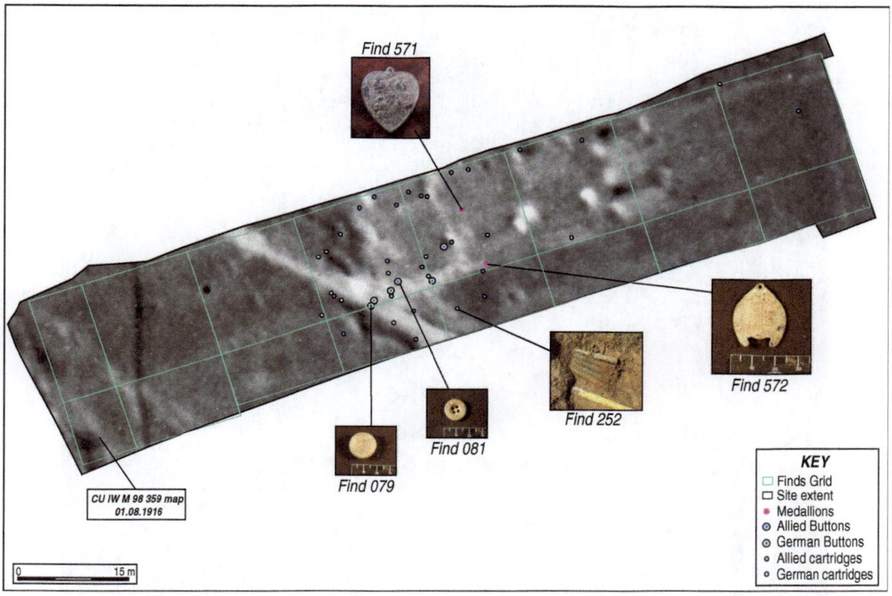

Fig. 6.3 Location of medallions and other selected finds overlain on aerial photograph (Courtesy of the Imperial War Museum)

Among the normal Western Front debris, which included shrapnel balls, shell casings and the like (Fig. 6.3), were two objects which have since gone on to attract a lot of attention, with one of them becoming associated with an individual soldier (Whitford and Pollard 2009). These were medallions which were presented by communities to Australian volunteers when they joined up – when pinned to a lapel they acted in lieu of a uniform and signified that duty had been done. Many of these may have been left behind in Australia, but two men had clearly brought theirs with them. The discovery of these objects removed any doubts that Australian soldiers had been buried on the site – the only reasonable explanation for the presence of these objects was that they had been brought in with bodies, as the Australians never got this close to the village during the assault. What the medallions did not tell us was whether the bodies were still in the pits.

There were other objects related to the 1916 burials. These included eyelets from the ground sheets or gas capes in which the already decaying bodies had been wrapped and carried to the grave site. Also present was a small collection of Allied and German buttons which may have fallen from the clothing of the living and the dead during the exertions of moving the bodies. The location of these finds was recorded with pinpoint accuracy with a total station, and it was not until they were plotted against one of the aerial photographs that their full relevance was realised.

When the photograph taken on 1 August 1916 (CU IWM 98 359 map) was enlarged (Fig. 6.4), it was found to contain some quite surprising detail. Firstly, there were footpaths worn by the German soldiers as they walked around the pits

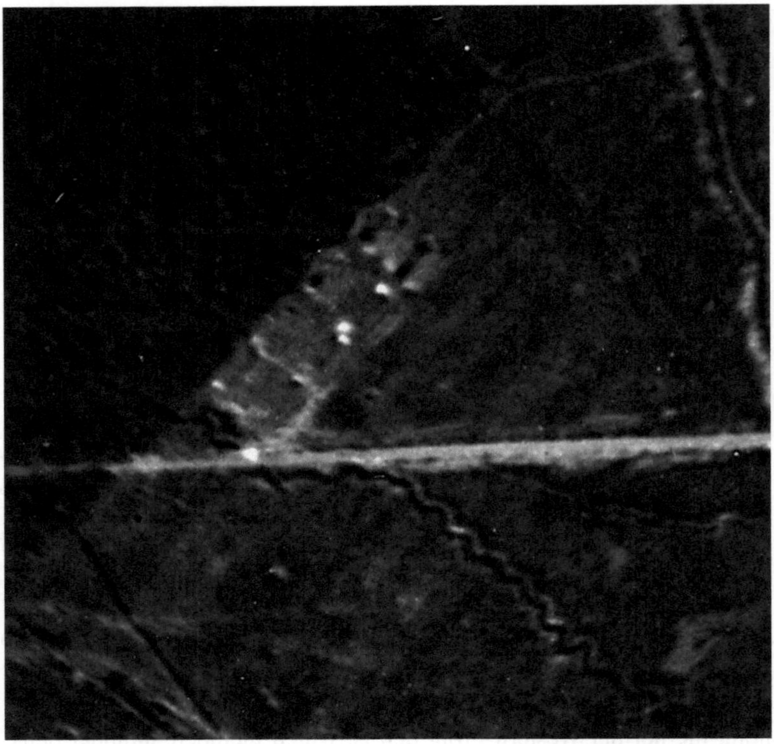

Fig. 6.4 Enlarged section of aerial photograph taken on 1 August 1916 showing differential backfills and the tracks worn into the ground surface by burial parties

and leading directly from the side of the railway where the bodies were unloaded. It is here that Law's realisation that aerial photographs were capable of capturing footprints comes most obviously to mind, with the trampled route-ways showing up as pale lines in the photograph. There is a dramatic correlation between these paths, the pits (5 of which have been backfilled) and the finds recovered during the metal detector survey. The buttons were all found on or close to the paths, with most of them clustered by the railway, where much movement would have been associated with lifting the bodies from the train. The medallions were located on either side of pit 4, though whether their locations mark the spot where the bodies were laid down prior to interment or points where they fell from bodies as they were being carried to the pits, we will never know.

6.7 The Evaluation

On the basis of the survey results, and the continued absence of any documentary evidence suggestive of a successful recovery operation in the immediate post-war period, it was decided by the Australian Army to commission a second programme of work, this time involving the invasive evaluation of the pits. Thus, in May–June 2008, all eight of the pits were subject to evaluation through the excavation of limited-sized sample trenches.

Given the less than clear nature of the geophysical survey, the aerial photographs were used to relocate the pits and guide the location of the trial trenches, which were marked out on the ground prior to digging. The first trench to be cut was a 1 m wide slot trench designed to expose the upper surface of pits 5 and 6. A mini digger equipped with a toothless ditching bucket was used to remove the 40–50 cm of heavy topsoil onto the heavy clay subsoil. Under the glare of the world's press, a nervous couple of hours passed as the careful removal of the topsoil failed to reveal any obvious sign of the pits. But then, to the authors' mutual relief, a tell tale soil change came to light. The first indicator was the presence of smears of blue clay intermixed with the browner clay soil beneath the topsoil. The blue clay horizon in an undisturbed profile should have appeared somewhere between 2 and 3 m below the surface, if not deeper, and its presence at the top of the subsoil horizon was therefore clearly indicative of digging to quite a considerable depth at some point in the past. As the newly exposed surface was cleaned, further straight edges became apparent, running across the width of the narrow trial trench. The pits were there, but were they still intact?

Certainly, the straight edges were a very good sign. It seemed highly unlikely that a burial recovery team, arriving on the site some years after the burials were made, would have used archaeological precision to re-excavate the pits. Much more likely, it was felt, the men charged with exhumation would have had a rather haphazard approach to excavation, resulting in the over-digging of the pit edges, leaving behind very ragged and irregular edges once the emptied pits had been backfilled for a second time. However, the presence or absence of bodies could only be established beyond doubt by digging down into the pits. Comparison with the pits as exposed in the base of the trial trench and as previously marked out on the surface showed some misalignment – the pits were in reality located immediately to the north of the pits as extrapolated from the aerial photograph plots, a drift of around 2.5 m on the north south axis, with the east west axis remaining consistent. The pits were around 10 m long and just over 2 m wide.

With the exposure of workable lengths of pits accomplished, the next task was to hand dig a small trial trench or *sondage* through the fill to check pit depth and presence or absence of human remains. The first pit to be investigated in this fashion was pit 4. Careful removal of the pit fill from an area of around 0.5 × 1 m revealed a number of zinc eyelets from ground sheets. These clearly related to a ground sheet which had been thrown back into the pit while it was being backfilled, though whether for the first or second time was as yet still unclear. Once the sondage was around a metre deep, the smell of decay became apparent, and white fragments of lime began to appear. Not long after, the first human bone was encountered – an ulna from the forearm.

By the end of day 2 of the project, the presence of human remains had been established in several of the pits, and so work progressed with extending the evaluation trenches, with the mechanical excavator in the hands of a skilled operator used to remove the pit fills down to a level just above the bones. The resulting trenches varied between 2×2 m and 1.5×2 m, depending on the width of the pit – with every attempt made to retain the original cut, where marks left by the German shovels were still preserved in the stiff clay.

All of the five pits known, from the aerial photographs, to have been backfilled shortly after the battle were found to be heavily populated with bodies. The remains were skeletonised, though decay products were evident as a black liquid deposit in the wettest of the pits. The pits closest to the wood were drier than those to the north as the roots assisted with drainage. The remains in these pits were also in a more advanced state of decay, with less in the way of clothing surviving. The most striking preservation was encountered in pit 2, which was the closest to the railway line. Among the bodies exposed in this pit were two Australians lying side by side on the same ground sheet. Elements of their clothing survived, including socks and trousers on one of them, and the canvas webbing belts were still in place on their torsos with the pouches still full of bullets which they never got the chance to fire. These men had clearly been deposited with some care, lifted into the pit in the ground sheet. The position of other bodies suggested a more hurried approach, with men lying on their fronts and one man in pit 5 lying spread-eagled with his arms outstretched.

In pit 4, possibly the wettest, the bodies were very densely packed and lay perpendicular to one another, given the impression of a raft of bodies. Preservation levels here were so good that a well preserved cardboard matchbox was recovered, the coloured graphic on it still perfectly legible. The majority of bodies appeared to be Australian, the first real indication being a metal 'rising sun' badge recovered from a body in pit 4. A photograph of this find was to appear in most Australian newspapers over the next few days, an indication of the interest in the operation in the country which remembers the battle of Fromelles as the worst 24 hours in its history. Other, less dramatic, indicators of nationality took the form of the ovoid buckles which were attached to the waist belts sewn into the Australian army tunic, a design feature not found on their British counterpart.

Evidence for a British presence in the graves took the form of General Service Buttons from pit 4 and one from pit 3. These brass buttons are of a distinctive design, with the lion and unicorn rampant on either side of a crest, and in this case with a Birmingham maker's mark on the reverse. These are distinctly British in character, as at the time the Australians were largely using plain buttons of resin, rubber or bone – their own General Service buttons were of a different design to the British types.

6.8 Benefits of Detailed Photographic Analysis

As described above, invasive evaluation served once and for all to establish the presence of the bodies of Australian and British soldiers within the pits at Pheasant Wood. A further suite of essential information was also recovered, including pit depths, variability of preservation levels from pit to pit, the presence of more than one layer of bodies in each pit, the nature of burial practice and so on. What also became apparent during this operation was the amount of useful information which had been gleaned from nothing more than an examination of the aerial photographs.

In addition to the photographs already discussed in relation to the medallion finds, two photographs in particular proved very informative (Fig. 6.2). The latest of these, taken on 22 October 1916, will be discussed first. This photograph captured the results of the burial operation in clear detail, with the nature of the mounded spoil still standing proud of the graves showing up very clearly. A natural result of returning soil to a pit in which bodies have been placed is that the soil will stand proud of the surface. In the case of pits 1–5, these mounds of backfilled soil have been landscaped in order to provide straight-edged platforms, probably of slightly larger dimensions in plan than the pits beneath them. Contrastingly, pits 6–8 have been left open, and the removed spoil can be seen as linear mounds heaped close to the northern and southern edges of the pits. Anomalous here is what appears to be the vestige of a spoil mound to the south of pit 5, which seems somewhat at odds with its backfilled status.

The obvious conclusion to draw from this was that pit 5 contains more bodies than the other four pits used for burial. Such a scenario would leave even less space for returned spoil and so ultimately lead to some of it being left behind as a vestigial spoil heap once backfilling was complete. The possibility of more bodies being present in pit 5 was noted in the 2007 survey report (Pollard et al. 2007), where it was suggested that pit 5, being the furthest from the rail line, was the last to be backfilled. If it became apparent to burial parties that with a little extra packing of the bodies, the need to utilise a sixth pit, with all the trouble that would entail in backfilling, could be circumvented, then the result may have been the lack of space for backfilled soil as indicated by the photograph.

This hypothesis seemed to be backed up by the results of the 2008 evaluation, which showed that bodies in pit 5, or at least that part excavated, lay not two deep as with the other pits, but three deep. However, when the graves were fully excavated and the bodies removed for reburial, by Oxford Archaeology in 2009, the pit was found to contain 50 bodies, the same as pits 1–4. The 2007 evaluation concluded that between 225 and 400 bodies were present at Pheasant Wood but with expectations leaning towards the higher end. This wide range gives some idea of the difficulty of making estimates from evaluation trenches – in any case something never tried before. The final figure was around 250, with the five full pits containing 50 bodies each.

The exposure of bodies three deep in the length of the pit examined in 2008 therefore appears to have been an anomaly within the pit. However, what does appear to have happened is that the relative care taken with bodies and their placement in pits 1–4 had broken down by the time they got to pit 5, and as a result, less space was left for backfilled spoil, which in any case may not have been replaced with the same zeal previously demonstrated – hence the excess left sitting beside the trench.

The second photograph to be discussed here was that taken on 29 July 1917 (CU IW 98 300 map). Whereas it had been apparent from the outset that pits 1–5 had been used for the burial of the dead in the days following the battle (their continued presence was established by the evaluation), some doubt still remained over pit 6. On first viewing of the photographs, this pit appeared to have been left open, along with pits 7 and 8, until at least 1918 when the last known photograph was taken. However, closer scrutiny of the photograph suggested that a small portion of the western end of the pit had actually been backfilled along with pits 1–5. This was suggested by nothing more than a smudge of grey at this end of the pit, between the two spoil heaps. It was enough, however, to merit the suggestion in the 2007 survey report (Pollard et al. 2007) that the western end of the pit may have been used for the interment of 'overspill' remains which had been left out of pit 5, perhaps because they were recovered after the backfilling of that pit. This partial use would certainly be in keeping with the desire to minimise effort during the tail end of the burial operation.

Once again, a suggestion drawn from photographic analysis was backed up by the subsequent evaluation. The excavation of a slot trench at a point around 2 m from the western end of pit 6 resulted in the exposure of a small amount of human remains lying in the bottom of the cut. These consisted of leg and foot bones, possibly from two individuals, partially wrapped in what appeared to be a sock. During the recovery operation in 2009, further human remains were found in the western end of the pit, which had been predicted on the basis of the earlier evaluation – during the evaluation, the sloping extent of the partial backfill had been apparent in both the east and west facing sections, a clear indication that burial-related backfilling had been limited to this end of the pit.

6.9 Conclusion

The Fromelles survey and evaluation was a unique project in several respects. It was the first time since 1921 that a concerted effort had been made to locate unmarked graves on the Western Front since post-war body recovery operations ceased in 1921. It was also the first time that the character and extent of mass graves had been established through invasive evaluation – such operations are certainly not associated with the recovery of bodies from more recent mass graves such as those in the Balkans.

On the basis of the 2007 survey and 2008 evaluation, a joint decision was made by the Australian and British governments to proceed with the full recovery of the human remains from the mass grave pits at Pheasant Wood. That operation was carried out in 2009 and was followed in January 2010 with the first reburials in individual graves located in a newly establish Commonwealth War Graves Commission cemetery very close to the Pheasant Wood site (the first such cemetery to be established in half a century). Accompanying the exhumation is a programme of DNA analysis which will hopefully allow at least some of these men to have their names carved into their gravestones, the rest will of course be 'Known Unto God'.

The forgoing has demonstrated how useful wartime aerial photographs were in the relocation, evaluation and interpretation of the Pheasant Wood Graves. This project was, however, just one of a growing number to have taken place on the Western Front over recent years, many of which have utilised wartime aerial photographs to some extent. These photographs represent a powerful tool which allows the modern researcher to view the battlefields as they appeared almost a century ago. Now, with digital technology, we are able to view these extraordinary photographs in an accessible form and scrutinise them as never before.

Bibliography

Barton, P. (2005). *The battlefields of the First World War: The unseen panoramas of the Western Front*. London: Constable.

Barton, P. (2007). *Fromelles: A report based on research in the Hauptstaatsarchiv Kriegsarchiv, Munich, November and December 2007*. Unpublished report on research carried out on behalf of the Australian Army History Unit.

Eastlake, L. E. (1857). Photography. *London Quarterly Review, 101*(202), 442–468.

Heiman, G. (1969). *Aerial photography: The story of mapping and reconnaissance*. New York: Macmillan.

Lundqvist, S. (2002). *A history of bombing*. London: Granta Books.

McMaster University Libraries. (2008). I Spy with my glass eye: Aerial photography and innovation in WWI. http://pw20c.mcmaster.ca/case-study/i-spy-my-glass-eye-aerial-photography-and-innovation-world-war-i. Accessed 5 Feb 2010.

Pollard, T. (2008). A view from the trenches: An introduction to the archaeology of the Western Front. In M. Howard (Ed.), *A part of history: Aspects of the British experience of the First World War* (pp. 198–209). London: Continuum.

Pollard, T., Barton, P., & Banks, I. (2007). *Pheasant wood Fromelles, evaluation of possible mass graves*. Unpublished GUARD report 12005, University of Glasgow.

Pollard, T., Barton, P., & Banks, I. (2008). *Pheasant wood, Fromelles, data structure report*. Unpublished GUARD report 12008, University of Glasgow.

Stichelbaut, B. (2005). The application of Great War aerial photography in battlefield archaeology: The example of Flanders. *Journal of Conflict Archaeology, 1*, 235–243.

Whitford, T., & Pollard, T. (2009). For duty done: A WWI military medallion recovered from the mass grave site at Fromelles, Northern France. *Journal of Conflict Archaeology, 5*, 201–229.

Chapter 7
Historic Vertical Photography and Cornwall's National Mapping Programme

Andrew Young

Abstract Between 1994 and 2006, a comprehensive programme of mapping and recording archaeological sites from aerial photographs in Cornwall was carried out as part of the National Mapping Programme, a project initiated and funded by English Heritage. Among the wide range of photographs consulted during the project, the most important were RAF verticals dating from the mid-twentieth century. These photographs cover the entire county, they were taken at propitious times of year and in favourable conditions both for cropmark and earthwork features, and they predate the widespread breaking-in of moorland in the later twentieth century, the post-war expansion of towns and the late twentieth-century move towards deep ploughing. For these reasons, the photographs were an indispensable source, and almost 12,000 sites, ranging from prehistoric settlements to post-medieval industrial remains, were transcribed from them.

Between 1994 and 2006, a comprehensive programme of mapping and recording archaeological sites from aerial photographs in Cornwall was carried out as part of the National Mapping Programme, a project initiated and funded by English Heritage. Among the wide range of photographs consulted during the project, the most important were RAF verticals dating from the mid-twentieth century. This chapter outlines the value of these historic photographs and highlights some of the principal areas of archaeological information for which they proved an indispensable source.

A. Young (✉)
Historic Environment, Cornwall Council, Kennall Building, Station Road, Truro, Cornwall, TR1 3AY, UK
e-mail: ayoung@cornwall.gov.uk

W.S. Hanson and I.A. Oltean (eds.), *Archaeology from Historical Aerial and Satellite Archives*, DOI 10.1007/978-1-4614-4505-0_7,

7.1 Aerial Reconnaissance in Cornwall

In common with many parts of Britain, Cornwall has been subject to a large amount of aerial photography, taken both for archaeological and non-archaeological purposes. The earliest archaeological aerial reconnaissance in the county was carried out in 1926 by O.G.S. Crawford and resulted in the now famous images of submerged late prehistoric or Romano-British field boundaries on Samson Flats, Isles of Scilly, which provide evidence of sea-level rise during the last two millennia (Thomas 1985; Ratcliffe and Straker 1996).

Intermittent flights were made during the 1950s and 1960s by Professor J.K. St. Joseph of Cambridge University and during the 1970s and 1980s by the Royal Commission on the Historical Monuments of England (RCHME). Since 1984, a more systematic programme of reconnaissance has been carried out by the Historic Environment Service of Cornwall County Council (formerly Cornwall Archaeological Unit), supplemented by occasional flights made by English Heritage. Collectively, these reconnaissance programmes have produced more than 22,000 oblique images of archaeological sites and historic monuments consisting mostly of black and white prints but including several thousand colour slides and, more recently, digital images.

Specialist archaeological photographs are generally taken at low altitudes and therefore offer detailed close-up views from oblique angles that use favourable conditions, such as low light and shadow, to great advantage. However, specialist archaeological reconnaissance is, by definition, site-specific, and flying time is always constrained by available resources. As a result, the pattern of archaeological aerial reconnaissance in Cornwall is uneven: some areas have been flown many times, others only rarely.

Non-archaeological aerial photography is taken for cartographic or military purposes (by the Ordnance Survey and the RAF, respectively) and for use in civil engineering projects by commercial companies (in Cornwall most notably by Meridian Airmaps Ltd during the 1960s and 1970s), and is commissioned by local authorities for census purposes (Cornwall County Council has commissioned three such surveys over the last 20 years). Non-archaeological aerial surveys invariably consist of vertical photographs often (but not always) at a scale of 1:10,000. Whilst these photographs are unlikely to provide the same level of detail as low-level oblique images, this is more than compensated for by the fact that vertical photographs provide comprehensive coverage of large swathes of the landscape.

In Cornwall, five vertical surveys provide coverage of the entire county. These surveys offer unique documentation of the Cornish landscape at a particular moment in time and are thus of great interest to landscape historians. The earliest complete survey was carried out in the years immediately after the Second World War and the most recent in 2005. Comparison of the two allows us to chart the fast-changing landscape in the latter half of the twentieth century, an era which witnessed the rapid expansion of towns, major road-building programmes and significant intensification in agricultural practice. These surveys, and indeed all the series of vertical photographs, also contain a wealth of archaeological information, and the study of Cornish archaeology has been greatly enhanced by the systematic mapping and recording of this information between 1994 and 2006 as part of English Heritage's National Mapping Programme (NMP).

The aim of the NMP is to enhance understanding about past human activity by providing information and syntheses for all archaeological sites and landscapes visible on aerial photographs from the Neolithic period to the twentieth century (Bewley 2001: 78). To achieve this aim, a methodology was developed from previous selective approaches to mapping from aerial photographs (e.g. Benson and Miles 1974). The guiding principle of the methodology is to map, describe and classify all archaeological sites recorded by aerial photography in England to a consistent standard (RCHME 1995). A great strength of this method is that it involves consulting all available and accessible photographs. For Cornwall, this amounted to more than 50,000 photographs, the majority of which were non-archaeological verticals. The photographs used during the project are held in three separate collections: English Heritage's National Monuments Record (NMR), Swindon; Cornwall County Council, Truro; and Cambridge University Unit for Landscape Modelling.

7.2 Historic RAF Photographs

The largest and most important archive is the NMR which loaned more than 33,000 photographs for consultation. A significant proportion of these (some 14,000 prints) were RAF verticals taken over the period from 1942 to 1964. Although the archive contains photographs from every year within this period, there are three main series: those of the mid-late 1940s covering the entire county, those from the early 1950s covering most of the county and those from 1964 again providing only partial coverage.

Each series differs in character and quality. The main sorties from the 1940s are the 3G/TUD series of April 1946, the 3G/TUD and 106G/UK series of July 1946, the CPE/UK series of October 1947 and the CPE/UK series dating from March 1948. The quality of photography from the 1940s is generally good, although weather conditions were somewhat variable over parts of the county on the day of the flights, with haze or patchy cloud cover in places. Many earthwork features are visible on these photographs, particularly the 1947 series which was flown in markedly low light. The July 1946 photography is also a prolific source of cropmarks, including some sites not seen on any subsequent photographs. The 540/483 series of April 1951, the 58/680 and 540/497 series of May 1951 and the 540/994 series from January 1953 constitute the bulk of the 1950s photography. On the whole, the quality of these photographs is inferior to the earlier series, in some cases very much so. Exposures are not always ideal, resulting in poor definition; this is a pity because the landscape visible on the 1951 springtime photographs is characterised by a large number of freshly ploughed fields which are frequently overexposed and appear bright white, with little chance of soilmark features showing. It is also the case that for some of the 1950s photographs, stereoscopic pairs do not produce a pronounced 3D effect when viewed through a pocket stereoscope. In contrast, the 58/6399 series of July 1964 contains photographs of excellent quality with very high definition and enhanced stereoscopic effect. The 1964 series are at a slightly small scale (1:12,000) than the norm, whereas the 1940s and 1950s photographs are nominally 1:10,000 (nominally because the pilots' ability to maintain a constant height was not always consistent). Broadly speaking, the quality of the actual prints is uneven, particularly those of the 1940s and 1950s; a minority are curled, creased or otherwise worn, and some have developed a sepia tone.

Notwithstanding these criticisms, there is no question that RAF vertical photography, in particular the historic photographs dating from the 1940s, constituted the most valuable photographic source available to Cornwall's National Mapping Programme. During the project, more than 30,000 individual archaeological features were mapped, and these formed the basis for 17,682 records either updated (in the case of previously recorded sites) or created (in the case of newly identified sites) in the county Historic Environment Record (HER). As a routine element of HER data entry, details of the photographs from which each site was mapped were noted. Analysis of this data shows that for 14,232 records, the mapping was derived from RAF photographs. In other words, 80% of all sites recorded during the project were transcribed from RAF photography.

7.3 Archaeological Sites on RAF Photography

One of the most significant outcomes of Cornwall's NMP was the large number of previously unknown or unrecorded sites which were newly identified from aerial photographs. As Table 7.1 shows, more than 10,000 of these new sites were mapped and recorded from RAF verticals, 72% of them from the 1946–1948 series.

Table 7.1 Statistical analysis of RAF aerial photography showing the number of sites transcribed and recorded from the various flights

	Form of remains					
Photo date	Cropmark	Extant	'Documentary' or 'Site of'	New site records	Existing site records	All site records
1946–1948	2,053	7,250	879	7,654	2,528	10,182
Other 1940s	96	308	44	368	80	448
1951–1953	339	851	140	927	403	1,330
Other 1950s	100	192	69	259	102	361
1964	444	1,260	113	1,340	477	1,817
Other 1960s	3	1	1	3	2	5
Undated[a]	32	54	3	43	46	89
Total	*3,067*	*9,916*	*1,249*	*10,594*	*3,638*	*14,232*

[a]For a small number of site records, the date of photography was erroneously omitted from the database

These figures can be qualified to an extent by pointing out that in some cases, the new records are for types of site which have only recently come to be included in the remit of Cornwall's HER. The most obvious example is the legacy of twentieth-century conflict. An important application of 1940s RAF photography lies in its documentation of Second World War defence structures, such as anti-aircraft batteries, barrage balloon mooring points, pillboxes and the like, many of which have been demolished over the last 60 years. Partly because of the rate of their destruction, these wartime installations are now recognised as a vulnerable and important element of the historic landscape and are included in local authorities' registers of monuments; indeed, some are designated as scheduled monuments. Prior to Cornwall's NMP, few or no Second World War sites were recorded in the HER. Almost 700 of these features have subsequently been mapped from historic RAF photography and virtually all of them are new site records.

Another example is small-scale quarrying. The Cornish stone industry has a long history; the main rocks quarried were granite and slate, and the industry has left a legacy of abandoned pits and quarries all over the county. Although the larger and more complex quarries were included in the HER, small-scale quarrying had not been systematically recorded prior to the NMP project. Mapping and recording of quarries accounted for 1,300 of the new sites identified from RAF vertical photography.

These observations in no way diminish either the results of NMP mapping or the value of historic RAF photography. In fact, the vast majority of sites mapped from RAF photographs are types such as prehistoric and medieval settlement and farming features, which have traditionally formed the core of the HER.

Of interest is the ratio of extant archaeological sites (with surviving stone or earthwork remains) to those visible only as cropmarks. For Cornwall's NMP as a whole, the majority of sites recorded had extant remains, and the predominance of earthworks is especially pronounced among features transcribed from historic RAF verticals. Almost 70% of these had extant remains, whilst only 20% were cropmarks; the remainder were classed as either 'site of' (extant features subsequently built over or otherwise destroyed) or 'documentary' (sites previously known from documentary evidence but where some form of physical evidence – usually cropmark evidence – was identified on the photographs).

This bias towards extant sites is not surprising given the nature of the Cornish landscape and the distinctive character of much of its archaeology. The upland areas of the county, especially Bodmin Moor, West Penwith and the Lizard Peninsula, are characterised by moorland and open pasture and in places contain extensive relict archaeological landscapes. Some other parts of Cornwall have, until quite recently, been relatively unaffected by intensive agricultural development, and throughout much of the county stone has been used for building since prehistory. As a result above-ground traces survive in a large percentage of Cornish archaeological and historic monuments compared with elsewhere in southern England. This fact is reflected by the unusually high number of sites – almost 2,000 – designated as scheduled monuments. Nowhere is this bias more apparent than in the remains of the tin and copper mining industries, which are among the most visually striking elements of Cornwall's rich and varied archaeological heritage. Only 5% of mining features transcribed from RAF photography were recorded as cropmarks.

7.4 The Cornish Mining Industries

Large mineral reserves coupled with technical advances in steam pumping, which made deep mining possible by the end of the eighteenth century, meant Cornwall dominated the world market in tin until the 1870s and by the early nineteenth century was the world's pre-eminent copper producer (Rowe 1953: 128; Barton 1969: 151). The impact of the industry on the landscape was large-scale, and the speed of its decline has left a well-preserved relict mining landscape. Its legacy includes thousands of mine shafts, numerous engine houses and the widespread remains of tin and arsenic processing (arsenic was a by-product of tin and copper mining and its production was pioneered in Cornwall). The international importance of the Cornish mining landscape was recognised when it was awarded World Heritage Site status (WHS) in July 2006.

Many of the more substantial remains, such as engine houses and the larger shafts, are recorded on Ordnance Survey maps, particularly the First and Second Edition maps (of 1880 and 1907, respectively) when some mines were still in operation. Archaeological field surveys and documentary research have added further detail to our knowledge of the mining landscape. Nonetheless, aerial photographs provide an abundance of additional information. In the main, this comprises the location of surface workings, shallow pits and their associated spoil heaps.

The development of the WHS nomination bid involved defining the nature and extent of the archaeological remains of the mining industry, and mapping from aerial photographs provided 40% of the data collated by the WHS team (A. Sharpe 2006, personal communication). The RAF verticals are the prime source for this information (Fig. 7.1); more than 2,000 of the 2,300 mine sites mapped during Cornwall's NMP were transcribed from RAF photographs. Although much of the mining landscape still survives today, it is clear from this photography that a considerable amount was lost during the latter part of the twentieth century. The land around some mine sites has been improved and landscaped, features in enclosed farmland have been levelled, and many structures have been obscured or damaged by scrub growth.

The period from 1700 to the early part of the twentieth century was the heyday of Cornish mining, but Cornwall's rich mineral resources have been exploited on a large scale since medieval times, and the tin industry was already internationally important by the thirteenth century (Hatcher 1973: 20–21) During this period, a substantial amount of tin produced in Cornwall came from tin streaming, a technique which involved washing away lighter sands and wastes from tin-rich gravels to leave the heavier tin ores which were then collected and smelted (Gerrard 1987: 7ff).

Accumulations of tin-rich gravels are known as tin streams. Essentially, these are geological deposits formed when erosion caused tin ore (cassiterite) to break away from exposed seams (known locally as 'lodes') and settle a short distance away or

Fig. 7.1 The Great Flat Lode, Carn Brea. RAF vertical photographs from the 1940s provide a uniquely graphic impression of the extensive relict landscape of Cornwall's tin and copper mining industries (Photo 106G/UK 1663/4152 (12 July 1946). © Cornwall County Council 2009)

to accumulate in the bottom of river valleys (Penhallurick 1986: 153ff and 164ff). After the formation of a tin stream, layers of sand and gravel settled on top of it; this overburden could be several metres deep and had to be removed to allow the tinners to reach the tin stream.

Streamers developed a range of methods to separate the cassiterite from the waste, depending on the nature of the tin stream and the elements constituting the overburden, and each method produced its own distinctive type of earthwork remains (Gerrard 2000: 63–79). Most commonly, water was brought through a channel into the working area, and the overburden was washed into the river or a drain where it flowed downstream in suspension. Heavy stones and gravel left behind were piled up to form steep linear banks. As the work progressed, the tinners produced a regular pattern of parallel banks of spoil (Fig. 7.2).

Fig. 7.2 Streamworks at Ennisworgey, Restormel. A diverted channel defines the southern limit of these workings. Waste has been dumped in a series of mounds forming a characteristic parallel pattern. To the north of the stream, an unusual amount of detail can be seen, including individual mounds of waste and the clearly defined squared-off end of one of the workings (*centre of the photograph*) (Photo RAF 543/2332/F22/0170 (26 July 1963). ©Crown copyright. MOD)

Streaming was carried out on a massive scale; the removal of millions of tonnes of overburden resulted in rivers and estuaries, such as the Fal, Fowey and Carnon, becoming heavily silted (Gerrard 2000: 61). Although tin streaming declined in importance after the seventeenth century, it continued as an industry up to the mid-1900s.

The tinners gradually turned their attention away from the increasingly exhausted tin streams towards the lodes themselves. The earliest underground mines in Cornwall were worked by pits or shallow shafts, often interconnected by tunnels and galleries underground, and usually dug in lines following the lodes (Fig. 7.5). Lode-back pits or shafts, as they are known, were in use by the thirteenth or fourteenth century (Herring 1997: 55–56). Unlike later deep mining, where very few shafts were used to extract the lode, shallow working required numerous pits. This is because the underground excavations were very narrow and it was easier to sink more shafts than to haul waste long distances back to the original shaft to take it to the surface.

These forms of early mining have left extensive remains which in places form spectacular 'other worldly' landscapes. Historic aerial photographs provide an ideal medium for viewing these landscapes, and more than 1,300 early mining sites have been mapped from RAF verticals. Of particular significance was the transcribing and recording during Cornwall's NMP of 259 new streamworking sites from RAF photography. Many of these were in the Hensbarrow area to the north of St Austell, the centre of the china clay industry in Cornwall, and here, a number of streamworks have been destroyed over the last few decades by clay extraction or have been buried beneath massive waste tips. Elsewhere, some streamworks have been drained and improved, and many have become overgrown by dense vegetation. The best preserved workings are found on Bodmin Moor, where the only extensive programme of field survey of Cornish streamworks has been undertaken (Gerrard 2000: 11ff).

7.5 Bodmin Moor

The granite massif of Bodmin Moor is of enormous archaeological significance nationally. As well as early tin workings, extensive relict landscapes comprising prehistoric settlements, fields and ceremonial sites, and medieval farms and field systems survive here. The importance of this landscape led to the development of the Bodmin Moor Survey, a ground-breaking initiative carried out jointly by the RCHME and Cornwall Archaeological Unit between 1980 and 1985, which was one of the first projects nationally to combine aerial survey with follow-up field survey. The results are contained in two separate publications (Johnson and Rose 1994; Herring et al. 2008).

For that project, a specially commissioned vertical aerial survey of Bodmin Moor was flown, at a scale of 1:7,500, by Cambridge University Committee for Aerial Photography in 1977. Archaeological sites visible on this photography were transcribed by RCHME staff at a scale of 1:2,500. Further detail was provided by a programme of specialist oblique photography carried out by RCHME. The resulting plots were then taken into the field by the survey team and amended where appropriate.

Many (if not all) of the features mapped during the Bodmin Moor Survey can be seen on RAF verticals, and in some instances, new features were identified from these photographs during Cornwall's NMP: for instance, 50 previously unrecorded round houses and 15 prehistoric field systems. Although these figures are small compared with the total number of similar features in the Bodmin Moor Survey area, it is testimony to the value of the historic RAF verticals that any new sites at all were identified in an area of open moorland which had been subjected to such intensive research in the past.

Many more new features were mapped from RAF verticals around the moorland edge and in farmland on the moor itself. The Bodmin Moor Survey was to a degree selective: firstly, only areas of moorland were surveyed and not the surrounding farmland; secondly, only photographs from the two sources mentioned above were consulted. As a result, there were gaps in the archaeological record, principally in areas that had been taken into agriculture during the post-war years but prior to 1977 when the aerial survey was commissioned. Archaeological features subsequently mapped from historic verticals filled in some of these gaps (Fig. 7.3). In the main,

Fig. 7.3 An extract of NMP mapping from the Siblyback area of Bodmin Moor, showing new sites transcribed from RAF vertical photography. For the most part, these comprise medieval field boundaries and ridge and furrow (the direction of the ridges is shown by *arrows*) (©Cornwall County Council 2009)

these features related to the medieval and post-medieval farming landscape, typically field systems (some quite extensive), removed field boundaries, ridge and furrow cultivation, enclosures of various types, holloways and trackways. In this respect, NMP mapping in Bodmin Moor exemplifies the mapping for Cornwall as a whole in that as many as a third of all new sites mapped from RAF verticals throughout the county were features of the medieval and early post-medieval farmed landscape.

7.6 The Medieval Farming Landscape

The nature of medieval settlement in Cornwall is most clearly seen on Bodmin Moor where more than 30 settlements were abandoned and survive in their entirety (Johnson and Rose 1994: 77–115). All of these deserted settlements can be seen on the 1940s RAF vertical photographs. They are made up of longhouses: some are single farmsteads but most are hamlets. Longhouse hamlets were the characteristic settlement type throughout medieval Cornwall. Typically, they would contain between two and six farmsteads, clustered around a small 'townplace'. Each farmstead comprised a main farmhouse and one or two smaller outbuildings serving as barns and animal houses. The houses are stone built and most in Cornwall date from the twelfth to fourteenth centuries, although some are earlier (Preston-Jones and Rose 1986: 146–147).

Each hamlet was surrounded by in-ground or 'townland' – an area of improved land sometimes enclosed by an irregular curvilinear boundary. The in-ground was divided into long narrow strips (often containing ridge and furrow) defined by low stony banks, and each farmstead within a hamlet had its own share of strips scattered through the fields. The in-ground was not permanently under arable or pasture but was worked in rotation. There is evidence that medieval agriculture in Cornwall was based on a long ley period, with crops grown for 2 or 3 years and grass for between 4 and 9 years (Fox 1971: 102–104). The whole cycle would take roughly 10 years to complete, and the fields were organised into ten cropping units or furlongs which were subdivided into strips (Herring 2006).

Earthwork remains of strip fields have been mapped from RAF vertical photography not just on Bodmin Moor, but throughout the upland areas of Cornwall (Fig. 7.4). In some places where the strips survive, they are likely to result from outfield cultivation involving the temporary cultivation of marginal land normally used as pasture. In Cornwall, outfield areas were only occasionally cultivated – maybe only every 50 or 60 years, when demand for produce was high or perhaps when prices were particularly favourable (Fox 1973).

Whilst few deserted longhouse settlements survive in lowland areas (most have been superseded by later buildings), the extent of the farming landscape in the medieval period is shown by the widespread occurrence of former strip fields which were enclosed with substantial boundary hedges at a later date. The enclosure of these strips often leaves distinctive traces in surviving field boundaries and field patterns, such as so-called reversed-s or reversed-j curves, and 'dog-legs', and the evidence

Fig. 7.4 The medieval landscape at Trerice, St Breock. In the south are the low earthwork remains of strip fields, abandoned in what is now uncultivated land. To the north, individual strips, or groups of two or three, have been enclosed by stock-proof hedges, whose reversed-j shape fossilises the pattern of the original open field (Photo: CPE/UK/1999/4049 (13 April 1947). © Cornwall County Council 2009)

suggests that the greater part of the farmed landscape of Cornwall was divided into strip fields (Herring 1998: 77–82).

The gradual enclosure of open strip fields took place mainly between the fourteenth and seventeenth centuries and transformed the landscape into that which survives to this day (Herring 1998: 77). The strips surrounding small hamlets were reorganised into block-shaped fields based on the enclosure of whole cropping units. These fields are distinctively irregular with very few straight boundaries. By contrast, fields of eighteenth or nineteenth century origin are predominantly straight-sided and rectilinear. Where hamlets contained many households and landholding arrangements were more complex, single strips or groups of two or three were enclosed; the resulting pattern closely resembles that of the original open field. Though less common than the larger block-shaped fields, there are good examples throughout Cornwall (Fig. 7.4).

In some parts of the county, the latter part of the twentieth century saw widespread field boundary removal, and in places, the character of the landscape has been altered. Over much of Cornwall, however, the present-day field pattern is clearly derived from medieval enclosure, and the mapping of field boundaries from RAF verticals has served to substantiate this interpretation. More than 2,300 newly recorded field boundaries and field systems originating from the late medieval and early post-medieval phase of enclosure have been mapped from RAF verticals, roughly half of them visible only as cropmarks. In some instances, the newly mapped field systems are relatively extensive and coherent. These represent medieval fields which were abandoned, like those at Trerice (Fig. 7.4), or which have been removed or overlain as a result of post-medieval reorganisation of the landscape. In many instances, however, the photographic evidence is for individual field boundaries or small groups of boundaries which fit into the present-day field pattern.

There are two important points to make regarding these features. Firstly, in accord with NMP methodology, field boundaries appearing on the First Edition OS map of 1880 (or on later maps) are not transcribed; thus, the mapping represents pre-1880 boundary removal. Secondly, these individual boundaries formed an integral part of more extensive field systems which are still in use today, but whose origins lie in the medieval period.

Using interpretations of field morphology in conjunction with place-name evidence and documentary sources (identifying settlements first recorded before 1540), it has been possible to define those areas in Cornwall which were enclosed and farmed in the medieval period (Herring 1998: 77). There is much circumstantial evidence to suggest that the medieval settlement and farming heartland corresponds to a large degree with the settlement and farming zone of the late prehistoric and Romano-British period (Johnson 1998). This evidence comes from archaeological field work, excavation, geophysical survey and mapping from aerial photographs. The general pattern becoming apparent from this work is that prehistoric settlement features are invariably found in areas of medieval settlement, but rarely from areas that were heathland or waste ground in medieval times.

7.7 The Prehistoric Farming Landscape

Many of the characteristic features of the upland prehistoric landscapes in Cornwall were mapped from RAF vertical photography during Cornwall's NMP. These include, for example, almost 500 stone built round houses (more than 100 of which were new sites identified from the historic photographs) and around 400 field systems. In the same way that the intact abandoned settlements and fields on Bodmin Moor serve as a model for the medieval settlement pattern for the rest of Cornwall, so the relict prehistoric landscapes in the Cornish uplands provide a likely template for the prehistoric settlement pattern in the county as a whole.

As a result, some models of prehistoric landscape development for Cornwall have been proposed, the most developed of which is set out as a narrative in a paper by Peter Herring (2008). Herring identifies, from different and superimposed settlement and field patterns, a series of key reorganisations of the Cornish farming system undertaken on a wide scale in response to changing pressures on land and resources.

The earliest definable patterns, from the Middle Bronze Age, can be traced on Bodmin Moor. Towards the fringes of the moor unenclosed round house settlements are set within curvilinear accreted field systems. Lanes lead through the fields to rough grazing land on the open moor beyond, which was probably shared with neighbouring groups as a form of common. In the heart of the moor are settlements consisting of round houses but with few or no associated field enclosures. These are best interpreted as the seasonal homes of people practising a pastoral economy, and it is possible that the permanent bases of these people were in lowland areas surrounding the moor.

A major reorganisation around the mid-second millennium involved the laying out of extensive coaxial field systems with round houses scattered within them. These have been recorded from coastal areas on the Lizard Peninsula and in West Penwith as well as on Bodmin Moor. The coaxial fields better organised the enclosed farmland (and in places extended its limits) and formalised access to the grazing land beyond.

Reorganisation in the Late Bronze Age/Early Iron Age saw the abandonment of coaxial fields and the development of dense grids of irregular brick-shaped fields (Fig. 7.5). There was an intensification of agriculture (evidenced by the formation of substantial lynchets) and an increase in settlement nucleation demonstrated in the Later Iron Age and Romano-British period by enclosed settlements (known in Cornwall as 'rounds') and courtyard houses (oval or irregular-shaped houses consisting of a central paved courtyard surrounded by several conjoined rooms all set within a massive stone-faced earthen enclosure). This model is clearest in West Penwith, where the layout of the main prehistoric boundaries has been encapsulated in the present-day field pattern.

A far-reaching but poorly understood reorganisation took place during the sixth or seventh century AD. Enclosed settlements were abandoned and replaced by open hamlets, many of which have Cornish names prefixed with *Tre* (Padel 1985: 223–224). Many of these early medieval settlements are situated close to abandoned rounds, and it is suggested that some may be overlying the site of former rounds (Rose and Preston-Jones 1995: 57; Johnson 1998).

These episodes of landscape reorganisation, derived from upland evidence, appear to have been on a wide scale so it is reasonable to suppose that similar models can be demonstrated in lowland Cornwall. However, the medieval strip fields were laid out apparently with little or no regard to the pre-existing Romano-British field systems, and parts of lowland Cornwall have been subjected to centuries of ploughing. For these reasons, the lowland prehistoric and Romano-British settlement and field pattern form a largely buried landscape. At present, we do not know the full extent of this landscape or how its various elements relate to each other in the way that we do for the uplands.

The most visible element of the lowland prehistoric landscape is the distribution of almost 400 rounds which survive as upstanding monuments. These include 64 previously unrecorded examples identified and mapped from RAF vertical photo-

Fig. 7.5 A multiphase landscape at Chysauster, Madron. In the *left centre* is the well-known courtyard-house settlement. To the right of this are the remains of associated irregular brick-shaped fields typical of Late Iron Age and Roman period settlements in West Penwith (these particular fields were badly damaged during agricultural improvement in the 1980s). Towards the *bottom right* and in the *top centre* of this photograph, lines of lode-back pits associated with the early tin industry are visible (Photo 3G/TUD/UK 209/5246 (13 May 1946). ©Cornwall County Council 2009)

graphs. Although there are hints that some rounds may have been in use during the Early Iron Age (Young and Quinnell 2001: 138–139), they are generally considered to have first appeared in the Later Iron Age. They were typical of the wider landscape between the fourth century BC and the sixth century AD and were a major part of the countryside during the Roman period, particularly during the second and third centuries AD (Quinnell 2004: 212–214).

Rounds were enclosed by substantial banks and ditches. When levelled by ploughing, these readily form cropmarks and a large number have been identified from aerial photographs (Fig. 7.6). In fact, the mapping of prehistoric enclosures is one of the most important aspects of Cornwall's National Mapping Programme. More than 1,000 new enclosures were identified, 550 of them from RAF vertical

Fig. 7.6 A cropmark round at Tregear, Ladock. This enclosure is bounded by a bank (visible as a *pale cropmark*) and an outer ditch (a *dark mark*). A secondary enclosure bounded by a single ditch is appended to the southern side of the main enclosure (Photo 3G/TUD/UK 222/5169 (11 July 1946). © Cornwall County Council 2009)

imagery. This more than doubles the number known prior to the NMP project (Johnson 1998), demonstrating that the density of rural settlement in Iron Age and Roman Cornwall was significantly greater than hitherto understood.

Of course, it should not be assumed that all of these enclosures were settlements. Current understanding of Cornish enclosures is based on a limited amount of evidence resulting mostly from small-scale investigations. Recent excavations have shown that the enclosures at Killigrew, St Erme (Cole and Nowakowski forthcoming) and possibly Little Quoit Farm, St Columb Major (Lawson-Jones 2010), were dedicated metalworking sites. Enclosures of this type were not previously recognised, nor were a considerable number of very small enclosures (less than 0.1 ha in extent) which were recorded from aerial photography (including the RAF verticals) during Cornwall's NMP.

Nonetheless, available evidence points to the likelihood that many of the new enclosures were settlements, and on a broad level, the distribution of enclosures provides a good indication of the currently definable Iron Age and Romano-British settlement pattern in lowland Cornwall. This settlement pattern is not uniform, but is marked by apparent 'hotspots' and by significant gaps. In general, the distribution of enclosures is weighted towards central and west Cornwall, and a large proportion of the new enclosures identified from RAF verticals are in western areas such as the Helford estuary.

Excavation and environmental evidence demonstrates that rounds were the homes of farmers who practised mixed agriculture (Quinnell 2004: 211–214 and 221–225). Whilst RAF verticals (and aerial photographs generally) reveal some elements of the buried landscape, such as field systems and trackways, the picture they provide is incomplete. So although 467 cropmark enclosures and rounds have been transcribed from RAF photography, only 87 cropmark field systems have been recorded. Furthermore, these field systems are invariably fragmentary, and the fields generally appear to be larger than the brick-shaped fields found, for example, in West Penwith, suggesting that not all the boundaries are visible (of course, it is possible they represent different patterns of field enclosure to those in West Penwith). Thus, even in locations where enclosure distribution testifies to a densely populated landscape, we do not have a full understanding of, for example, the extent of arable land between the settlements. In this respect, information mapped from RAF verticals, by identifying 'hotspots' and gaps, indicates those locations where geophysical survey and other remote sensing techniques might best be targeted to elucidate the ways in which the landscape was used.

7.8 Conclusion

This brief review of the role of historic RAF photography in Cornwall's National Mapping Programme has been necessarily selective. There has been no mention, for example, of Bronze Age barrows and other prehistoric ceremonial monuments (more than 1,000 barrows were transcribed from RAF photographs). Even so, the account presented above underlines the extent to which historic vertical aerial photographs were an indispensable source for the mapping.

A small range of factors combine to make RAF photography such a valuable archive. The photographs cover the entire county, they were taken at propitious times of year and in favourable conditions both for cropmark and earthwork features, and they predate the widespread breaking in of moorland in the later twentieth century (most notably during the 1980s), the post-war expansion of towns and the late twentieth-century move towards deep ploughing. These benefits are likely to be replicated elsewhere in the UK, and RAF photography should be regarded as an essential source for programmes of mapping archaeological sites from aerial photographs.

Bibliography

Barton, D. B. (1969). *A history of tin mining and smelting in Cornwall*. Exeter: Cornwall Books.

Benson, D., & Miles, D. (1974). *The Upper Thames Valley: An archaeological survey of the river gravels*. Oxford: Oxford Archaeological Unit.

Bewley, R. (2001). Understanding England's historic landscapes: An aerial perspective. *Landscapes, 2*, 74–84.

Cole, R., & Nowakowski, J. (forthcoming). Killigrew Round. *Cornish Archaeology.*

Fox, H. S. A. (1971). *A geographical study of the field systems of Devon and Cornwall*. Unpublished PhD thesis, University of Cambridge.
Fox, H. S. A. (1973). Outfield cultivation in Devon and Cornwall: A reinterpretation. In M. Havinden (Ed.), *Husbandry and marketing in the South-West* (Papers in economic history, Vol. 8, pp. 19–38). Exeter: University of Exeter.
Gerrard, S. (1987). Streamworking in Medieval Cornwall. *Journal of the Trevithick Society, 14*, 7–31.
Gerrard, S. (2000). *The early British tin industry*. Stroud: Tempus.
Hatcher, J. (1973). *English tin production and trade before 1550*. Oxford: Clarendon.
Herring, P. (1997). *Godolphin, Breage: An archaeological and historical assessment* (HES Report 1997 R042). Truro: Cornwall County Council.
Herring, P. (1998). *Cornwall's historic landscape: Presenting a method of historic landscape character assessment*. Truro: Cornwall County Council.
Herring, P. (2006). Cornish strip fields. In S. Turner (Ed.), *Medieval Devon and Cornwall* (pp. 44–77). Macclesfield: Windgather Press.
Herring, P. C. (2008). Commons, fields and communities in prehistoric Cornwall. In A. Chadwick (Ed.), *Recent approaches to the archaeology of land allotment* (British Archaeological reports, international series 1875, pp. 70–95). Oxford: Archaeopress.
Herring, P., Sharp, A., Smith, J. R., & Giles, C. (2008). *Bodmin Moor an archaeological survey, Vol. 2: The industrial and Post-Medieval landscapes*. Swindon: English Heritage.
Johnson, N. (1998). Cornish farms in prehistoric farmyards. *British Archaeology, 31*, 12–13.
Johnson, N., & Rose, P. (1994). *Bodmin Moor an archaeological survey, Vol. 1: The human landscape to c1800*. London: English Heritage/RCHME.
Lawson-Jones, A., & Kirkham, G. (2010). Smithing in the round: excavations at Little Quoit Farm, St Columb Major, Cornwall. *Cornish Archaeology 48–49*(2009–10), 173–227.
Padel, O. J. (1985). *Cornish place-name elements*. Nottingham: English Place Name Society LVI/II.
Penhallurick, R. D. (1986). *Tin in antiquity*. London: The Institute of Metals.
Preston-Jones, A., & Rose, P. (1986). Medieval Cornwall. *Cornish Archaeology, 25*, 135–186.
Quinnell, H. (2004). *Excavations at Trethurgy Round, St Austell: Community and status in Roman and post-Roman Cornwall*. Truro: Cornwall County Council.
Ratcliffe, J., & Straker, V. (1996). *The early environment of Scilly, palaeoenvironmental assessment of cliff-face and intertidal deposits, 1989–1993*. Truro: Cornwall County Council.
RCHME. (1995). *Guidelines for the production of project specifications*. Royal Commission on the Historical Monuments of England internal document. Swindon.
Rose, P., & Preston Jones, A. (1995). Changes in the Cornish countryside AD 400–1100. In D. Hook & S. Burnell (Eds.), *Landscape and settlement in Britain AD 400–1066* (pp. 51–69). Exeter: University of Exeter Press.
Rowe, J. (1953). *Cornwall in the age of the industrial revolution*. St Austell: Cornish Hillside Publications.
Thomas, C. (1985). *Exploration of a drowned landscape: Archaeology and history of the Isles of Scilly*. London: Batsford.
Young, A. R., & Quinnell, H. (2001). Time team at Boleigh fogou, St Buryan. *Cornish Archaeology, 39–40*(2000–01), 129–145.

Chapter 8
The Use of Historical Aerial Photographs in Italy: Some Case Studies

Patrizia Tartara

Abstract Aerial photography should be considered one of the essential instruments for the analysis and the reconstruction of human activity in the landscape. Because it was acquired before the onset of mechanised farming, intensive industrial activity and major urban development, historical aerial photography reveals a landscape that is both closer to that of antiquity and more legible in detail. This chapter outlines the main sources of archival imagery available for Italy, summarises the photogrammetric methodology applied and illustrates the approach with a series of illustrated site and area case studies.

Acronyms and Abbreviations

AM	Aeronautica Militare
ICCD	Istituto Centrale del Catalogo e della Documentazione
IGM	Istituto Geografico Militare, Florence
SARA Nistri	Società per Azioni Rilevamenti Aerofotogrammetrici, Rome
CGR	Compagnia Generale Ripreseaeree, Parma
EIRA	Ente Italiano Rilievi Aerofotogrammetrici, Florence
ESACTA (Rome)	Ente SpecializzatoAerofotogrammetria Cartografia Topografia Aerofotogeologia
ETA Nistri (Rome)	Ente Topografico Aerofotogrammetrico
I-BUGA (Milan)	di Annalena Bugamelli
IRTA (Milan)	Istituto Rilievi Terrestri Aerei
SIAT (Rome)	Studio Italiano Aerofotogrammetria e Topografia
UTE (Florence)	Ufficio Tecnico Erariale

P. Tartara (✉)
Consiglio Nazionale delle Ricerche – Istituto Beni Archeologici e Monumentali, Campus Universitario, via dei Taurini 19, 00185, Roma, Italy
e-mail: patrizia.tartara@cnr.it

W.S. Hanson and I.A. Oltean (eds.), *Archaeology from Historical Aerial and Satellite Archives*, DOI 10.1007/978-1-4614-4505-0_8,

8.1 Introduction

For those disciplines concerned with study of the landscape, or ancient topography, aerial photography, and especially historical aerial photography, must be considered one of the essential instruments for the analysis and the reconstruction of human activity in the landscape, whether urban or rural (Lugli 1939; Castagnoli and Schmiedt 1955: 10ff.; Bradford 1957; Adamesteanu 1964; Castagnoli 1969: 7ff.; Schmiedt 1970). Insofar as it constitutes an objective record of a given situation at a particular moment in time, aerial photography is far too frequently undervalued or little used for the enormous volume of traces of the past that it can reveal, whether archaeological, geological, historical or simply agricultural. It is in fact an indispensable instrument for our knowledge of the landscape and environment, both present and past.

It is worth recalling that it is not always possible to gain information on buried ancient evidence from the most recent aerial photographs, or from satellite images, or from different sensory devices, such as multispectral imagery. Of equal or even greater utility in the vast majority of cases are historical images, such as the photographs taken by the Royal Air Force (RAF), the United States Army Air Force (USAAF), and the German Luftwaffe in the Second World War and the Istituto Geografico Militare (IGM) photographic coverage beginning in 1929–1930. These can show the landscape and environment in a state of preservation that is both closer to that of antiquity and more legible in detail. When this imagery was acquired, the use of machinery for cultivation was virtually non-existent, trees and bushes covered less terrain, and the numerous changes and damage brought about by the use of mechanised means of farming and earth moving, as well as by intensive industrial activity (such as quarries and illegal dump sites) and urban development that is neither planned nor supervised, had yet to occur.

8.2 Italian Aerial Photographic Archives

A specific photo-interpretation linked to the topographical analysis of a given territory, whether archaeological, geological or pedological, must be done not with individual photographs or stereoscopic pairs but on the basis of the totality of the documentation for the area in question. This requires a systematic checking of the various archives and the acquisition of all useful images. For Italy, apart from the documentation from the Second World War cited above, there exists a vast amount of historical material that is available in public archives and in the possession of private businesses. Access to this material varies correspondingly. First, there is the archive of the IGM, which constitutes the most extensive aerial photographic collection in Italy. It is possible to consult material at their offices in Florence, where location of the photographs is depicted cartographically at a scale of 1:100,000, or obtain them on-line (www.igmi.org). This photographic material comes primarily from flights performed by the military or by private entities or was acquired in relation to the production and updating of national territorial maps. A fundamental instrument of research for our sector is the *Volo Base*, produced in 1954–1955 to provide complete and systematic coverage of the country at a scale of 1:25,000 for updating maps. In the historical archive of the IGM, moreover, material is present from European and North African countries where the Institute had a role in producing maps for the former Italian colonies, for occupied territories and for military campaigns.

Second only to the IGM, in terms of its wealth of documentation, is the collection of the Aerofototeca of the Ministry for Cultural Heritage and Environment, which is located at the office of the Istituto Centrale del Catalogo e della Documentazione (ICCD). This is indubitably the largest civilian photographic archive in Italy. Also known since 1973 as the 'Laboratory for Interpretation of Photographs and the Measurement of Aerial Photographs', this structure was created in 1958. It was a response to the need to study the cultural heritage of Italy, especially with regard to archaeological sites. The scholar who promoted, and served as the first director of, the aerial photographic collection was the Romanian archaeologist Dinu Adamesteanu, one of the pioneers in the use of aerial photography for archaeological research in Italy.[1] Over the course of 50 years, the aerial photographic collection has acquired a great many other collections of aerial photographs. These include some of the wartime photographs made by the RAF (largely over the central and southern portions of the Italian peninsula) which were formerly maintained at the British School at Rome. Other photographs of Italian territory – taken by active service aircraft from North African bases – are preserved in the TARA archive now housed in RCAHMS in Edinburgh (see Chap. 2 by Cowley et al., this volume). Similar collections of the USAAF, which largely focused on the central and northern sections of the Italian peninsula and which were formerly maintained at the American Academy, are also included. In addition, there is some coverage effected by the Luftwaffe, and there are numerous collections of different origins, such as the Aeronautica Militare Italiana (AM) and a variety of private companies specialising in photographic survey in Italy, but which are no longer active today. In the meantime, especially in the 1960s and 1970s, many vertical and perspective shots of specific archaeological sites were taken. The aerial photographic collection, aside from providing assistance in reproduction, is also organised to facilitate study of its material as though it were a library for the history of the landscape of Italy.

[1] Together with Ferdinando Castagnoli, Professor of Topography of Rome and ancient Italy at the University of Rome 'La Sapienza', and Giulio Schmiedt, Director of the Istituto Geografico Militare and subsequently Professor of Archaeological Aerial Topography at the University of Florence, Dinu Adamesteanu began his career as an archaeologist in Romania, where he was one of the first to make use of aerial photo-interpretation. He participated in missions to the Middle East and thereafter moved to Italy, where he worked above all in Sicily from 1959 to 1964. He directed the Aerofototeca, subsequently transferred to the Archaeological Superintendency of Basilicata, where he was Superintendent, and lastly taught at the University of Lecce.

Finally, there are many archives in private photogrammetry companies in Italy.[2] Some of these have been organised and can be consulted. There are also the archives of the Regions and Comuni, which can be consulted at the technical offices that preserve aerial material created for the production and updating of technical maps. Other collections, tied to public works, are maintained by public organisations, such as the state railways, motorway companies, Azienda Nazionale Autonoma delle Strade (ANAS), companies for landscape reclamation, Azienda di Stato per gli Interventi nel Mercato Agricolo (AIMA), civilian protection and so on. There are also specialist archives created by research organisations and universities that work on different aspects of the landscape (archaeology, geology, environment, agronomy), such as the Institute for Ancient Topography at the University of Rome 'La Sapienza' and the Laboratory for Ancient Topography and Aerial Photogrammetry of the University of Lecce.

8.3 Methodology of Study

For a proper historical understanding and a diachronic interpretation of the data, it is necessary to transfer the results of the air-photo interpretation onto an adequate cartographic base, with the best definition and updating possible, by means of the methodologies and instruments that are currently available, as the needs of the occasion and the cartographic base require. For positioning with greatest precision, such as for single structures within an urban centre, it will prove useful to make use of photogrammetric transcription which is analytical when possible. The individual elements restored need to be codified; otherwise, use must be made of modern digital photogrammetric instruments. In those instances where less precision is required, or when the original documents are not available, or where images are not photogrammetric, use can be made of suitable commercially available geo-referencing software. In the case of oblique images, use can be made of similar software with satisfactory results, even if they obviously possess less geometrical precision.

The first example of the aerial photogrammetric mapping of an ancient city was a response to the stimulus that Ferdinando Castagnoli gave to archaeological cartography (Castagnoli and Schmiedt 1957: 125–148; Piccarreta and Ceraudo 2000: 88–89). The principles of cartography focused ('finalizzata') on archaeology (Piccarreta 1989, 2003a; Ceraudo 1999, 2003; Guaitoli 2003b) have been

[2] There are nearly 100 companies that have worked or are still active in Italy. Foremost amongst these are Aerofoto Consult (Rome), Aerotop (Rome), CGR (Parma), EIRA (Florence), ESACTA (Rome), ETA Nistri (Rome), Aerofotogrammetrica Nistri (Rome), Fotocielo (Rome), BUGA (Milan), Luigi Rossi (Florence), IRTA (Milan), Rossi (Brescia), SARA Nistri (Rome), SIAT (Rome) and UTE (Florence). For a detailed listing of the main companies, see Ceraudo (2000): 189–203, Boemi (2003): 37–42.

successively calibrated on the basis of subsequent developments of this prototype, even with the evolution of instruments enabling analytical and digital photographic transcription. Today it is also systematised for the codification of the individual elements of the territory and its archaeological heritage[3] and employed upon an ever wider scale for the study and topographic reconstruction of ancient urban centres and sections of territories. Especially when entirely or partially abandoned urban centres are being studied, 'finalizzata' analytical (i.e. numerical) photogrammetric maps (with appropriate codification) must be produced *ex novo*. In more recent work, a further element has been introduced: the transcription of the archaeological traces visible from previous or historical flights, sometimes referring to levels that have since been lost, and the merging of these traces in a single archaeological file upon a representation of the present terrain.

8.4 Case Studies

8.4.1 Caere

An in-depth study of the territory of Caere, between the urban centre of the ancient city and the coastline, has been based in effect upon the photo-interpretation and transcription of traces visible on aerial photographs. Attention has been paid to detailed topographic accuracy and the correct positioning of the individual archaeological features, particularly the extraordinarily extensive necropolises, the ancient road network and various structures. The territorial analysis, which is still in progress, started with the plains of Banditaccia and Monte Abbadone that lie on two sides of the ancient city and has gradually been extended to the coast, south and north-north-west (Pyrgi-Santa Severa).

Historical and more recent photographic images from both national public archives and private organisations have been used from flights of various types and dates. Particularly useful were IGM 1929–30; RAF 1943; IGM 1955, 1958, 1960 and 1968; AM 1959–60; SARA Nistri 1986, 2004; IGM 2004; '*Volo base*' IGM

[3] A modern vector cartographic base makes work in our sector easier, with its codification of the individual elements. Apart from eliminating the obligation to work with scale, the innovative contribution of analytical photogrammetry permits the assignment of a code to each and every element that is included. There is a general codification in use today for commercial cartography. By structuring a specific system of codes with a section given over to archaeological elements and others to all other aspects of the terrain (geomorphology, hydrography, viability, buildings, infrastructures, etc.), we obtain a graphic database of what is represented. This constitutes the first information system for a specific territory. Each element included consists of geographical coordinates: the altitude and encoded information that is available to all non-specialised users. By providing numerical data and the grid of codes employed, the system can easily be transferred to anyone who has access to even a simple CAD or graphic design program.

2002; *Volo* 2004 (this last was expressly made for the Laboratory for Ancient Topography and Photogrammetry of the University of Lecce) and oblique aerial photographs taken in collaboration with the Nucleo Tutela Patrimonio Culturale dei Carabinieri and the Carabinieri Helicopter Groups of Pratica di Mare and Bari in the years 1997–2008. Together they document both the hundreds of traces of completely unknown or unpublished structures and complexes (largely funerary in nature) and the, unfortunately, very high number of clandestine excavations that have taken place.

The oldest images available for the territory are the aerial photographs of 1930, which are held at the IGM in Florence. These photographs constitute the first example of photogrammetric relief for cartographic and cadastral purposes and consist of a series of pairs of 13 × 18 cm negatives on glass slides, which can be laterally superimposed only for a few millimetres (e.g. Fig. 8.1). The search for and identification of useful individual images required a great deal of effort. Each pair of slide negatives covers only a modest portion of territory that is extremely difficult to recognise, for they are almost completely without identifiable references to recent maps. Furthermore, significant environmental changes (such as modified watercourses) and human interventions (such as roads, buildings and the parcelling of fields) have been frequent. However, the photographs, taken at a relatively low altitude and at a moment when the landscape was still largely untouched by major disturbances, allow the reading, interpretation and mapping of a large number of archaeological traces that are especially visible. Consequently, they document a situation that in many instances was not visible in later photography. The 1943 RAF photographs published by Bradford, although of especial importance, do not achieve the same level of information because of problems of definition, scale and quality of photographic emulsion (cf. Bradford 1947: 74–80, tab. I, 1957: pl. 34).

Frequently distorted on their long sides and with light reflections caused by the sun's rays striking the photographic objective (e.g. Fig. 8.1), the 1930 photographs were taken from planes that did not yet possess the required stability, and they obviously do not achieve the geometrical characteristics of a modern photogrammetric series. In many instances, as a result of sharp changes in altitude and the failure to maintain a level flight path, the strips are irregular and sometimes the images do not even overlap. Thus, even determining the geographical positioning of these images has proved difficult. To achieve this, the best available cartographic resources have been employed, especially the 1:10,000 maps of the Regione Lazio, which are of medium quality. But the best coverage, at the same scale, has been that of the Comune of Cerveteri. A special aerial photographic strip was acquired on which the finalised photogrammetric transcription of the enormous complex of monumental necropolises was superimposed. The archaeology of the urban area is excluded from the present research since it has for some time been the object of study by the Istituto ISCIMA of the CNR.

Once a mosaic had been created from more than 40 photographs (Tartara 2003a: pl. III), a first attempt was made at geo-referencing the whole ensemble. However, since the precision obtained was not considered satisfactory, it was necessary to deal

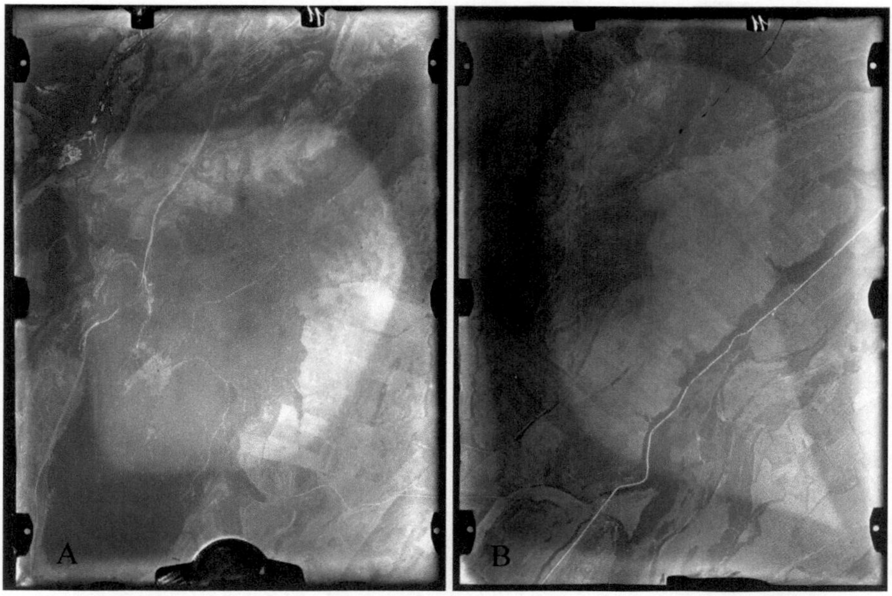

Fig. 8.1 Cerveteri. A 1930 IGM glass slide negative pair (IGM photograph, 1930 flight, strip 4, neg. 11, authorization IGM nr. 6467 of 16 Jan. 2009)

with smaller portions of the territory in turn. Each photograph was subdivided to allow for work upon individual sectors, and the results were deemed sufficiently satisfactory.[4]

The transcription of the vast number of traces evident on the historical aerial images of 1930 onto the map, and subsequently onto the three-dimensional model, allows a good overview of the archaeological heritage of the area and a partial reconstruction of the ancient landscape (Fig. 8.2). The complex consisted of the urban area and very extensive monumental necropolis flanking, or rather surrounding, it on both sides of the large plain. The necropolis extends along the major axes of communication as far as the coastline and ports, with occasional nuclei or large isolated tombs.

The excellent definition of the images also allows the identification and transcription of even the entrances (dromoi) and sometimes the burial chambers of the individual tumuli, as well as the primary and secondary road system. In other sectors of the necropolis, traces of funerary complexes can be seen with a series of entrances to burial spaces at regular intervals along the same frontage (Fig. 8.2d). These might reasonably be interpreted as tombs of 'aedicula' or 'dado' type laid out in rows. Similarly, the

[4] With a mean square error that varies between 1.90 and 3.90 m; at other points, the precision achieved can be measured in terms of centimetres.

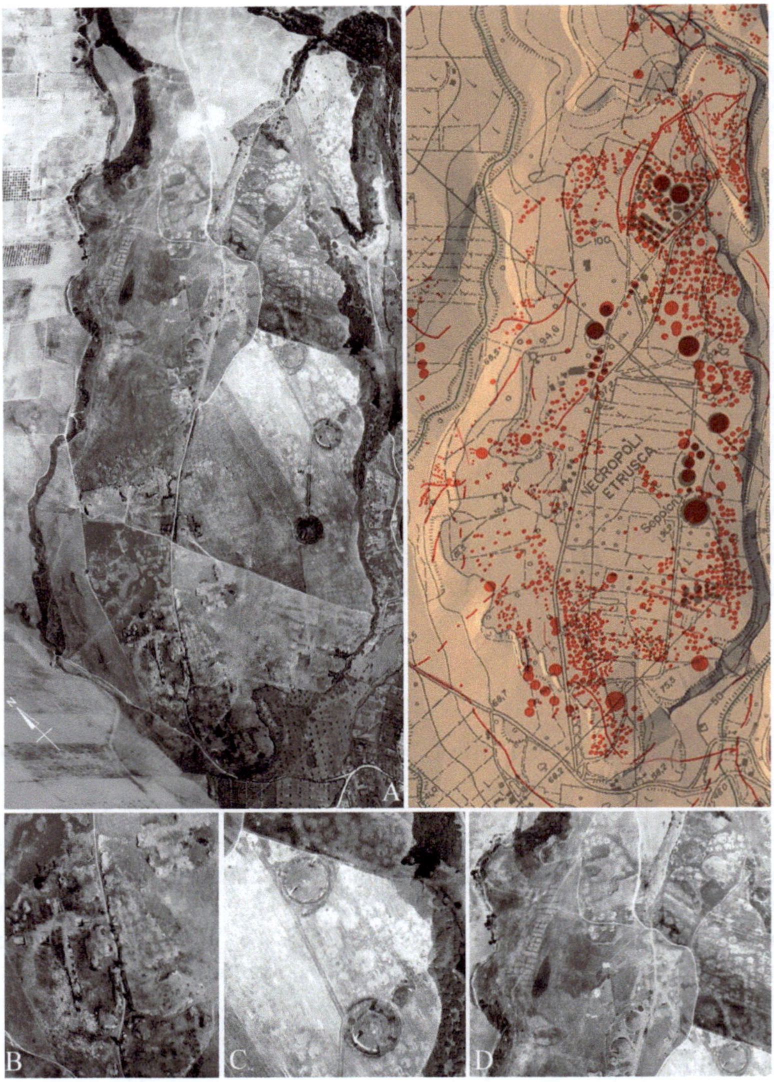

Fig. 8.2 Cerveteri: (**a**) A section of the Banditaccia plateaux showing, on the *left*, a mosaic of aerial photographs (IGM 1930) and, on the *right*, the partial transcription of visible traces on 3D cartography (After Tartara 2010). (**b**, **c**, and **d**) show details of different kinds of tombs on the plateau

imagery has made it possible to plot much of the ancient road system, both primary links between the city and the coast and with the neighbouring centres, and service roads for the numerous agricultural settlements scattered over the territory.

The overall result of the graphic reconstruction (Tartara 2003a), based upon objective data derived from photographic interpretation and systematic geographical positioning, presents a situation quite different from the limited impression of the archaeological area that is available to the visitor to the ancient city today. The monumental necropolis of Banditaccia that people visit represents but a minute fraction (roughly 10%) of the ancient burial zone, most of which still survives and is measurable. This zone has a high density of tombs of various types, primarily tumuli of varying dimensions, sectors with aligned tombs of the 'dado' and 'chamber' type, and areas with very dense clusters of 'fossa' tombs, extending along the very extensive plain of Banditaccia covering an area of roughly 3,000 by 700 m. Various parts of the complex cannot easily be reached or understood, such as the necropolis of Monte Abbadone, Polledrara, and the less known example of Macchia della Signora, Monte Abbadoncino, the vast burial area at the base of the acropolis where systematic excavations and chance discoveries have documented even the earliest phases of the necropolis, such as the Villanovan cemetery of Sorbo, and where a masterpiece of the Orientalising culture of Etruria, the Regolini Galassi Tomb, can still be visited (Tartara 2010). Consequently, the visitor does not appreciate that they are parts of a single, enormous monumental complex that is exceptional both for its extent and for its structural evidence.

In view of the quantity and precision of the sometimes quite detailed information discernible, a further development of the transcription of the archaeological evidence onto a three-dimensional model of the terrain might be to allow a reconstruction of the ancient appearance of the plain, of the Etruscan metropolis, of its cemeteries and of the surrounding territory. This would be useful for scientific purposes as well as providing a basis for multimedia products aimed at informing the public and enhancing the value of the heritage. The observation of the overall image that results from piecing together the information gathered, even if still largely provisional, allows some insight into the present state of the archaeological heritage, which has been included among the UNESCO World Heritage sites for some years.

In the flat plain at the base of the acropolis, where the modern town of Cerveteri has developed over the past 30 years, numerous tumuli of sizeable dimensions are attested on the historical photographs relating to a series of road lines. The greater part of the evidence has been obliterated by urban development, and the density of buildings has rendered it impossible to appreciate the few monuments that survive. A good knowledge and detailed census of the archaeological heritage would have facilitated better planning of the urban development, but the work of 'recovery' done from the 1930 photographs does allow a form of environmental restoration of this portion of the territory.

8.4.2 Arpi

Of particular interest for reading the numerous archaeological and geological traces for the reconstruction of the landscape are the aerial photogrammetric strips already mentioned as having been made in the *Volo base* of 1954–1955, taken to update the cartography of the whole country. In an IGM photograph taken in September 1954, used by Schmiedt (1970), the large Daunian settlement of Arpi (Foggia), the most extensive indigenous centre of pre-Roman Italy covering an area of c. 1,000 ha, is clearly visible. All visible traces have been restored photogrammetrically[5] and then added to a detailed overall map based upon a recent flight and combined with data from flights of various periods (Guaitoli 2003c) (Fig. 8.3). The settlement area is D-shaped, the straight side defined by the course of the river Celone, which ancient sources claim was partially navigable. The perimeter of the inhabited area is defined by a defensive wall, c. 11 km in length, which consisted of either an earthen *agger* or unbaked bricks partially faced with stones. Two phases have been identified, the *terminus post quem* of the earlier provided by the rediscovery of a child's tomb, which indicates a date in the sixth century BC (Mazzei 2003: 186). From photogrammetric measurements, it emerges that the *agger* was preserved to a height of c. 6 m in 1954, but today is reduced to 1.5–2 m. Detailed study by the Laboratory of Ancient Topography of Lecce indicates that the *agger* was c. 17–18 m wide. Farming activities have continually damaged the structure, which is nonetheless visible on the ground as a light brown coloured line c. 40–60 m in width, with large fragments of unbaked bricks that crumble at the first rains and mix with the terrain. The line of the fortification is cut at various points by dark traces running from outside towards the centre of the urban area. These are probably to be interpreted as ancient watercourses or prehistoric riverbeds. Reaching a width of as much as 40 m, these were narrowed and used in the historical period as routes providing a communications network for the inhabited area.

The perimeter of the city was first identified in the aerial photographic strips of the RAF for the area of Foggia by J. Bradford (1957: 167–169), who was acting as aerial observer for the airport of San Severo a few kilometres away. The first traces of occupation of the area have been dated to the Neolithic; while during the Iron Age, the Daunian settlement developed, perhaps in scattered concentrations. A decisive break with the past occurred in the sixth century BC, when the defensive circuit was built. The city was closely linked to all the historical events of the region, which determine both its extraordinary development and its progressive impoverishment from the second century BC down to its almost complete abandonment in the Imperial period (Mazzei 2003).

8.4.3 Roman Centuriation

The contribution of the initial IGM flight of 1955 was extremely useful for the reconstruction of Roman centuriation in the territory between the rivers Vulgano, Celone and Cervaro. This centuriation was included in the leaf IGM 164, between Arpi and the present inhabited centre of Troia (Martin 1990: 175ff.),

[5] The transcription was undertaken with the analogue instrument Galileo V, which was turned into an analytical instrument. The special characteristics of the 'base flight' (*Volo Base*) did not allow the employment of more recent analytical or digital instruments because they cannot be used to create a model of the terrain, which normally serves as a basis for the transcription.

Fig. 8.3 Arpi (Foggia). (**a**) An IGM flight in September 1954 shows the Daunian settlement perimeter clearly visible, outlined by defensive rampart circuit variously cut by fossil riverbeds and roads (After Guaitoli 2003c: 187; photo in Aerofototeca Nazionale, ICCD-Roma). (**b**). Analytical transcription of the archaeological traces of the settlement (After Guaitoli 2003c: 190)

which came into being after the destruction of the Daunian and Roman city of Aecae (Livy, 24.20; Polyb. 3. 88; Pliny, *Nat. Hist.* 3.105). The regular division of the fields into *centuriae* of 20 *actus* in length has been known for some time (*Lib. Col.* I, 210, 7–9; *Lib. Col.* II, 261, 3; Bradford 1949: 67ff.; Schmiedt 1985: 272ff., 1989: pls. XII–XIV) and is connected to the reference in the *Liber Coloniarum* to the assignment of land in the *Ager Aecanus* at the time of the Gracchi.

The photo-interpretation of this aerial coverage has revealed an exceptional visibility and wealth of traces of ancient and medieval structures, which permit their precise location upon the terrain or on maps. There are clear traces of roads, identified by the drainage ditches that run alongside them, especially at intersections. These allow for satisfactory precision in interaxial measurements and indicate the exact dimensions of the *limites* and related elements, without having to rely, as is generally the case, on the partial survival of such elements in modern land divisions. The measurement of the *centuriae* varies from 705 to 707 m, just under the theoretical norm of 710 m which is frequently documented. These figures are consistent over large portions of the land and must, therefore, be considered reliable. Insofar as the photographs provide the coverage and precision that permits this,[6] an analytical photogrammetric transcription of the traces that can be identified through stereoscopic analysis has been undertaked. This system must considerably enlarge these traces,[7] but the data can be drawn with sufficient precision on the available IGM maps at a scale of 1:25,000.[8] A further step forward has been obtained through the analysis of the photographs themselves in a digital photogrammetric system, after the digital format of the slides had been acquired by means of a photogrammetric scanner with a very high definition.[9]

---

[6] Apart from the problems due to scale, which is close to 1:30,000, the 1954–1955 IGM photographs are not always of high quality. They possess photogrammetric characteristics only in part and present a series of technical difficulties including anomalies in the creation of the model, type of marks and inadequate elements for the optical correction of the camera with which they were acquired. Transcription has been performed with the analytical system Galileo Stereosimplex and the analogue system Galileo V (transformed into analytical), both of which have a precision of 1 μm. Consequently, the data gathered have relatively large tolerances, but nonetheless they seem more than adequate for the study of large-scale land divisions.

[7] When enlarged, the stereoscopic model resulting from slides illuminated from behind allows for the identification of details that are invisible in photographic enlargements or else fuzzy in traditional photo-interpretation done with table stereoscopes.

[8] Several IGM maps were acquired in raster format with the highest definition of a cartographic scanner, corrected for various deformations (of support, printing, paper) by means of a grid calibrated on the basis of geographical coordinates, assembled and inserted in the GIS modules of the territorial information system of the CNR, Ufficio Sistemi Informativi Territoriali per i Beni Culturali (GIS for Cultural Heritage Office). The archaeological data gathered and encoded were subsequently put back into the system as part of the cataloguing of the archaeological data of the territory.

[9] Thanks to the CIGA of the Italian Military Air Force for use of the photogrammetric scanner.

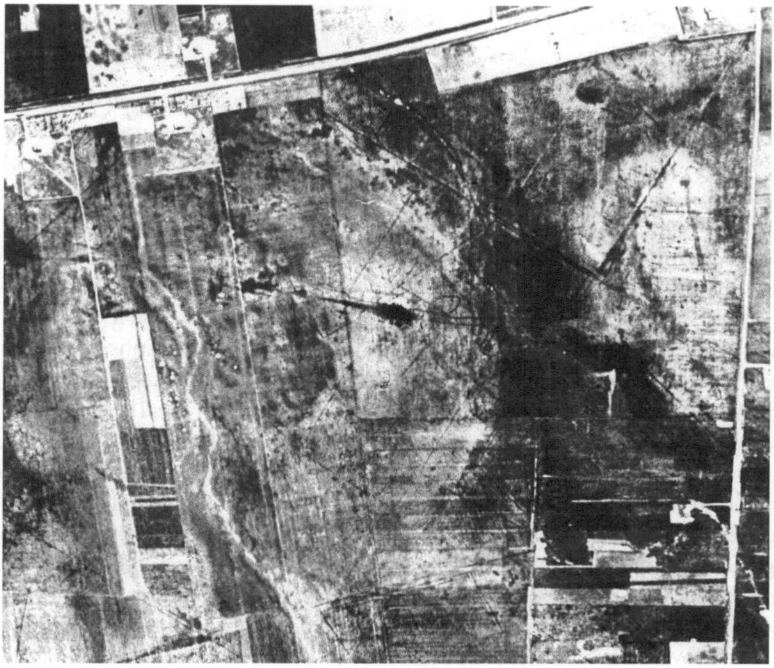

Fig. 8.4 Arpi and Aecae territory: traces of Roman *limitatio* axes, settlements, field allotments and stretches of road system in a detail from the IGM May 1955 flight (After Guaitoli 2003e: 475; photo in Aerofototeca Nazionale, ICCD-Roma)

Within the *centuriae,* there are clear traces of settlements, even complex ones, aligned with roads and large parcels of land or fields. Some of the latter display parallel trenches for vines or holes laid out in parallel rows for olive or fruit trees, as well as a noteworthy number of components of land divisions that are apparently older or more recent. In one detail (Fig. 8.4), it is possible to see clearly the traces of the axes of centuriation, settlements from different periods, allotments that are in part aligned with the field divisions and stretches of road systems (Guaitoli 2003e).

The assignment of this *limitatio* to the *Ager Aecanus* cannot be viewed as certain. The *agrimensores* also mention the *Ager Arpanus* (Arpi lies roughly 4 km northeast of Foggia), as well as the *Collatinus*, *Herdonitanus*, *Ausculinus*, *Sipontinus* and *Salpinus* (*Lib. Col.* I, 210, 10–13 L). Traces of centuriation appear to the east of Aecae and continue as far as the current periphery of Foggia, where, even on older cartography, they appear to be obliterated by the disposition of plots of land in radial format around the modern city. However, while not ruling out the attribution to Aecae, we should consider the possibility that the centuriation is to be referred to the *Ager Arpanus*, whose urban centre does not seem to have been continued in use later than the early second century BC, unless, on the basis of analogies elsewhere, we envisage a single subregional *limitatio* not tied to any individual urban centre.

8.4.4 *Abruzzo*

The RAF photographs used in the topographical analysis of a portion of the territory of Abruzzo, focusing on the section between L'Aquila and Capestrano and along the route of the Royal Sheep Track (Tratturo Regio), provide another detailed example of this approach. This work involved systematic analyses of a portion of the vast territory, beginning with test zones and gradually extending the area of research; on-site inspections; updating and documenting of previous data, both bibliographical and archival, as well as on-site information; and a detailed review of the existing aerial photographic documentation, with especial attention given in the various archives to the recovery and interpretation of historical photographs.

The historical photographs in this research, primarily those taken during the Second World War and to a lesser extent IGM images from the 1960s, have shown themselves especially useful because of the fact that at the time they were taken, there was an almost complete absence of trees and woodland, unlike today. This was the result of the extensive fires related to long-term, large-scale pastoral use. Consequently, it has proved possible clearly to discern fortified settlements on the high ground,[10] some necropolises not presently visible, and stretches of road, particularly the course of the Tratturo Regio and the routes of minor sheep-tracks that can no longer be made out on the ground (Fig. 8.6a). The monitoring flights over areas at risk, which have been conducted since 2003 with the Reparto Elicotteri del Raggruppamento Aeromobili Carabinieri di Pratica di Mare, coordinated by the Nucleo Tutela Patrimonio Culturale dei Carabinieri,[11] and since 2007 with the Sezione Aerea della Guardia di Finanza di Pescara, have been particularly helpful.

These flights have provided previously unknown information concerning individual complexes, necropolises, roads and traces of the cultivation of various crops that is extremely important for our knowledge of the evolution of human occupation in the area. Subsequently, some of these remains have been studied and extensively excavated by the Soprintendenza per i Beni Archeologici dell'Abruzzo so as to preserve and assign value to them. A detail from the RAF photograph of 13 May 1944 (Fig. 8.5) has made it possible to detect the remains of the fortified medieval settlement of Leporanica, which had at least two phases, built upon a previous, more extensive Vestine settlement of Iron Age date (Tartara

[10] Some of which are otherwise known only from scaled sketches drawn up during occasional surveys. On-site surveys were done by dott. E. Mattiocco, a dedicated amateur whose work has contributed in a meaningful fashion to our current knowledge of the archaeological evidence on the uplands of the Abruzzo interior.

[11] Within the context of collaboration between Prof. M. Guaitoli, director of the Laboratory of Ancient Topography at the University of Lecce – CNR, the Nucleo Tutela Patrimonio Culturale dell'Arma dei Carabinieri and the author, this work has shown itself to be productive over many years of monitoring particular areas of the country at risk, with flights in Abruzzo undertaken since 2003. These have focused upon the mountainous interior areas of the province of L'Aquila, with particular attention dedicated to the Vestine area along the Royal Sheep Track (Tratturo Regio), and the Marsican area along the territory bordering the Fucino basin.

Fig. 8.5 Abruzzo. Medieval settlement of Leporanica in an RAF photograph of 13 May 1944 (After Tartara 2003b: 205; photo in Aerofototeca Nazionale, ICCD-Roma)

2003b: 204–205, 2007: 518–520). The external edge of the ditch (C) that served to defend the castle's circuit of defensive walls on the northern and eastern sides can clearly be seen. Nearly triangular in shape, a first circuit occupied the summit of the hill (A). This was partially reused in a second phase, to which belongs the clearly visible circuit of defensive walls (B), towers, access gate and numerous remains of structures connected with the life of the village. The settlement was built upon a previous, more extensive Vestine settlement of Iron Age date.

In a photograph taken by the RAF in September 1943 (Fig. 8.6b), an archaeological complex can be seen for the first time (Tartara 2003b: 202–204, 2007: 545–547), though the two settlements were partially documented in a scaled sketch made during surveys undertaken by a dedicated amateur (Mattiocco 1986: 114ff.). Not unusual in the area under consideration, the complex consists of two fortified settlements (Monte Boria and the so-called Colle Campo di Monte, at a and b, respectively) and diverse clusters of tumuli (c) that make up the necropolis. In particular, it is clearly visible from the photograph that even the settlement at Colle Campo di Monte (b) had two defensive walls, where previously only a single circuit had been noted. The image also shows the various groups of tumuli still retaining their domed coverings, which have since been destroyed.

The enlargement of an AM image from 1955 has enabled us to observe at Colle Sinizzo, to the east of the small lake of the same name, the traces of a necropolis consisting of inhumation burials (Fig. 8.7). Previously unknown, this necropolis cannot now be seen on the ground (Tartara 2003b: 205–206, 2007: 494–495). The numerous burials are arranged in two distinct groups, aligned east–west and northwest-southeast, respectively. It is quite likely that these are to be linked to the neighbouring settlement of Colle Sinizzo (Tartara 2007: 492–493), a small, isolated

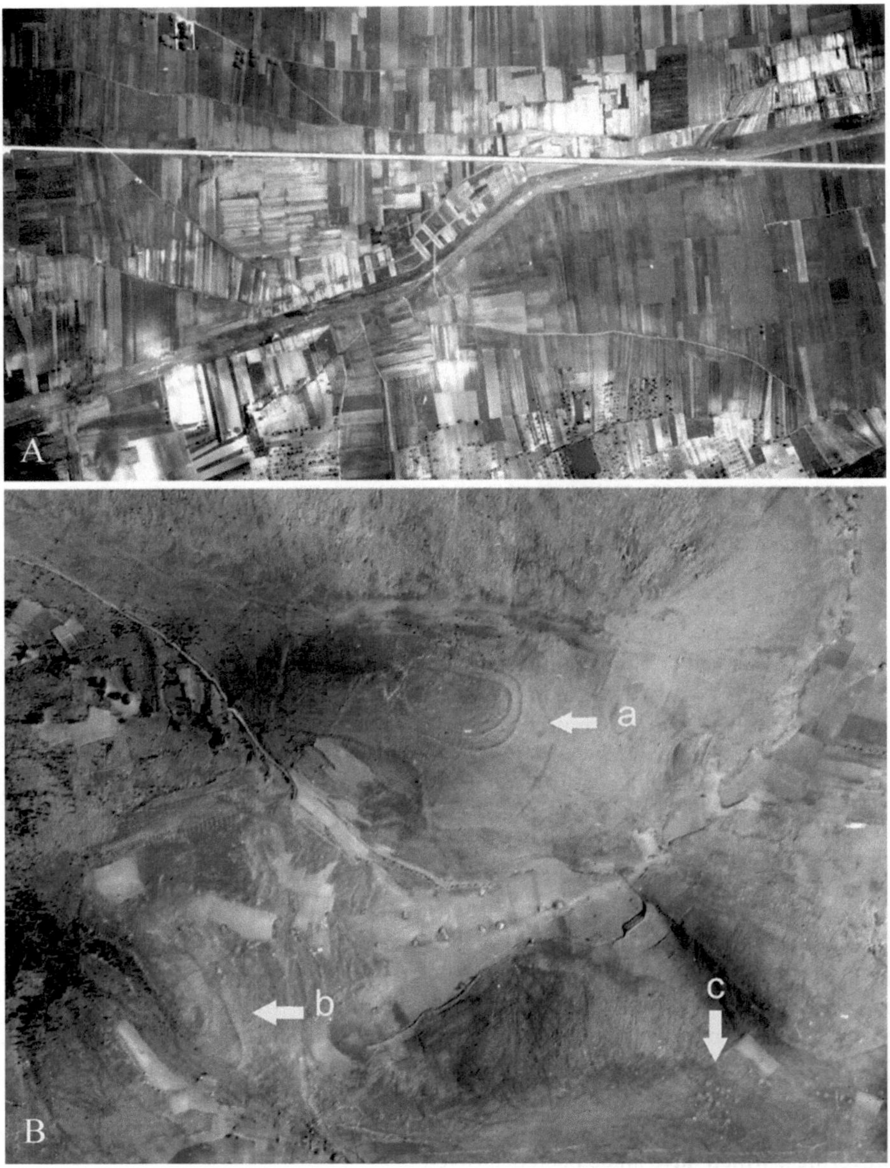

Fig. 8.6 Abruzzo. (**A**). Surviving remains of the Royal Sheep Track (Tratturo Regio), the route visible in the continuous strip alignment of fields along the boundary of the track (After Tartara 2003d: 455; photos in Aerofototeca Nazionale, ICCD-Roma from 1955). (**B**). Two fortified hilltop settlements (*a* Monte Boria and *b* the so-called Colle Campo di Monte) and *c* clusters of tumuli with their dome coverings still intact, in an RAF photograph of September 1943 (After Tartara 2003b: 202–204)

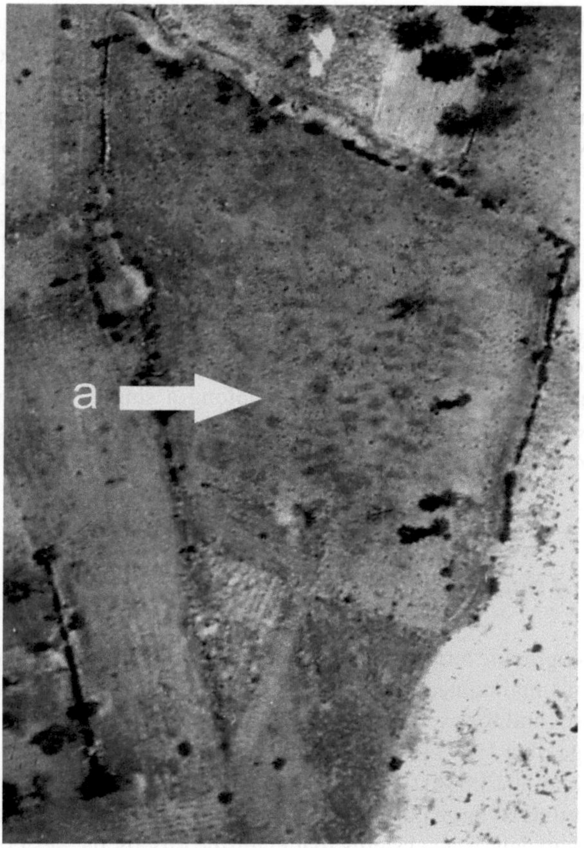

Fig. 8.7 Abruzzo. Detail from an RAF photograph of 1955, just above the small Sinizzo Lake, showing traces of a necropolis consisting of inhumation burials (After Tartara 2003b: 205–206; photo in Aerofototeca Nazionale, ICCD-Roma)

eminence, roughly 230 m to the west of the necropolis, which exhibits traces of structures and materials belonging to the Bronze Age and Middle Ages.

For the territory of the L'Aquila district that has been examined, despite limited information and often accidental recovery, it is possible to outline in broad terms a picture of human activity from at least the Neolithic onwards, whether enduring or simply transitory, scattered along a primary natural route. This route was subsequently defined and formalised in historical times and then institutionalised in the Aragonese period as the Royal Sheep Track (Tratturo Regio).[12] The archaeological evidence is located along this axis, even in those areas that subsequently constitute

[12] It was connected to a complex network of minor or service routes, e.g. the sheep-route that breaks off at the church of the Madonna of Centurelli in the direction of Capestrano and then Forca di Penne, or that which runs SE-NW towards to the valley of Capestrano from the plain of Collepietro to the E of Serra di Navelli, (see Tartara 2007: 551).

the cores of three large, stable concentrations of population – perhaps related to three different groups – at the present Fossa (*Aveia*) and Capestrano (*Aufinum*) and at *Peltuinum.*

Running for 243,257 km, the Tratturo Regio is the principal Sheep Track of the Dogana della Mena delle Pecore, which was established by Alfonso of Aragon. The visible details have since been largely obliterated by roadworks and associated infrastructure (see also Tartara 2003d, 2007: 534–538). The Tratturo Regio L'Aquila-Foggia, which was the most important route employed for transhumance, cut across three regions: Abruzzo, Molise and Puglia. With a width varying between roughly 75 and 120 m, it followed ancient pastoral routes of communication and exchange. There are numerous settlements from prehistoric and proto-historic times identified along this route, both temporary and more established. A continuous strip of campsites is visible in the aerial images situated along a single, unvarying alignment constituting the boundary of the land belonging to the state, which was clearly occupied over the years and given over to cultivation (Fig. 8.6a).

8.4.5 City of *Satricum*

Numerous photographic strips of the Aeronautica Militare Italiana can be consulted in the collections maintained at the Aerofototeca Nazionale. These were taken for military purposes and for exercises, with flights focused occasionally on the archaeological heritage. In a 1941 AM photograph of the territory of the ancient Latin city of *Satricum* (Guaitoli 2003d), it is easy to make out the trace of the rampart and of the ditch that outlines a large portion of the perimeter of the inhabited area (Fig. 8.8). Traces of some ancient road systems that enter the city by its western gate are also clear. Some tumuli, excavated in the 1800s, are visible along the road coming from the north-northwest. In this instance, the historical aerial photographs are especially important as the area of the city was so heavily damaged by agricultural changes wrought by powerful mechanised means in the early 1960s that it was effectively obliterated. Archaeological studies and research conducted in the 1800s (see Guaitoli 2003d), before 1960 (Castagnoli 1963: 505ff.) and during the 1990s by the Dutch Institute of Rome in collaboration with the University of Groningen (Heldring and Stibbe 1991; see especially Maaskant Kleinbrink 1987, 1992; Knoop and Stibbe 1989) have shown occupation of the inhabited centre from the ninth century BC until the mid-Republic, confirmed by materials recovered from the necropolises which have been investigated. Knowledge of the city has been greatly enhanced by detailed study of the aerial images, especially the oldest photographic strips for 1936, 1939, 1941 and 1943, as well as of the historical cartographic documentation, which was realised with extraordinary care by the Opera Nazionale Combattenti in 1934 for the reclamation of the Pontine Marshes. The innumerable black stains of charcoal pits, which recur on the photographs and can be seen even decades later (see Guaitoli 2003a), constitute proof of the deforestation which occurred in an area previously covered with dense forest.

Fig. 8.8 An AM photograph of 1941 showing the ancient Latin city of Satricum. The traces of the rampart and the ditch surrounding the city are visible at *A* and *B*, and a road approaching the city at *C* (After Guaitoli 2003d; photo in Aerofototeca Nazionale, ICCD-Roma)

8.4.6 Cavallino

The inhabited area of the ancient Messapic settlement of Cavallino occupies almost 68 ha, located a few kilometres to the southeast of Lecce. Systematic excavations conducted between 1964 and 1967 by the Universities of Lecce and Pisa indicated occupation during the Bronze Age and Iron Age. The wall circuit was built in the second half of the sixth century BC, its destruction occurring during the fifth. Thanks to a 24×48 cm AM aerial photograph taken in 1968, it has proved possible to delineate a previously unknown part of the circuit of the external defensive walls which, where visible (Fig. 8.9, solid arrows), consisted of a wall made of blocks of local limestone and a shallow external ditch (dark line) (Tartara 2003c). The south-south-western sector of the city has been progressively obliterated by uncontrolled urban development. The dark line seen in the photograph (Fig. 8.9, clear arrows) shows the ditch as a soil mark, thus revealing the route taken by the wall circuit that is no longer extant. Recently, the area of the ancient city was purchased by the University of Salento, which has created an archaeological park and, within it, the Museo Diffuso.

Fig. 8.9 A detail from an AM 1968 aerial photograph of the ancient inhabited area of the Messapic centre of Cavallino, partially occupied by the modern town. *Solid white arrows* indicate the line of the external defensive walls; *clear arrows* indicate the dark line of the ditch, where the accompanying city-wall circuit has disappeared (After Tartara 2003c: 334; photo in Aerofototeca Nazionale, ICCD-Roma)

8.4.7 Rocavecchia

A 1968 AM aerial photograph (Fig. 8.10) of the territory of the Comune di Melendugno, along the Adriatic coast to the south of Lecce, shows the outline of the ancient inhabited centre at Rocavecchia with great clarity. Occupied from the Bronze Age to the fourteenth century AD, when it was used by the Aragonese in their struggle with the Turks after the occupation of Otranto (Guglielmino and Pagliara 2004: 527–548; Pagliara and Guglielmino 2005; Pagliara et al. 2007, 2008), the site includes the Grotta della Poesia (Cavern of Poetry), which is remarkable for its hundreds of Messapic, Greek and Latin inscriptions (Pagliara 1987). This image has facilitated the recognition and mapping of a double ditch outside the wall circuit where nothing is visible on the ground (Piccarreta 2003b), while in older photographs, a third ditch closer to the wall circuit has been identified.

Fig. 8.10 Rocavecchia (Lecce): a detail from an AM 1968 aerial photograph shows the area of the inhabited Messapic centre defined by the line of a double ditch outside the wall circuit (After Piccarreta 2003b: 236; photo in Aerofototeca Nazionale, ICCD-Roma)

8.5 Conclusion

Aerial photography, especially historical aerial photography, has made an immense contribution to our knowledge of the past landscape, as the case studies outlined above demonstrate very clearly. Each example is different in character, giving us information about the past by means of different kinds of archaeological traces. I would like to emphasise the importance of the transcription of those traces onto maps in order to provide a better understanding of the evidence as a whole and particularly to facilitate better government care for and protection of the monuments. Nonetheless, I must underline how necessary it is not to detach the process of aerial photographic analysis from field survey; otherwise, we may lose a substantial portion of the information that can be read but rarely precisely contextualised, either chronologically or typologically.

Bibliography

Adamesteanu, D. (1964). Contributo dell'Aerofototeca Archeologica del Ministero P. I. alla soluzione dei problemi di Topografia antica in Italia. In *Atti del X Congresso Internazionale di Fotogrammetria* (pp. 1–76). Lisbon: Consiglio del X Congresso Internazionale di Fotogrammetria.

Boemi, M. F. (2003). L'Aerofototeca Nazionale – Appendice. In M. Guaitoli (Ed.), *Lo Sguardo di Icaro. Le collezioni dell'Aerofototeca Nazionale per la conoscenza del territorio* (pp. 17–42). Roma: Campisano.

Bradford, J. S. P. (1947). Etruria from the air. *Antiquity, 21*(82), 74–83.

Bradford, J. S. P. (1949). Buried landscapes in southern Italy. *Antiquity, 23*(90), 58–72.

Bradford, J. S. P. (1957). *Ancient landscapes. Studies in field archaeology*. London: Bell.

Castagnoli, F. (1963). Satricum. *L'Universo, 43*, 504–518.

Castagnoli, F. (1969). La prospezione aerea negli studi di Topografia antica. *Quaderni de 'La Ricerca Scientifica', 60*, 7–13.

Castagnoli, F., & Schmiedt, G. (1955). Fotografia aerea e ricerche archeologiche. *L'Universo, 35*, 3–14.

Castagnoli, F., & Schmiedt, G. (1957). L'antica città di Norba. *L'Universo, 37*, 128–148.

Ceraudo, G. (1999). *Introduzione all'aerofotogrammetria applicata all'archeologia*. Formia: Il Grande Blu.

Ceraudo, G. (2000). Reperibilità delle foto aeree. In F. Piccarreta & G. Ceraudo (Eds.), *Manuale di Aerofotografia archeologica – Metodologia, tecniche e applicazioni* (pp. 7–218). Bari: Edipuglia.

Ceraudo, G. (2003). Fotografia aerea: tecniche, applicazioni, foto interpretazione. In M. Guaitoli (Ed.), *Lo Sguardo di Icaro. Le collezioni dell'Aerofototeca Nazionale per la conoscenza del territorio* (pp. 75–85). Roma: Campisano.

Guaitoli, M. (2003a). Carbonaie. In M. Guaitoli (Ed.), *Lo Sguardo di Icaro. Le collezioni dell'Aerofototeca Nazionale per la conoscenza del territorio* (pp. 86–88). Roma: Campisano.

Guaitoli, M. (2003b). Dalla cartografia finalizzata ai sistemi informativi territoriali. In M. Guaitoli (Ed.), *Lo Sguardo di Icaro. Le collezioni dell'Aerofototeca Nazionale per la conoscenza del territorio* (pp. 101–102). Roma: Campisano.

Guaitoli, M. (2003c). Arpi. In M. Guaitoli (Ed.), *Lo Sguardo di Icaro. Le collezioni dell'Aerofototeca Nazionale per la conoscenza del territorio* (pp. 187–193). Roma: Campisano.

Guaitoli, M. (2003d). Satricum. In M. Guaitoli (Ed.), *Lo Sguardo di Icaro. Le collezioni dell'Aerofototeca Nazionale per la conoscenza del territorio* (pp. 283–289). Roma: Campisano.

Guaitoli, M. (2003e). Centuriazione tra Aecae ed Arpi. In M. Guaitoli (Ed.), *Lo Sguardo di Icaro. Le collezioni dell'Aerofototeca Nazionale per la conoscenza del territorio* (pp. 470–474). Roma: Campisano.

Guglielmino, R., & Pagliara, C. (2004). Nuove ricerche a Roca. *Annali della Scuola Normale Superiore di Pisa, 2*, 527–548.

Heldring, B. H. M., & Stibbe, C. M. (1991). Scavi a Satricum – Campagne 1988–89. *Archeologia laziale, 10*, 229–233.

Knoop, R. R., & Stibbe, C. M. (1989). s.v. Satricum. In *Enciclopedia dell'arte antica* (suppl. II). Rome: Treccani.

Lugli, G. (1939). *Saggi di esplorazione archeologica a mezzo della fotografia aerea*. Roma: Istituto di studi romani.

Maaskant Kleinbrink, M. (Ed.). (1987). *Settlement excavation at Borgo le Ferriere 'Satricum', I*. Groningen: Egbert Forsten.

Maaskant Kleinbrink, M. (Ed.). (1992). *Settlement Excavation at Borgo le Ferriere 'Satricum', II*. Groningen: Egbert Forsten.

Martin, J. M. (1990). Troia et son territoire au XI siècle. *Vetera Christianorum, 27*, 175–201.

Mattiocco, E. (1986). *Centri fortificati vestini*. Sulmona: Museo civico di Sulmona.

Mazzei, M. (2003). Arpi. In M. Guaitoli (Ed.), *Lo Sguardo di Icaro. Le collezioni dell'Aerofototeca Nazionale per la conoscenza del territorio* (pp. 185–186). Roma: Campisano.

Pagliara, C. (1987). La Grotta Poesia di Roca (Melendugno, Lecce). *Note preliminari. Annali della Scuola Normale Superiore di Pisa, 17*, 267–328.

Pagliara, C., & Guglielmino, R. (2005). Roca: dalle curiosità antiquarie allo scavo stratigrafico. In S. Settis & M. C. Parra (Eds.), *Magna Grecia. Archeologia di un sapere* (pp. 298–321). Milan: Electa.

Pagliara, C., Maggiulli, G., Scarano, T., Pino, C., Guglielmino, R., De Grossi Mazzorin, J., Rugge, M., Fiorentino, G., Primavera, M., Calcagnile, L., D'Elia, M., & Quarta, G. (2007). La sequenza cronostratigrafica delle fasi di occupazione dell'insediamento protostorico di Roca (Melendugno, Lecce). Relazione preliminare della campagna di scavo 2005 – Saggio X. *Rivista di Scienze Preistoriche, 57*, 311–362.

Pagliara, C., Guglielmino, R., Coluccia, L., Malorgio, I., Merico, M., Palmisano, D., Rugge, M., & Minonne, F. (2008). Roca Vecchia (Melendugno, Lecce), SAS IX: relazione stratigrafica preliminare sui livelli di occupazione protostorici (campagne di scavo 2005–2006). *Rivista di Scienze Preistoriche, 58*, 239–280.

Piccarreta, F. (1989). Fotogrammetria finalizzata alla cartografia archeologica. In *Atti del Convegno 'La cartografia dei Beni Storici, Archeologici e Paesistici'* (pp. 143–149). Rome.

Piccarreta, F. (2003a). Aerofotogrammetria finalizzata all'archeologia. In M. Guaitoli (Ed.), *Lo Sguardo di Icaro. Le collezioni dell'Aerofototeca Nazionale per la conoscenza del territorio* (pp. 96–98). Roma: Campisano.

Piccarreta, F. (2003b). Rocavecchia. In M. Guaitoli (Ed.), *Lo Sguardo di Icaro. Le collezioni dell'Aerofototeca Nazionale per la conoscenza del territorio* (pp. 236–237). Roma: Campisano.

Piccarreta, F., & Ceraudo, G. (2000). *Manuale di Aerofotografia archeologica – Metodologia, tecniche e applicazioni*. Bari: Edipuglia.

Schmiedt, G. (1970). *Atlante aerofotografico delle sedi umane in Italia*. Florence: Istituto Geografico Militare.

Schmiedt, G. (1985). Le centuriazioni di Lucera e di Aecae. *L'Universo, 65*, 260–277.

Schmiedt, G. (1989). *Atlante aerofotografico delle sedi umane in Italia. III: la centuriazione romana*. Florence: Istituto Geografico Militare.

Tartara, P. (2003a). Ortofotopiano storico IGM 1930 del territorio tra Cerveteri e la costa. In M. Guaitoli (Ed.), *Lo Sguardo di Icaro. Le collezioni dell'Aerofototeca Nazionale per la conoscenza del territorio* (pp. 157–166). Roma: Campisano.

Tartara, P. (2003b). Insediamenti d'altura dell'Abruzzo. In M. Guaitoli (Ed.), *Lo Sguardo di Icaro. Le collezioni dell'Aerofototeca Nazionale per la conoscenza del territorio* (pp. 201–209). Roma: Campisano.

Tartara, P. (2003c). Cavallino. In M. Guaitoli (Ed.), *Lo Sguardo di Icaro. Le collezioni dell'Aerofototeca Nazionale per la conoscenza del territorio* (pp. 333–335). Roma: Campisano.

Tartara, P. (2003d). Insediamenti lungo il Tratturo Regio nel settore aquilano. In M. Guaitoli (Ed.), *Lo Sguardo di Icaro. Le collezioni dell'Aerofototeca Nazionale per la conoscenza del territorio* (pp. 454–456). Roma: Campisano.

Tartara, P. (2007). Il territorio aquilano lungo il Tratturo Regio: primi dati per una carta archeologica sistematica. (Area tra Bazzano e Capestrano). In A. Clementi (Ed.), *I Campi Aperti di Peltuinum dove tramonta il sole ...- Saggi sulla terra di Prata d'Ansidonia dalla Protostoria all'età moderna* (pp. 449–565). L'Aquila: Deputazione Abruzzese di Storia Patria.

Tartara, P. (2010). Fotografie aeree storiche e recenti – Alcuni risultati dal territorio ceretano. In F. D'Andria, D. Malfitana, N. Masini, & G. Scerdozzi (Eds.), *Il Dialogo dei Saperi – Metodologie integrate per i Beni Archeologici e Monumentali, Tomo I* (pp. 427–432). Napoli: Edizioni Scientifiche Italiane.

Chapter 9
A Lost Archaeological Landscape on the Lower Danube Roman *Limes*: The Contribution of Second World War Aerial Photographs

Ioana A. Oltean

Abstract Extensive landscape developments in the Lower Danube region over the second half of the twentieth century have resulted in the loss of numerous archaeological features before appropriate rescue and recording work could be performed. The discovery of a number of historical aerial photographs in one of the uncatalogued sections of The Aerial Reconnaissance Archive has enabled the recovery of a range of landscape features of various dates in the Galați area on the Lower Danube even after their destruction, including unparalleled elements of the Roman frontier defences. This chapter considers the way in which the historical context of acquisition of these photographs impacts on their quality and quantity and how such contextual information can be used to help identify imagery in sections of the TARA archive which presently lack finding aids.

Archaeological research must often build interpretations based on incomplete evidence, heavily biased by the rate of survival and recovery of archaeological information. Extensive landscape developments in the Lower Danube region over the second half of the twentieth century have resulted in the loss of numerous archaeological features before appropriate rescue and recording work could be performed. This situation not only affects our level of knowledge of local landscapes but can have much wider impact, such as on our appreciation of the nature and evolution of the defensive strategies of the Roman Empire. Aerial photographs taken before modern developments took place can provide important information and help reduce some of the biases affecting our understanding of past landscapes and cultures.

I.A. Oltean (✉)
Department of Archaeology, University of Exeter, Laver Building,
North Park Road, Exeter, EX4 4QE, UK
e-mail: I.A.Oltean@exeter.ac.uk

W.S. Hanson and I.A. Oltean (eds.), *Archaeology from Historical Aerial and Satellite Archives*, DOI 10.1007/978-1-4614-4505-0_9,

The discovery of a few historical aerial photographs in the uncatalogued collections of TARA (The Aerial Reconnaissance Archive) enabled the identification of unparalleled elements of the Roman *limes* after their destruction. These photographs provide an opportunity to discuss the way in which the context of acquisition impacts on the quality and quantity of various types of Second World War archival material and how such information can be used to help the identification of imagery in areas of this archive where finding aids have not been yet developed.

9.1 Galaţi: Geographical and Historical Context

The modern town of Galaţi in eastern Romania is located on the left bank of the river Danube at its confluence with River Siret, overlooking the last major bend of the great European river before it reaches the Black Sea (Fig. 9.1(1)). Much of the Danube valley cutting across the steppe from south to north is, with few exceptions, spread across a marshy landscape with multiple river channels which changed and evolved over time. However, until the twentieth century, seagoing vessels were able to reach its harbour and, with the resources of cereals and timber for the international markets just emerging in the nineteenth century, Galaţi developed rapidly into the most important shipping base of the Lower Danube. Indeed, between 1856 and 1938, it was the seat of the European Commission of the Danube which oversaw the free navigation on the river. Moreover, the growth of Romania's oil industry after 1857 in the Ploieşti region (155 km W of Galaţi), mostly at the hands of Western companies, rapidly developed the network needed to establish efficient rail transportation to harbours, not just on the Black Sea shore at Constanţa but also on the Danube (Jensen and Rosegger 1978; Turnock 1974: 37).

Given its ideal location for fishing, farming and transport/communications, the area of Galaţi has been inhabited since at least the early Bronze Age (Dragomir 1996: 662). Much as during its later history, the area has been a sensitive point of border defence since the Roman Empire expanded into Dobrogea and established its boundary along the Danube in the first century AD. But even though a number of military bases were established along the right bank of the Danube in Dobrogea at *Troesmis*, *Arrubium*, *Dinogetia* Luncaviţa and *Noviodunum*, the large marshes in front of them would have provided little comfort if faced with eventual attacks from the north-west without additional fortifications on the raised plateau on the opposite shore (Fig. 9.1(3)). The Roman *limes* installations in the area of Galaţi (Romania) on the left bank of the Danube were known by Romanian archaeologists since the late nineteenth century to have included a rampart with an outer ditch to the west and north, which extended for 20 km between the modern villages of Traian on the Siret to the west and Tuluceşti overlooking the Prut river valley to the north. This position was therefore extremely favourable, allowing control of the problematic marshes with minimal effort. Also, it provided a foothold inside *Barbaricum* overlooking movement along the most important river valleys of Moldavia, the Siret and the Prut. Behind this linear rampart, previous field-based research had already

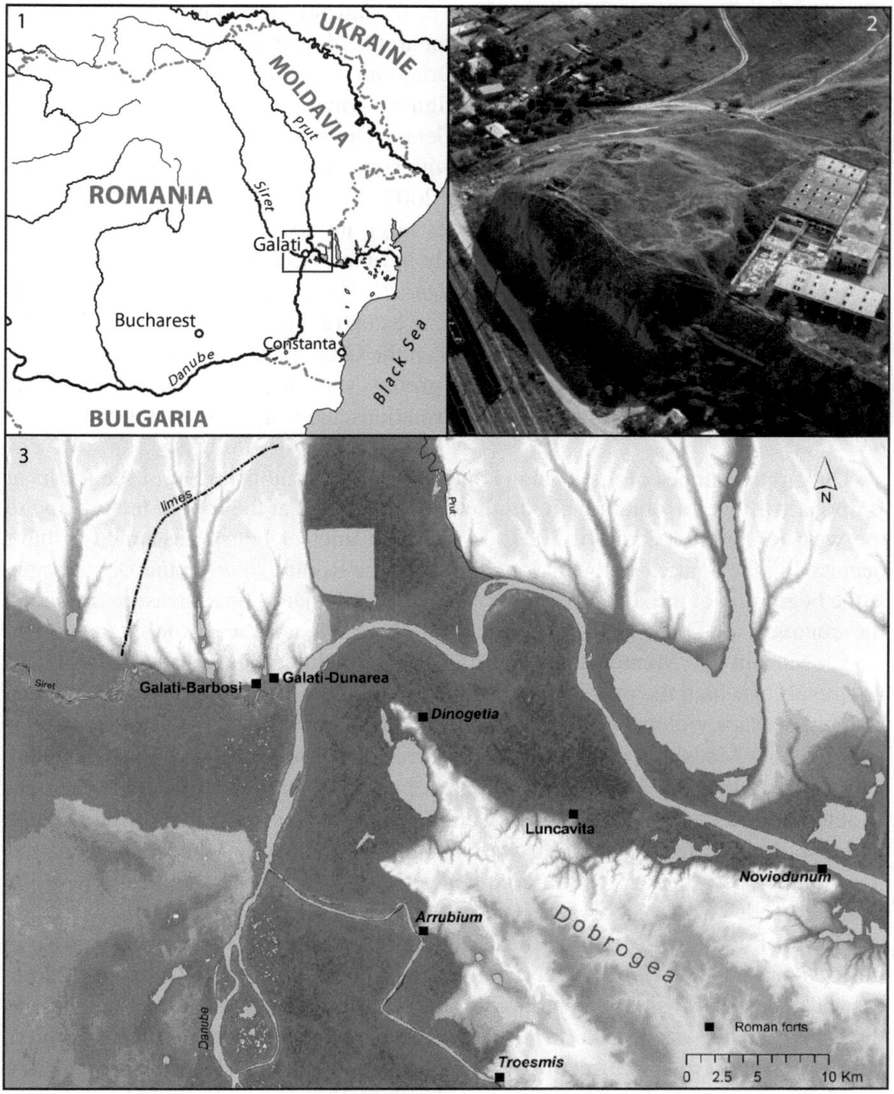

Fig. 9.1 **3**. The topographic location of Roman *limes* forts along the Danube in north-western Dobrogea (**1**: location of the study area; **2**: oblique aerial photograph of Galați-Bărboşi fort in July 2008; topographic and hydrographical raster data ©USGS)

located a number of military and non-military sites, the latter consisting mainly of funerary remains, including stray urn burials and barrows, with a few settlements (e.g. Dragomir 1996: 661–695, with bibliography; Croitoru 2007).

The site that attracted most specialist attention for over 170 years is the stronghold at Galați-Bărboşi on the Tirighina promontory overlooking the Siret

(Fig. 9.1(2)). Though the full chronology of the site still remains to be established in detail, various excavations by Romanian archaeologists, including G. Săulescu, V. Pârvan, N. Gostar, S. Sanie and I.T. Dragomir, seem to indicate that a Roman fortlet had been built there during the reign of Trajan on a former Iron Age site, the residence of the local leader of the Getae. The hill fort was built probably at some point in the second century BC but with some traces of an earlier, Bronze Age phase of occupation (Sîrbu and Croitoru 2007). The pre-conquest archaeological evidence recovered from the site includes large numbers of Greek amphorae or black- and red-figured pottery found mostly in funerary contexts in the vicinity and a large hoard of 517 Roman Republican *denarii*. These serve to indicate its importance from the sixth to the first century BC as a central place within the wider social and economic networks in operation between the Greco-Roman world and the local population (mainly Getae alongside other groups, such as the Scythians early in the sixth to fifth centuries BC and later the Sarmatians and Carpi – see Dragomir 1996: 663–664).

The circumstances of the Roman conquest and the establishment of the fortlet at Bărboşi, with its surrounding civilian settlement located at the foot of the hill and to the west (Sîrbu and Croitoru 2007: 14), remain unclear but are assumed to have occurred in the context of the major changes to the Roman *limes* in the Danube area at the beginning of the second century AD made by Trajan. These measures included the conquest of Dacia and the reorganisation of Moesia, with the expansion of Lower Moesia into Muntenia and south Moldova, including the area of Galați itself, as a result of which the pre-Roman occupation on the Tirighina hill may have met a sudden, perhaps violent end (Dragomir 1996: 665). However, if the Trajanic boundary of Lower Moesia in this sector was located further north, as is usually assumed, it is unlikely that the *limes* fortifications between Traian and Tuluceşti were constructed before the boundaries of Lower Moesia reverted to the line of the Danube under Hadrian. A few strongholds were kept on the left bank of the Danube, including Galați where control of the confluences of the rivers Siret and Prut with the Danube needed a Roman presence on both banks of the great river.

However, the fortification on the Tirighina hill was not the only military base in the area of the Galați *limes*. A second fortlet was located at Galați-Dunărea 1.25 km east of that at Bărboşi, also overlooking the Siret and closer to its confluence with the Danube. This was little known until the recent rescue excavations of 2004 which, despite severe erosion affecting the recovery of internal features, revealed a fortification some 40 by 40 m square within its wide surrounding ditch, giving an estimated internal area, with due allowance for a rampart, of some 0.09 ha (Țentea and Oltean 2009: 1515. See also below). The dimensions of the fortlet indicate that it was never intended to host more than a small detachment garrisoning this strategic outpost. Analogies are encountered elsewhere in various fortlets in Britannia (e.g. Martinhoe or Barburgh Mill) or those overlooking the Iron Gates of the Danube in Moesia Superior or those along communication routes in Dacia (e.g. Tibiscum – the early phase; Abrud; Boița) and along the *limes Transalutanus* in southern Dacia. The lack of constant maintenance work on the ditches, as apparent from their stratigraphy, and the nature of the archaeological material recovered from three separate layers of ditch infill, consisting mainly of fragments of wine amphorae, indicate that the fortlet had been used sporadically during the first half of the second century AD. This contradicts previous interpretations of its continuous use over almost two centuries (Ţentea and Oltean 2009: 1515–1518).

Unfortunately, indications of settlement in the immediate vicinity of these two Roman fortifications come mainly from funerary activity, rather than the identification of the settlements themselves. Burials of Roman date continue the earlier monumental fashion of tumuli, but with additional elements, including stone sarcophagi (see Ţentea 2007 with bibliography).

The archaeological discoveries reported so far were made predominantly in the immediate vicinity of Galaţi, either to the west along the Siret, around the villages of Şendreni and Bărboşi, or in the northern part of the town and along Lake Brateş, while much of the rest of the area enclosed behind the *limes* has remained insufficiently explored. The landscape has undergone great transformations in modern times as the town has developed and, therefore, the reported discoveries provide only a very fragmentary picture of the archaeological landscape. Most of the area to the west and the north of the town suffered dramatic intervention and destruction in the decades following the Second World War. Development was particularly extensive between the lakes of Cătuşa and Mălina, to the north of the known ancient fortification and cemetery at Bărboşi, with a huge steel factory covering some 14 km^2 (now Mittal Steel-Galaţi) being constructed there from 1960 to 1968 (Fig. 9.2). Moreover, the town itself has expanded greatly with new residential areas needed as a result of the massive industrial development (Turnock 1974: 135–136).

Given the scale of the areas affected by development around Galaţi, it is now difficult to place in their appropriate site and landscape context previously reported archaeological discoveries, which consist mainly of material culture. Therefore, the extent to which they accurately represent the nature and full extent of past human activity in the region remains highly questionable. In this situation, historical aerial photographs from the first half of the twentieth century which document the landscape prior to its destruction, such as those made during the Second World War, now provide probably the only chance of recovering plans of lost sites and filling in the gaps in the archaeological maps of the area.

9.2 Second World War Aerial Photographs of Romania: Acquisition, Coverage and Distribution

The few Second World War photographs utilised in the present study were acquired in 2005 from The Aerial Reconnaissance Archive (TARA – at that time based in the University of Keele, UK; now part of the RCAHMS, Edinburgh. See Chap. 2 Cowley et al., this volume) in the context of the author's postdoctoral British Academy-funded research project entitled *Contextualizing change on the Lower Danube: Roman impact on Daco-Getic landscapes* (see Chap. 18 by Oltean and Hanson, this volume). They all represent medium altitude vertical photographs

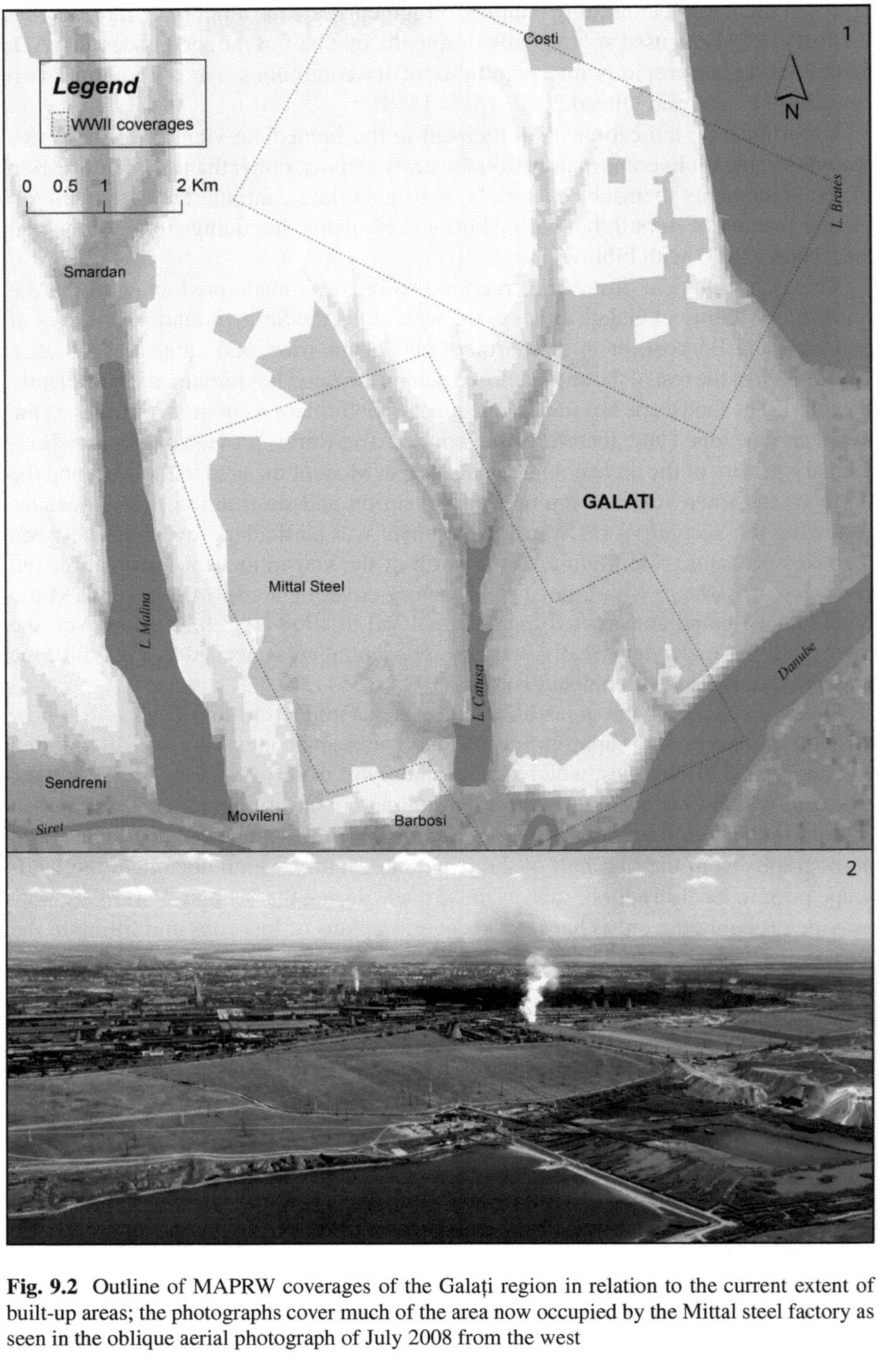

Fig. 9.2 Outline of MAPRW coverages of the Galați region in relation to the current extent of built-up areas; the photographs cover much of the area now occupied by the Mittal steel factory as seen in the oblique aerial photograph of July 2008 from the west

acquired during a reconnaissance flight gathering military intelligence on 31 May 1944 by the RAF (sortie no: 60.PR.460; height 28,500 ft). Unfortunately, the original films have not been located and may be lost, but high resolution digital copies of the original prints were purchased from TARA.

Given the size of the archive and the scarcity of finding aids available, locating any useful photographs may seem to be a stroke of luck (see also Chap. 2 by Cowley et al. this volume). However, prior knowledge of the historical context that stimulated the need for aerial photographs to be acquired in the study area provides useful information for archaeological users, helping them to assess the quantity and the quality of the photographic material which may be found in the archive. As stated elsewhere in this volume (Chap. 18 by Oltean and Hanson; Chap. 2 by Cowley et al.), military reconnaissance photography was acquired for very specific strategic reasons, depending on the needs of the opposing armies. Also, depending on the positions of those armies within the area under reconnaissance, the acquisition flights themselves would have been undertaken under very different conditions. The research programme allowed access through TARA to Second World War aerial imagery from different parts of Romania, in the north-west, centre and the south-east, produced by both sides in the conflict. Thus, it provided ideal conditions for illustrating the significant differences in the extent of coverage, acquisition dates, media and state of survival between the photographs taken by the RAF and the USAF and those taken by the Luftwaffe (see Chap. 18 by Oltean and Hanson, this volume).

The geographical position of Romania and its proximity to the Russian front line made the British and American armies less interested in acquiring extensive coverage to help the advancement of ground troops. Invariably, the intelligence gathered by them was aimed at narrower areas around points of interest for strategic bombing operations. Aerial photographs were taken of targets before and after attacks in an effort to improve bombing accuracy and assess the significance of the damages inflicted (Meilinger 2007: 148; McCloskey 1987: 916–918). Early in the war, the production and supply bases of German forces and of their allies (among them supply depots for oil and refined products), along with the necessary transport infrastructure (such as rail marshalling yards, harbours and bridges), were defined as among the most important objectives of strategic bombing; in particular, oil supply became a major priority after the Casablanca Conference in January 1943 (Meilinger 2007; BBSU 1998: 1–12; Schaffer 1985: 32; Werrell 1986). During the war the Ploieşti area was the most important European producer of crude and refined oil products, which were at that time directed exclusively to Romania's German ally. This made it a primary bombing target of the American 15th Air Force, with the occasional participation of the RAF (Meilinger 2007: 146–147; Stout 2003: 2–4; 7–20). Naturally, targets included the extraction fields and refineries in Ploieşti, Câmpina and Brazi, but facilities further afield at Bucharest, Braşov, Târgovişte and Doiceşti Teleajenul were also affected. Particular attention was dedicated to processing and storage facilities at shipping points along the Danube (e.g. Giurgiu and Cernavodă) or the Black Sea key outlet at Constanța (Fig. 9.3). The Ploieşti area was heavily defended not only by the numerous local anti-aircraft ground installations but also by a series of aerial bases near Bucharest (Băneasa and Otopeni) and

Fig. 9.3 Contemporary hand-annotated request for an enlargement of the areas of strategic interest at Constanţa, Romania: the large oil storage facilities, the railway terminal and the harbour (MAPRW sortie 60PR460 frame 4017. Licensor NCAP/aerial.rcahms.gov.uk)

elsewhere, such as Zilıştea (near Focşani), Tecuci, Piteşti, Buzău or, indeed, Galaţi. The latter town was an important node for rail transportation to Moldavia, where the front line had moved by early 1944. It was an important Danube harbour and ship-maintenance centre and a vital part of the Galaţi-Nămoloasa-Focşani defence line, an extensive system of fortifications built by the Romanian army at the end of the nineteenth century, which was rehabilitated and operational in the Second World

War (see below). However, it was Galaţi's small defensive airfield that constituted the most likely target for the reconnaissance flight of 31 May which produced the photographs under study here. The 15th Air Force attack which eventually followed on 6 June, at the same time as the D-Day operations in Normandy were under way, targeted the airfield with its 104 B-17s and 42 P-51s which had arrived via Ukraine (Carter and Mueller 1991: 247; Stout 2003: 145–146). The same reconnaissance sortie provided intelligence for a separate attack on the 11th June on the oil installations by the sea terminal at Constanţa (Fig. 9.3) (Carter and Mueller 1991: 251), which have provided equally important information for the reconstruction and analysis of the archaeological landscape in southern Dobrogea (see Chap. 18 by Oltean and Hanson, this volume, with earlier bibliography). It is to be expected, therefore, that coverage of other areas in the general vicinity of strategic targets in Romania or elsewhere should be found in the archive.

Another factor which needs to be considered when estimating the overall quantity and likely chronological distribution of the USAAF/RAF imagery of Romania is that this was acquired with great difficulty as operations in the region started to take place. Given the distance of the targets from the bases in the Mediterranean, flying over Romania was difficult for the Allies until 1944; indeed, the massive attack on Ploieşti of 1 August 1943, under the operation entitled 'Tidal Wave', was a landmark in military aviation history because of the numerous technical perils (over and above that of enemy fire) faced as a result of the long flight at low altitude from Benghazi over the occupied Balkans. Few aircraft returned to base without difficulty, if at all (Stout 2003). The situation improved in the spring of 1944 after the establishment of new aerial bases in Italy and after the cooperation agreements with the USSR allowed bases in the Ukraine to be used by 'shuttle' missions (Operation Frantic, June to September 1944).

In contrast, German aerial imagery of Romania consulted as part of this research programme reflects different strategic needs and possibilities from those of the British and American armies. German interest in strategic objectives in Romania was manifested quite early on, with the important railway node at Simeria in western Transylvania photographed as early as 7 June 1939 (GX mosaic no. 14331), a few months before the invasion of Poland had started. Similar photographs continued to be made in the country throughout the war, even though Romania was Germany's ally between October 1940 and August 1944, some of which provide valuable archaeological information (e.g. Medgidia, see Chap. 18 by Oltean and Hanson, this volume and Fig. 18.7). However, extensive block coverages which document in detail the topography, land use and transport facilities were also made. Such an example is that of the entire region of southern Dobrogea, extending well into north Bulgaria along the Danube, made by the Luftwaffe in April 1944 over a number of separate sorties. At first sight, this may come as a surprise, given that at the time both Romania and Bulgaria were allies of the Axis. But, apart from the data already mentioned, the coverage documents an extensive network of defensive trenches built in the first decades of the twentieth century as a result of the divergent positions of Romania and Bulgaria over the status of Dobrogea in the Balkan War (1912–1913) and in the First World War. With the area to the south of the existing modern boundary lost to Bulgaria in September 1940, the defences were renovated

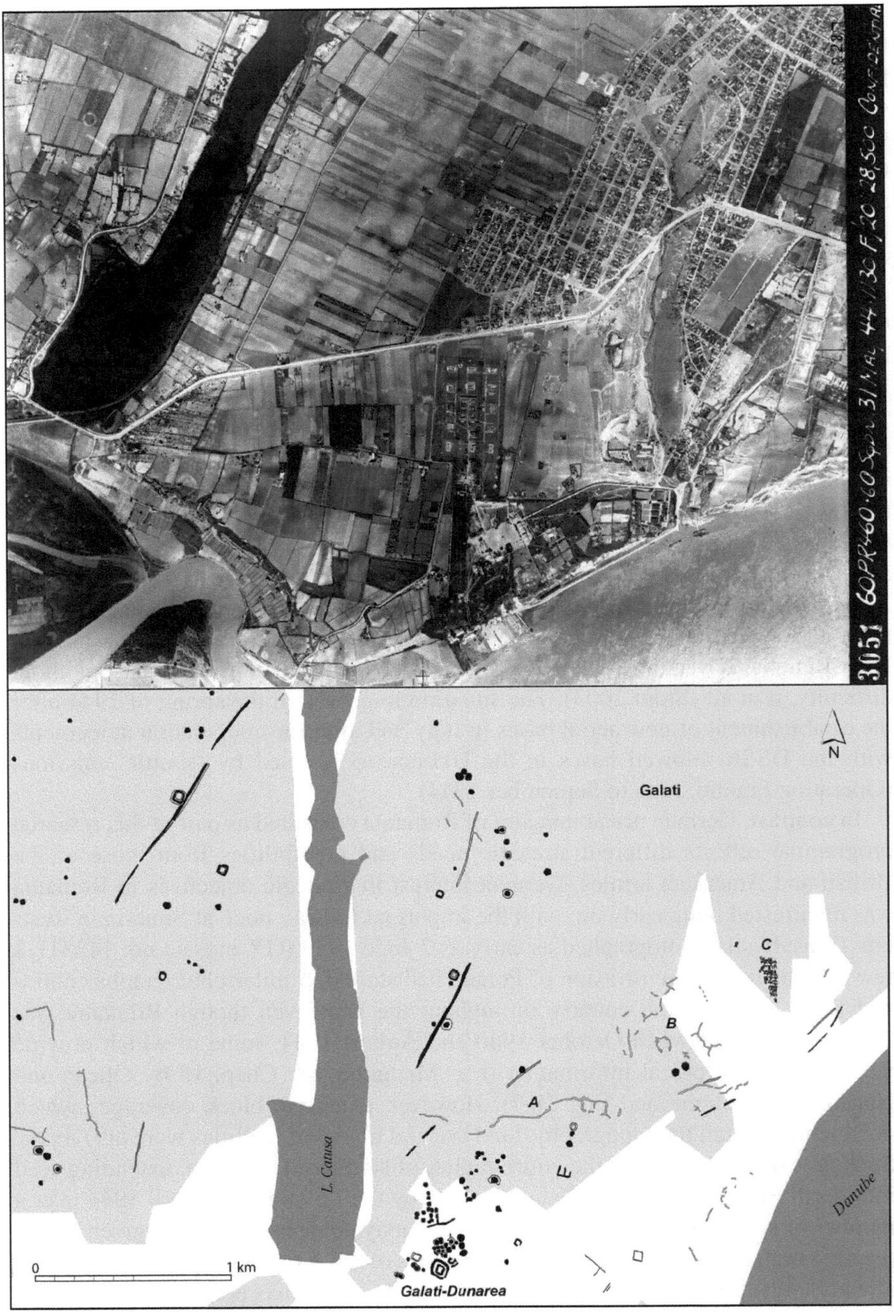

Fig. 9.4 *Upper*: 1944 aerial photograph (MAPRW sortie 60PR460 frame 3051. Licensor NCAP/aerial.rcahms.gov.uk) of Galați indicating the presence of archaeological features of Roman, medieval and early 20th century date lost under post-war development as transcribed (*lower*)

to ensure their usability if needed. In the spring of 1944, political developments in the Balkans raised the possibility of Bulgarian capitulation in March 1944 (see BBSU 1998: 18). This eventuality alongside the generally fragile basis of Romanian loyalty towards the Axis were potentially significant threats to the vital supply of Romanian oil resources to the German battle lines. Thus, the intelligence-gathering interest here is unequivocally aimed at providing the German army with the most up-to-date information for an eventual defence on the ground of the crucial system for the transport of oil in the area of the Lower Danube and across Dobrogea to Constanța on the Black Sea in the eventuality of an attack from the south. Alongside strategic information, however, the photo coverage also provides a huge amount of archaeological information which has enabled the first large-scale archaeological landscape reconstruction work ever to be performed in Romania (see Chap. 18 by Oltean and Hanson, this volume). As this dataset also covers a good part of the north of Bulgaria, it is easy to envisage its potential importance for reconstructing the ancient landscape also in that country, where aerial archaeology has yet to be introduced (see Chap. 1 by Hanson and Oltean, this volume). Moreover, given the spatial and chronological scale of German military operations and strategic interests in the Second World War, similar material of great archaeological potential would no doubt have been produced for other regions in occupied territories, allied countries and, possibly, even neutral countries (see also Going 2002).

9.3 Transcription and Mapping Issues

The vertical aerial photographs taken on 31 May 1944 provide incomplete coverage of the area surrounding Galați to the west and the north, particularly in relation to the area later affected by the construction of the steel factory (Fig. 9.2). Moreover, as noted above, the photographs were acquired with different priorities in terms of timing compared to those favoured by archaeological photo-interpreters. The best time for cropmark formation in the area of Galați is from early-mid June to early-mid July, so the photographic coverage was taken slightly too early in the growing season to best reveal buried archaeological remains through cropmarks. But elsewhere, even imagery taken earlier in the season was able to reveal larger extant or soil-marked archaeological features, such as ancient roads, linear defensive systems and tumular cemeteries (see Chap. 18 by Oltean and Hanson, this volume). In the area around Galați, a combination of extant features and early cropmarks, perhaps occurring as differential germination, give sufficient information to document a wide range of modern fortifications from the First and the Second World Wars, along with a number of ancient fortlets, roads and field systems (Fig. 9.4). Virtually all of them are new to the archaeological record and thus required immediate mapping and interpretation.

However, there are a number of problems which can affect the process of transcription and mapping of archaeological information from Second World War imagery onto modern maps. Physical and chemical deterioration of the photographs and films due to poor archival conditions may create additional problems. Stains and scratches can add false 'features' (spots or lines), while breakage, fading prints or chemical decomposition of films can all affect severely the clarity of the images at our disposal across whole landscapes or in more confined locations. It is, therefore, recommended that at least once during the process the archaeological photo-interpreter should have access to the original photographs, rather than the scanned copies which can be purchased. Furthermore, the original use of the photographic material by photo-interpreters and archivists during the war sometimes resulted in annotations being added to the original prints and films. These vary from occasional frame numbers or schematic interpretations (e.g. Fig. 18.7a) to more extensive handwritten or stamped texts which, although themselves a source of historical information, may result in additional obstruction to the visibility of archaeological features (Fig. 9.3). The printed photographs of Galați luckily survived quite well, with only occasional horizontal scratches and fingerprints transferred from the original film onto the photographs.

When dealing with aerial photographic material accumulated during the Second World War, it is not always easy for the archaeological photo-interpreter to identify the places recorded. Given the scarcity of finding aids surviving from the war period and the development of the landscape over the last 65 years, it is often difficult to relate the historical photographs to modern maps and imagery. The task of identifying control points is easier when individual datasets cover larger landscapes, such as higher altitude photographs and block coverage over extensive areas (such as the Luftwaffe coverage of southern Dobrogea), but can be difficult or even impossible when the coverage of one particular area is limited and the details observable on the ground are too generic or have since suffered radical transformation. The difficulty of finding control points also affects the accuracy of the geo-referencing of the photographs and the transfer of archaeological information onto site plans and area maps. The general wartime destruction and the post-war development in and around Galați mean that numerous possible control points have disappeared (particularly in the area now covered by the steel factory and new post-war residential areas) or have suffered transformations (such as road-widening or new buildings replacing older ones). This may cause location errors for transcribed features which are occasionally higher than is usually achievable, but since this problem relates primarily to areas where such transformations have erased all earlier traces of human activity, the importance of recovering these lost sites at all far supersedes concerns over the level of imprecision.

9.4 The Archaeological Landscape of Galați: Lost Features and Their Significance in Wider Context

The photographs studied here provided the basis for mapping a large number of features which at the date of photography had gone out of use and which may have been built at various dates from late prehistory onwards. The large majority of these are likely to be of ancient date, probably late pre-Roman and Roman. These include some 200 funerary barrows (tumuli), 6 linear features, probably indicating ancient roads, and 16 rectangular or square enclosures. Among the latter, the remains of the fortlet at Galați-Dunărea are clearly visible as at least partially extant where it is not overlain by modern cultivation, located on the edge of the higher ground overlooking the confluence of the Siret and Danube with good visual contact with both the fortlet at Bărboşi nearby and with the forts bordering the Danube in Dobrogea. The aerial photographs provide further details about the fortlet at Galați-Dunărea to augment those obtained by excavation undertaken in 2004, noted above (Țentea and Oltean 2009). Given the better preserved state of the site, a rampart some 6–7 m wide can be seen enclosing an area approximately 26 by 30 m, somewhat less than estimated from the recent excavations. The surrounding ditch also appears wider, up to 11 m, beyond which was an outer bank some 12 m wide, created by the upcast from the ditch, providing an additional defensive obstacle. Thus, the fortlet can be seen to have been more extensive overall than is apparent from the recent excavations, occupying an area across its defences of some 89 by 85 m. Slightly wider marks at the corners of the rampart may perhaps indicate the original provision of towers. Despite its better state of preservation in 1944, no buildings or subdivisions could be identified in the interior, but such detail is rarely obtained from this type of imagery, especially if the internal buildings were of timber (which may well have been the case here).

The aerial photographs also provide considerable additional data confirming that there is significant evidence of further associated activity in the immediate vicinity, with as many as 54 larger agglomerations of circular and sub-circular negative cropmarks, probably barrows of Roman date, clustered within 500–600 m of the fortlet (mainly to the north and north-west). This indicates the presence of one the largest concentrations of funerary activity and, by implication, of occupation so far recorded in the area. As elsewhere on the Lower Danube, tumuli have been previously documented around Galați, but, again as elsewhere, a full appreciation of the phenomenon is only now beginning to be grasped with the recent application of aerial remote sensing facilitating the identification and mapping of thousands of such monuments (see e.g. Chap. 18 by Oltean and Hanson, this volume, with earlier bibliography). The historical photographs allowed the recovery of a further 229 barrows (out of 883 recorded in total in the wider region from additional aerial and satellite imagery), helping to reveal the character and structure of the settlement pattern and the use of space behind and beyond the Galați *limes* (Fig. 9.5). Most of these barrows have been destroyed by subsequent development which, unfortunately, does not allow more precise chronological and cultural interpretation to be undertaken. Lines of communications articulating movement through the landscape also constitute new additions to the archaeological record, with 6 road lines and 3 more alignments of tumuli, probably also reflecting routeways, identified in the areas affected by development. Important access routes have been recovered leading to the north along Lake Brateş and the Danube east of Galați-Dunărea, linking this with a civilian settlement which, according to the national gazetteer (RAN -75105.01; http://ran.cimec.ro/sel.asp?Omod=1&nr=8&ids=708), was located under the modern town. Other roads to the north-east of the Bărboşi fortlet would have ensured its connection to sites north of modern Galați. Funerary barrows emerged along at least two of these roads, while other alignments on elevated positions border topographic

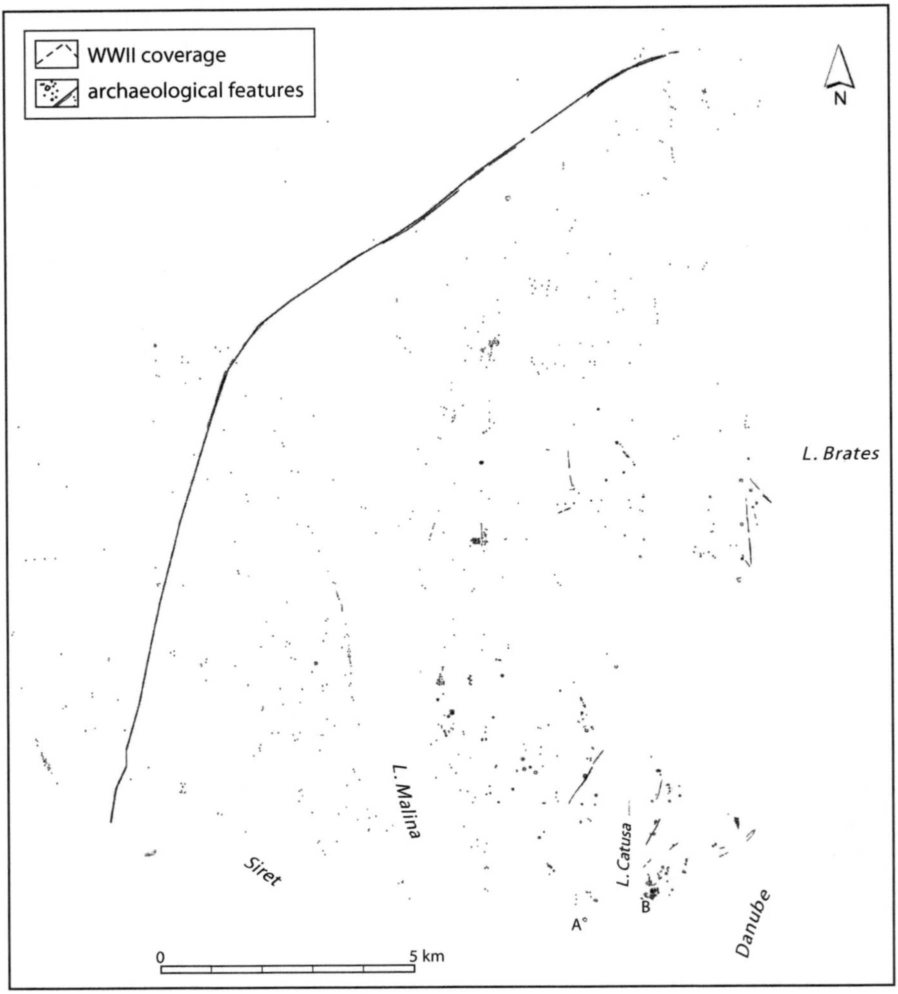

Fig. 9.5 Distribution of ancient archaeological remains recovered from the air in the area of Galați (**A** – Bărboşi-Tirighina; **B** – Galați-Dunărea)

features (e.g. along Lake Brateş and Lake Cătuşa or overlooking deeper valleys cutting through the higher plateau at the Siret-Danube-Prut confluence).

On the basis of experience further south in Dobrogea (Chap. 18 by Oltean and Hanson, this volume), the presence of ancient funerary barrows and access routes in the area was anticipated, even though their number and complexity exceeded these expectations. However, the most important and entirely unexpected discovery was the identification on the aerial photographs of a further 15 rectangular enclosures of broadly similar dimensions (see examples on Fig. 9.4). Based on the transcriptions alone, it is difficult to interpret all of them as fortifications, even though this possibility should not be entirely dismissed. However, two of them located some 2.8 km apart to the NW of Galați-Dunărea, one measuring 73 by 76 m and the other 98 by 99 m across their defences, were provided with a surrounding ditch in front of the rampart and are likely to have been Roman fortlets. Another probable fortlet was located further to the NE in the northern area of the modern town. Most of the enclosures, however, do not display an outer ditch; indeed, in four cases internal ditches may have been present, thus considerably reducing the occupied area. Though there would still have been sufficient space to accommodate a watchtower, the military nature of the remaining examples, defined only by a surrounding bank, is less obvious, so they might instead represent settlement or funerary enclosures. Recent surveys (including aerial reconnaissance in 2008) in the area confirmed the presence of additional such enclosures, at least one of them producing scattered Roman material during field walking.

The archaeological interpretation and mapping of archived RAF photographic material from the Second World War helped both to reconstruct the lost archaeological landscape of the area and to bring more light on the complex structure of this sector of the Roman Danubian *limes*. The installations around the modern town of Galați would have included at least four or five small fortifications (fortlets) along with a number of watchtowers. Although the line of the *limes* itself remains surprisingly clear of any known forts or towers, control extended as far as 6.5 km beyond the line of the Danube within the territory enclosed by the Traian-Tulucești *vallum*. Based on these observations, it is difficult to estimate the full extent and potential efficiency of Roman garrisoning of the area, but these installations were sufficient at least to facilitate the control and surveillance of the territory and provide vital communication links to military bases across the Danube.

Additional features observed on the 1944 aerial photographs also include some five rectangular enclosures and various field boundaries of uncertain date, along with smaller enclosures and internal features which could relate to late medieval or early modern occupation, the latter overlooking a small inlet on the left bank of the Danube used as a harbour until later in the nineteenth century but now dried out (Dragomir 1996: 667) (Fig. 9.4). Extensive information is also provided on part of the nineteenth- to twentieth-century defensive works, particularly on and around the so-called Focșani-Nămoloasa-Galați line (Fig. 9.6(3)). Now destroyed by development and agriculture and partly visible as cropmarks, this defensive system for heavy artillery was originally built in 1888–1893 by the Romanian army based on German plans. It was intended to defend the core of the country against eventual invasion from the north-east by Russia, but eventually ended up being used to stop the advance of enemy troops into Moldavia from the opposite direction during the First World War in 1916 when all of the southern part of Romania (then an ally of the Entente Powers) was occupied by German, Austro-Hungarian and Bulgarian troops (Georgescu 2007: 95–96; Stevenson 2004: 113–114, 171–173). Earlier maps, such as the Austro-Hungarian maps of the area (e.g. the 3rd Military Mapping Survey of Austria-Hungary, Sheet 46-45 Galați, General Map of Central Europe, scale 1:200,000, 1901), describe these fortifications as consisting essentially of two fortified lines with 27 casemates on their northern side. The 1944 aerial photographs help us to reconstruct the complexity of the Galați fortifications which would have included over 50 casemates built along three separate fortified lines (Fig. 9.6).

Fig. 9.6 (**1**) Eastern end of the Focşani-Nămoloasa-Galaţi fortifications in 1944 (MAPRW sortie 60PR460 frame 3055. Licensor NCAP/aerial.rcahms.gov.uk) indicating that the use of some of the earlier casemates from 1918 has been changed, as at **A** (on map **3**). The casemate at **B** (on inset **2** and map **3**) was still visible as cropmark in 2008

They provide significant additional information on the location and plan of 23 casemates, on their network of communications (roads) and information on further defensive trenches in various locations. At least one such set, fragmentarily recovered and now completely disappeared underneath the modern town, appears to be a consistent barrier system extended between the Danube and Lake Cătuşa (see Fig. 9.4). Of uncertain date, it undoubtedly preceded the late nineteenth-century fortifications. Additionally, the photographs allow identification of the features in use during the First World War and the extent of their reuse and adaptation during the Second World War, providing ultimately excellent information for analysing, probably for the first time, the development of the major conflicts of the twentieth century on the ground in this area.

9.5 Conclusions

The photographic imagery acquired during the Second World War to gather strategic information is important in its own right as a document of that developing conflict. However, the above discussion highlights the considerable contribution that even a very small amount of such imagery can make to reconstructing the development of the archaeological landscape in a particular area over a very long period. In the case discussed here, the utilisation of vertical photography from the Second World War has enabled the identification and recognition of a large number of archaeological features of diverse character and date in an area now permanently and irreversibly damaged by modern urban and industrial expansion.

The discoveries discussed above clearly indicate that the system of Roman surveillance and control of the left bank of the Danube in this area was far more complex than previously estimated and, through the positioning of fortlets at a significant distance behind the linear earthworks, reflects a strategic approach different from those documented elsewhere along the Roman *limes*. These sites now constitute the basis for a thorough reassessment of the Roman landscape of Galaţi and its surroundings through an intensive programme of survey on the ground and from the air (the latter involving the author), under the aegis of the STRATEG programme of the National Museum in Bucharest (http://www.strateg.org.ro/). The main research priority now is to date the few elements still surviving in order to inform interpretations of strategic networking and communication with Roman bases across the Danube in Dobrogea and to better understand the wider defensive strategy employed along the Lower Danube Roman *limes*.

The new information from historical photographs may contribute in the future to the successful scheduling of the Roman *limes* in Romania as part of the Frontiers of the Roman Empire World Heritage Site. In a similar fashion, the imagery could help the development of studies of more recent archaeological remains from this area, such as modern conflicts, which are currently not covered by Romanian archaeologists. Finally, the present study will hopefully encourage more archaeologists to use similar early aerial photographs in their work, whether from TARA or elsewhere.

Acknowledgments I am grateful to Sean Goddard (Exeter) for help with the illustrations and to Ovidiu Ţentea (Bucharest) and the STRATEG programme for facilitating further aerial research and providing ground truthing information in the area of Galaţi.

Bibliography

British Bombing Survey Unit. (1998). *The Strategic Air War against Germany 1939–1945: Report of the British Bombing Survey Unit*. London/Portland: Frank Cass.

Carter, K. C., & Mueller, R. (1991). *U.S. Army Air Forces in World War II: Combat chronology, 1941–1945*. Washington, DC: Center for Air Force History.

Croitoru, C. (2007). Some fortuitous funerary discoveries at Bărboşi, Galaţi County. *Istros, 14*, 53–59.

Dragomir, I. T. (1996). *Monografia arheologică a Moldovei de Sud* 1. *Danubius*, 16 (Special edition). Galaţi: Muzeul Judeţean de Istorie.

Georgescu, M. (2007). Dimensiunile reformei armatei române între Războiul de Independenţă şi intrarea în Primul Război Mondial. In P. Oţu (Ed.), *Reforma militară şi societatea în România, de la Carol I la a doua conflagraţie mondială* (pp. 77–100). Occasional Papers of MAPN, 6(8). Bucharest: Editura Militara.

Going, C. J. (2002). A neglected asset. German aerial photography of the Second World War period. In R. H. Bewley & W. Raczkowski (Eds.), *Aerial archaeology – Developing future practice* (Nato science series, pp. 23–30). Amsterdam: Ios Press.

Jensen, J. H., & Rosegger, G. (1978). Transferring technology to a peripheral economy: The case of Lower Danube transport development, 1856–1928. *Technology and Culture, 19*(4), 675–702.

McCloskey, J. F. (1987). U.S. operations research in World War II. *Operations Research, 35*(6), 910–925.

Meilinger, P. S. (2007). A history of effects-based air operations. *Journal of Military History, 71*, 139–178.

Schaffer, R. (1985). *Wings of judgement: American bombing in World War II*. New York/Oxford: Oxford University Press.

Sîrbu, V., & Croitoru, C. (2007). Bărboşi – 170 ans de récherches archéologiques. *Istros, 14*, 13–25.

Stevenson, D. (2004). 1914–1918*: The history of the First World War*. London: Allen Lane

Stout, J. A. (2003). *Fortress Ploesti: The campaign to destroy Hitler's Oil*. Havertown: Casemate.

Ţentea, O. (2007). The Roman Necropolis at Bărboşi – Several remarks. *Istros, 14*, 217–225.

Ţentea, O., & Oltean, I. A. (2009). The Lower Danube Roman *limes* at Galaţi (Romania). Recent results from excavation and aerial photographic interpretation. In Á. Morillo, N. Hanel, & E. Martín (Eds.), *Limes XX. XX Congreso Internacional De Estudios Sobre La Frontera Romana – XXth International Congress of Roman Frontier Studies León (España), Septiembre, 2006* (pp. 1515–1523). Madrid: Ediciones Polifemo.

Turnock, D. (1974). *An economic geography of Romania*. London: Bell & Sons.

Werrell, K. P. (1986). The strategic bombing of Germany in World War II: Costs and accomplishments. *Journal of American History, 73*(3), 702–713.

Chapter 10
The Value and Significance of Historical Air Photographs for Archaeological Research: Some Examples from Central and Eastern Europe

Zsolt Visy

Abstract Because many countries of Eastern Europe were forced to join the communist totalitarianism of the Soviet Empire, when aerial reconnaissance was relegated to the category of military secrets, they were unable to follow the developments in archaeological aerial reconnaissance experienced in Western Europe in the period after the Second World War. This chapter outlines some of the difficulties archaeologists experienced in acquiring aerial photographs in Eastern Europe, traces the development of aerial archaeology there and provides examples of the value of historical imagery taken for purposes other than archaeology which could be accessed. Examples from Hungary, mainly from the Roman period, serve as a particular case study.

10.1 Introduction

The invention and use of Zeppelins and airplanes was one of the major advances of the late nineteenth century, fulfilling a many-thousand-year-old dream of mankind. Until then, more distant view observations could be made only from mountain peaks, hills and higher elevations. The invention of flying machines meant that such observations could now be made from any desired location and from any altitude. Maps of various regions and towns, formerly drawn only from imagination, could now be based on personal experience. With the spread of photography, the number of pictures taken from higher elevations increased, and photographs made from balloons, the antecedents of real aerial photographs, appeared at the end of the nineteenth century. As often happens in the case of major inventions, the pioneering work in this field was done by the military. The advantages of aerial reconnaissance

Zs. Visy (✉)
Department of Archaeology, University of Pécs, Rókus utca 2, Pécs, Hungary 7624
e-mail: visy.zsolt@pte.hu

W.S. Hanson and I.A. Oltean (eds.), *Archaeology from Historical Aerial and Satellite Archives*, DOI 10.1007/978-1-4614-4505-0_10,

and the military potential of recording observations on a photograph were quickly realised during the First World War. Most pictures were taken at an altitude of about 3,000 m, and it was at this time that overlapping photographs were first made that also allowed the production of mosaic maps.[1]

Observation from a high altitude and the accompanying photographs opened new perspectives for scientific research. The specialists of this new method soon determined the optimal conditions for its application and worked out when, from what altitude and at what time of day the best results could be obtained. They soon realised that many features that remained undetected on the ground became visible from the air and, also, that phenomena that appeared as random features on the ground formed a coherent pattern if viewed from above, revealing a number of points that could never otherwise have been detected.

A number of partially or totally buried remains and other relics of bygone ages could be identified in this way. Aerial photography was, thus, one of the positive accomplishments of the First World War; many pilots fighting in the war were the first to observe and register archaeological remains. After returning to civil life, they began to organise systematic aerial reconnaissance, documentation and evaluation of archaeological features. The pioneers of aerial archaeology elaborated the methods of this discipline in the 1920s and 1930s. In addition to work in Europe, they were also interested in the exploration of buried ruins in the desert areas of Africa and the Near East. The doyens of the field, Theodor Wiegand, Antoine Poidebard and, later, Osbert Guy Stanhope Crawford, were joined by Aurél Stein, who in 1938, at the age of 76, began the aerial exploration of the Roman *limes* and other archaeological monuments in Iraq, where he succeeded in identifying several military sites.

10.2 Aerial Archaeology in Central Europe Before and After the Second World War[2]

The beginnings of aerial archaeology in the countries of Central Europe go back to the period after the First World War and followed a course similar to that in other European countries. The new method had been developed during the war, and many pilots were ready to apply their knowledge for science and for archaeology. They were the first who noticed and photographed archaeological remains from the air (e.g. Böhm 1939; Mencl 1937, Taf. 1; Ondrouch 1941; Radnai 1939, 1940). Some archaeologists adopted the methodology and started to evaluate aerial photographs (Radnóti 1945).

---

[1] The McMaster University collection contains almost 600 examples of First World War aerial photographs (http://library.mcmaster.ca/maps/ww1/home.htm). See also Chap. 5 by Stichelbaut et al.; Chap. 6 by Pollard and Barton, this volume.

[2] Based on Visy 1997a.

The Second World War led to the rapid development of the method and technology of aerial observation and investigation, and an immense number of aerial photographs were taken for military purposes. A significant number of these pictures also contain archaeological features, and the potential of this material has still not been fully exhausted. On the other hand, because these countries were forced to join the communist totalitarianism of the Soviet Empire, this material also represents the end of active aerial archaeology in their territory. Thereafter, there followed a long caesura in the aerial archaeology of many countries of Europe. The most difficult situation developed in the German Democratic Republic, but the regulations for observations made from aircraft were almost the same in Czechoslovakia, Hungary and in the other socialist countries under Soviet control. These strong and restrictive regulations were in force exactly in that 40-year period when aerial archaeology was undergoing rapid development and came into general use in Western Europe.

The strict regulations introduced in the name of the vigilance allowed the investigation of existing aerial photographs only in specific circumstances. One needed special permission from the ministry, which was given only for a specific territory within a given time. It was not allowed to make notes or to use a map while working with the aerial photographs and, of course, it was forbidden to make any kind of copies. A copy could be requested, but in these cases, it was made from the print, not the negative, so that it often showed the notes and signs of the military cartographers who had used the vertical photographs for correcting military maps. The copy could be given only to an institute which owned a safe for keeping maps and other politically 'dangerous' material. It did not mean that these pictures could be published, because they retained their categorisation as secret documents. Only an institute could ask for this category to be changed to the designation 'free' to allow publication. Two identical pictures had to be forwarded to the authorities, and if permission was given, one went to the press, while the other was retained for later control.

Another and more complicated issue was how to organise archaeological flights. For a long time this could not even be mentioned, because anybody who wanted to make observations from the air could became suspicious politically. This privilege available to the military forces was extended only slowly in the late 1960s as aerial observation was more widely used for economic purposes. At the same time, the door was opened slightly for the archaeologists also, but the high costs made it almost impossible to undertake such work. The main problem lay, however, in the methodology which it was necessary to apply. As was standard practice for cartographic or economic flights, archaeologists were required to specify in advance the exact location to be photographed as well as its interpretation. That meant that the only archaeological sites which were allowed to be photographed were those previously identified, rather than any unexplored ones. As a result, therefore, it was almost impossible to undertake any real aerial archaeological investigation. Only by extending the specified territory around identified archaeological sites was there a slight possibility to search for new sites in its vicinity. Accordingly, the investigation of existing aerial photographs and the careful examination of earlier flights were of considerable importance for increasing the number of the known sites or at least those suspected of having some archaeological relevance. The biggest problem was that one could not choose the optimal time for archaeological flights, although this is the most important precondition to ensure a successful outcome. Permission to fly was given with 1–3 months notice, and it specified the exact day or days on which the photographs were to be taken. As a result, it was often the case that the conditions were totally wrong for archaeological investigation, but having been given permission, it was necessary to undertake the flight and hope that something would be found, even under unfavourable conditions. Under such conditions, aerial archaeology could not become a successful discovery technique (Sedlácek and Vencl 1975; Plesl 1983).

Furthermore, the opportunity to fly in socialist countries became available again only in the 1960s. Scientists had learnt about the new method of aerial archaeology being practised in Western countries, which had achieved significant results. One of the first steps in this process was the presentation of a French aerial archaeological exhibition in Prague in 1967 (Hásek 1968).

After the Second World War, the first intentional aerial archaeological flight in Moravia was organised in 1949. It was made by J. Poulík over the rich Langobardian king's grave in Zuran at Podoli (Poulík 1995). Successful research flights were made in Moravia also over the Slavic fortified sites at Mikulcice, Pohansko bei Breclav and Sady bei Uherské Hradište at the end of the 1950s, and soon after over the grave of the Bell-Beaker culture at Prosimerice (Pernicka 1961) and later at some other sites (Pavelcik 1976; Cerny 1978). In 1972–1973, a radio-controlled model plane was used to document the important archaeological site in Tešetice-Kyjovice. These investigations continued to contribute to interdisciplinary results until the middle of the 1980s (Bálek et al. 1986).

Change also occurred slowly in the Slovakian region. The excavations at Nitriansky Hrádok (Šurany) in 1967 (Tocik and Vladár 1971: 383, Abb. 16), in Bína (Habovštiak 1985: 181), and the earthwork at Stary Tekov in 1963 (Visy 1997a: 24, fig. 2) were investigated and documented through aerial photography. The Slavic earthwork at Majcichov was also identified by this method in 1967, and the 'Roman Station' at Cífer-Pác also documented from the air in 1972 (Visy 1997a: 25, fig. 4). Beginning in 1978, a series of aerial photographs were taken both in Czechoslovakia and Hungary in the territory of the planned dam on the Danube at Gabcikovo-Nagymaros. The first pictures were taken without any permission, but no punishment followed. It is a pity, however, that most of the photographs were taken under conditions which were unfavourable for the recovery of archaeological information. It should be noted that after the democratic change in these countries, the construction work on this huge dam has been stopped.

Aerial archaeology in Hungary began more or less simultaneously with international experiments in this field. In 1938, Lóránd Radnai published a paper in which he described the archaeological uses of aerial photography and the basic requirements for successful observation. The first aerial photographs were published in *Archaeologiai Értesítő* 2 years later. In his discussion of these photographs, Radnai convincingly proved his point and demonstrated that aerial photography could be successfully applied under Hungarian conditions too.

Archaeologists soon became acquainted with this important new research technique. A few years later, Aladár Radnóti (1945) published high quality photographs that could be evaluated archaeologically in his study of the Dacian *limes* along the ridge of the Meszes mountains. Wartime conditions greatly contributed to advances in archaeological aerial photography, but they also brought a number of restrictions. While planned reconnaissance flights could rarely be made, there were no objections to the archaeological analysis of the high number of excellent aerial photographs made by the army. Sándor Neógrády spent long years studying these photographs. He accumulated an impressive collection of aerial photographs, publishing a part of his collection at the last possible moment before the clamp down (1950). Indeed, it is pure chance that he was allowed to publish many aerial photographs in Hungary, for thereafter it was restricted for more than two decades. His photographs were taken during the Second World War and earlier. We can only hypothesise what else there may have been in his collection that never became generally accessible because of the changes in the political climate. The all-pervasive atmosphere of suspicion characterising the communist system did not allow the complex mapping of Hungary's territory, and aerial reconnaissance was relegated to the category of military secrets.

Thereafter in Hungary, the National Museum was allowed to make a series of photographs of the archaeological investigations and of the immediate area around the second dam on the Tisza (Patay 1968, 1971), and also, the Archaeological Institute of the Academy obtained some aerial photographs for its topographical work. The Savaria Museum at Szombathely succeeded in making a series of aerial photographs of the county of Vas. A theoretical work was also published on the subject providing a context for the topographical works in Hungary (Eöry and Szabó 1972). The first opportunity to present aerial photographs to an international audience came in 1976 at the XIth International Congress of Frontier Studies in Dunaújváros. The material for this exhibition was made up of earlier aerial photographs from the Institute of the Military History and the Budapest Cartographical Institute assembled by the current author and focussed on the remains of the Roman frontier of Pannonia (*Ripa Pannonica*) in Hungary (Visy 1978, 1981, 2003a). The first paper about the methods and history of the aerial archaeology in the region was written by B. Erdélyi (1982).

In these decades, almost the only possibility to apply the principles of aerial archaeology was the interpretation of photographs made for different purposes at different times. They could be investigated according to the strict rules referred to above. The bulk of these photographs were vertical and topographical ones, sometimes made from a great height. Their scale varied between 1:5,000 and 1:20,000, so they could be used only within certain limits. Apart from the military photographs, a great number of civil photographs made for economic, meteorological or other purposes were taken in the 1970s and 1980s, while pictures taken in order to better understand settlements and their environment enriched the material. The scale of the latter was often about 1:2,000, which is better for archaeological interpretation than the small-scale photogrammetric coverage. On the other hand, it could be demonstrated that a

Fig. 10.1 The Roman legionary fortress at Brigetio indicated by the large *black arrow* on a vertical aerial photograph taken in 1940 (Hungarian Institute for Military History 69396). The semicircular earthen structure in the SE corner of the fortress is a bastion of the huge Komárom fortification built in the 1880s

great number of such photographs preserved archaeological features and archaeologists could make use of them and identify many archaeological sites.

The political thaw in the 1970s at last made possible the application of aerial photography for purposes other than military reconnaissance, but obviously still with strict observance of the regulations. Aerial photography for archaeological research could at last begin, although the photographs made during this period often had little scientific value since they were not always made at the optimal time and under optimal conditions, but only when the flight was permitted. It then became possible systematically to study the photographs made for topographic or economic purposes on which archaeological features could be clearly made out. Most important among these was a series from the early 1940s that showed the entire Hungarian section of the Danube and other territories. A number of features that now lie concealed under buildings and factories built subsequently, or have been destroyed by intensive cultivation, are still visible on these photographs. The Roman legionary fortress at Brigetio was clearly visible in 1940 entirely free of modern buildings (Fig. 10.1). However, an oil refinery was built there in 1942, which was bombed in 1944.

Fig. 10.2 The Roman legionary fortress at Brigetio indicated by the large *white arrow* on a vertical aerial photograph taken in 1951 (Hungarian Institute for Military History 22924)

Both the craters of the bombardments and the rebuilt refinery are visible on a photograph made in 1951 (Fig. 10.2). In the 1950s and early 1960s, some parts in the rear front of the fortress remained undeveloped, but soon thereafter, both the southern sector of the fortress and that of the surrounding *canabae* (military settlement) were covered by new industrial buildings. Similarly, the well-preserved platform of the Roman auxiliary fort of Azaum was clearly visible in 1940, but now, tailings (red mud waste products) from an aluminium factory cover the site completely (Fig. 10.3). The plateau at Százhalombatta is full of Iron Age tumuli. In 1955, they were clearly visible, but now, only a few of them survive. Near Kulcs the Roman *limes* road was readily visible from the air in 1940 (Fig. 10.4), but in 1949, a major road was built on top of it removing all traces. The situation is the same in Dunaújváros. This territory was ploughed land in 1940, but in 1950, a new town and the iron works were built, and now, almost the whole area is built over (Fig. 10.5).

The photographs made in the 1950s and later also contained much useful information, and their study can still yield new data, since the careful inspection of these photographs can lead to new discoveries. The restrictions on aerial photography were gradually lifted, first by easing the strict regulations and, after the political changes, by declassifying certain maps and photograph types. The earlier strict regulations allowed only photography of already known archaeological sites, meaning that archaeological reconnaissance flights aimed at discovering and documenting

Fig. 10.3 (**a**) The extant platform of the Roman auxiliary fort of Azaum clearly visible in the centre of a vertical aerial photograph taken in 1940 (Hungarian Institute for Military History 69397). (**b**) The site of the fort on Google Earth (©2011 Google Earth; ©2011 Map Link/Tele Atlas; ©2011 Geoeye)

new archaeological features and remains were not permitted. The new regulations allow flights over larger areas and the unrestricted photography of assumed and identified features. Thus, the current Hungarian system more or less conforms to the regulations in most European countries.

Fig. 10.4 The Roman *limes* road near Kulcs (*arrowed*) on vertical aerial photograph taken in 1940 (Hungarian Institute for Military History 69425)

It is evident that only the active practice of the aerial archaeology brings high quality results. However, during the work on earlier historical photographs in central and eastern European countries, it has become apparent that such photographs are potentially of great value not only because they preserve older remains and features but also because much archaeological information has been destroyed since they were taken. It is easy to understand how much damage agriculture has caused in the fields only through the application of deep ploughing. Archaeological features, which are highly sensitive to agricultural activity, can often be identified in old photographs much better than in more recent ones. Archaeological remains have been damaged through other human activity as well. Building development covers much larger areas now than 50 or 80 years ago, settlements are twice or three times bigger, and these newly built-up areas are no longer suitable for aerial

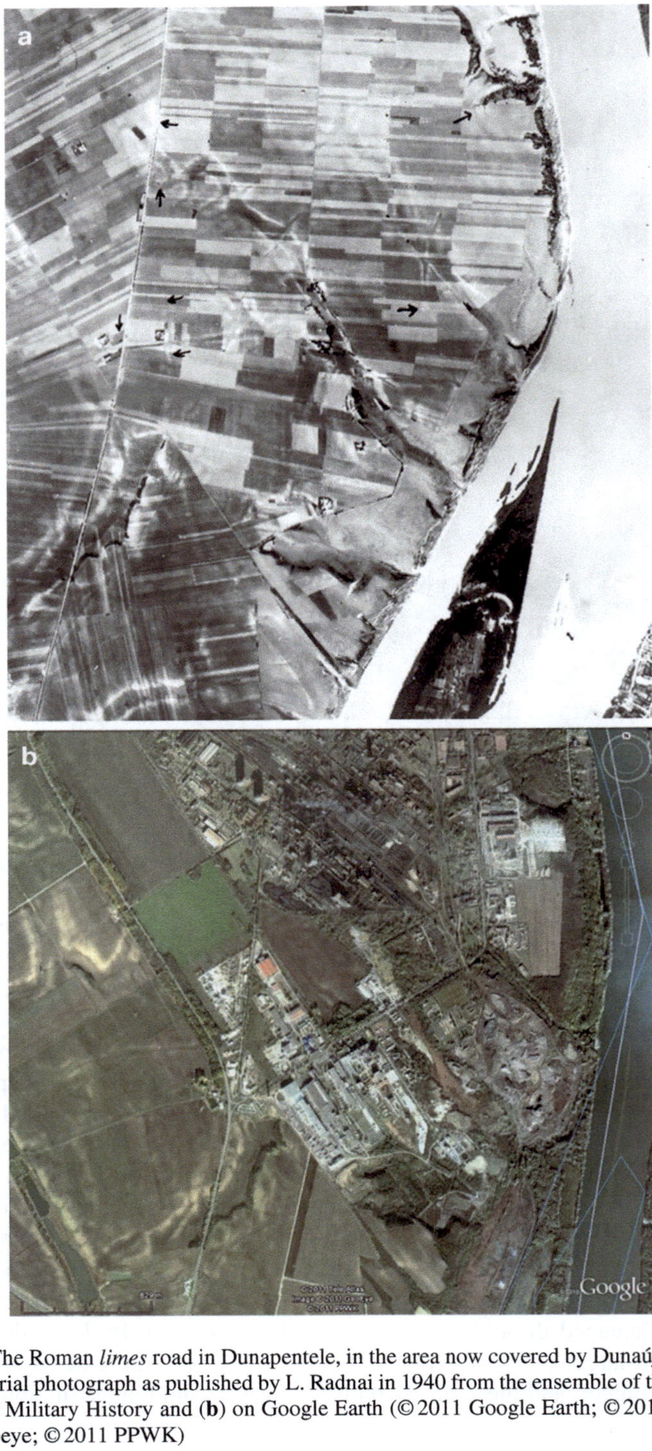

Fig. 10.5 The Roman *limes* road in Dunapentele, in the area now covered by Dunaújváros, (**a**) on a vertical aerial photograph as published by L. Radnai in 1940 from the ensemble of the Hungarian Institute for Military History and (**b**) on Google Earth (©2011 Google Earth; ©2011 Tele Atlas; ©2011 Geoeye; ©2011 PPWK)

Fig. 10.6 The *limes* road in Ercsi on a vertical aerial photograph taken in 1940 (Hungarian Institute for Military History 69422). The Roman road line is still partly used today, while other parts are visible as tracks or cropmarks in the field

archaeology. Again in such cases historical photographs can often reveal archaeological remains much more clearly.

Older photographs contribute much archaeological information. The excavations of the aeneolithic fortified settlement in Stehelcemes-Homolka could be photographed in 1963 (Visy 1997a: 25, Fig. 5). In another picture made at Klucov, Kolin district, the remains of prehistoric and early medieval walls could be seen, which are hardly visible today (Visy 1997a: 26, Fig. 6). A systematic investigation was made in the different institutes in Hungary which preserve aerial photographs of the area of the *Ripa Pannonica* in the 1970s and 1980s (Visy 2003a), and about

200 photographs out of more than a 1,000 were found which contained some archaeological features. Apart from Roman forts, watch towers and other military buildings, the most striking result was that lines of the *limes* road tens of kilometres long could be identified (Fig. 10.6).

Bibliography

Bálek, M., Hašek, V., Merínsky, Z., & Segeth, K. (1986). Metodicky prínos kombinace letecké prospekce a geofyzikálních metod pri archeologickém vyzkumu na Morave. *Archeologické rozhledy, 38*, 550–574.

Böhm, J. (1939). Letecká fotografie ve sluzbách archeologie. *Zprávy památkové péce, 3*, 63–65.

Cerny, E. (1978). Typy zaniklych stredovekych vesn(ickych sídlišt' na Drahanské vrchovine z historicko-geografického hlediska. *Archaeologica Historica, 3*, 31–40.

Eöry, K., & Szabó, I. E. (1972). *Aerial photo interpretation for regional field research in archaeological topography*. Bratislava: Geodéziai és Kartográfiai Egyesület Közleményei.

Erdélyi, B. (1982). Régészeti légifényképezés és légifénykép-értelmezés. *Hermann O. Múzeum Évkönyve, 21*, 81–88.

Habovštiak, A. (1985). *Stredoveká dedina na Slovensku*. Bratislava: Obzor.

Hásek, I. (1968). Archeologie z letadla. *Archaeologické rozhledy, 20*, 94–95.

Mencl, V. (1937). *Stredoveká architektúra na Slovensku*. Praha-Prešov: Československá grafická únie.

Neógrády, S. (1950). A légifénykép és az archeológiai kutatások. *Térképészeti Közlöny, 7*, 283–332.

Ondrouch, V. (1941). *Rímska stanica v Stupave a rímske stavebné stopy v Pajštúne* (pp. 1–2). Bratislava: Historica Slovaca.

Patay, P. (1968). *Rechérches d'archéologie aérienne en Hongrie*. Budapest: Geodéziai és Kartográfiai Egyesület Kiadványa.

Patay, P. (1971). Neue Ergebnisse der Luftbildinterpretation in der ungarischen archäologischen Forschung. *Internationales Archiv für Photogrammetrie., 18*, 519–529.

Pavelcík, J. (1976). Letecká archeologie v severomoravském kraji. *Vlasivedné listy, 3*(1), 17–18.

Pernicka, R. M. (1961). Eine unikate Grabanlage der Glockenbecherkultur bei Prosimerice, Südwest-Mähren. *Sborník praci filozofické fakulty Brnenské univerzity, E 6*, 9–54.

Plesl, E. (1983). K vyuzití leteckych snimku pro potreby archeologie v Cechách. *Geofyzika a archeologie, 48*, 239–242.

Poulík, J. (1995). Zurán in der Geschichte Mitteleuropas. *Slovenská archeológia, 43*, 27–109.

Radnai, L. (1939). Légi fényképezés a régészeti kutatás szolgálatában. *Magyar Fotogrammetriai Társaság Évkönyve, 1938–1939*, 141–142.

Radnai, L. (1940). Újabb archaeologiai nyomok Dunapentele környékén. *Archaeol. Értesítő*, 3rd Ser. 1, 62–65.

Radnóti, A. (1945). A dáciai limes a Meszesen. *Archaeol. Értesítő*, 3rd Ser. 5, 137–151.

Sedlácek, Z., & Vencl, S. (1975). Zpráva o leteckém snímkování na Kolínku. *Archeologické rozhledy, 27*, 151–158.

Tocik, A., & Vladár, J. (1971). Prehl'ad bádania vyvoja Slovenska v dobe bronzovej. *Slovenská archeológia, 19*(2), 365–416.

Visy, Zs. (1978). Pannoniai limes-szakaszok légifényképeken. *Archaeol Értesítő, 105*, 235–259.

Visy, Zs. (1981). Pannonische Limesstrecken auf Luftaufnahmen. *Antike Welt, 12*(4), 39–52.

Visy, Zs. (1997a). Stand und Entwicklung der archäologischen Luftprospektionen in der DDR, der Tschechoslowakei und Ungarn in den Jahren 1945 bis 1990. In J. Oexle (Ed.), *Aus der Luft – Bilder unserer Geschichte: Luftbildarchäologie in Zentraleuropa* (pp. 23–27). Dresden: Landesamt für Archäologie Sachsen mit Landesmuseum für Vorgeschichte.

Visy, Zs. (2003a). *The ripa Pannonica in Hungary*. Budapest: Akadémiai Kiadó.

Additional Bibliography for Historic Aerial Photography in Central and Eastern Europe

Bewley, R., Braasch, O., & Palmer, R. (1996). An aerial archaeology training week, 15–22 June 1996, held near Siófok, Lake Balaton, Hungary. *Antiquity, 70*, 745–750.

Erdélyi, B. (1998). Régészeti célú légifényképezés. *Panniculus, B3*, 37–50.

Erdélyi, B., & Sági, K. (1984). A magyarországi régészeti légifényképezés története és a Szent György-hegyi kolostorrom. *Veszprém Megyei Múzeumok Közleményei, 17*, 272–280.

Gojda, M. (2011). Remote sensing for the integrated study and management of sites and monuments – A Central European perspective and Czech case study. In: D. C. Cowley (Ed.), *Remote sensing for archaeological heritage management* (EAC Occasional Paper No. 5, pp. 215–227). Budapest: Archaeolingua.

Kis Papp, L. (1982). Fotogrammetria az építészetben és a régészetben. *Hermann O. Múzeum Évkönyve, 21*, 89–100.

Visy, Zs. (1988). *Der pannonische Limes in Ungarn*. Budapest: Theiss.

Visy, Zs. (1990). Légifelvételeken megfigyelt halom-sírok a Dunántúlon. In S. Palágyi (Ed.), *Noricum-pannoniai halomsírok, Várpalota 1988* (pp. 23–45). Budapest: Veszprém.

Visy, Zs. (1995). Luftbildarchäologie am römischen Limes in Ungarn. In J. Kunow, (Ed.), *Luftbildarchäologie in Ost- und Mitteleuropa. Forschungen zur Archäologie im Land Brandenburg* 3 (pp. 213–218). Potsdam: Brandenburgisches Landesamt für Denkmalpflege.

Visy, Zs. (1997b). Erdwälle und Burgen in Transdanubien, Ungarn. In J. Oexle (Ed.), *Aus der Luft – Bilder unserer Geschichte: Luftbildarchäologie in Zentraleuropa* (pp. 77–81). Dresden: Landesamt für Archäologie Sachsen mit Landesmuseum für Vorgeschichte.

Visy, Zs. (2003b). Aerial archaeology in Hungary. In Zs Visy (Ed.), *Hungarian archaeology at the turn of the millennium* (pp. 25–28). Budapest: Ministry of National Cultural Heritage.

Chapter 11
Archaeology from Aerial Archives in Spain and Portugal: Two Examples from the Atlantic Seaboard

Iván Fumadó Ortega and José Carlos Sánchez-Pardo

Abstract This chapter seeks to illustrate the largely underestimated potential for the use of available historical aerial photographs in archaeological research projects conducted in the Iberian Peninsula. The richest Spanish and Portuguese aerial photographic archives are described, before presenting two different examples of their use from the Atlantic seaboard. While the first case study takes a wide perspective on a Galician landscape, the second is a more local analysis from the Sado Estuary. Though both are preliminary assessments, they are intended to encourage the directors of research projects in the Iberian Peninsula to take an interest in these useful aerial photographic archives, even though they were compiled without any archaeological purpose in mind.

11.1 Introduction

In this brief chapter, we want to illustrate and emphasise the interesting and comprehensive possibilities of the use of available aerial photographs in archaeological research projects in the Iberian Peninsula, even when they were not taken for

I. Fumadó Ortega (✉)
Escuela Española de Historia y Arqueología en Roma – Spanish Research Council, Via di Torre Argentina 18, 00186 Rome, Italy
e-mail: i.fumado.ortega@gmail.com

J.C. Sánchez-Pardo
Fundación Española para la Ciencia y Tecnología, Ministerio de Ciencia e Innovación, Madrid, Spain
e-mail: jsp1980@hotmail.com

W.S. Hanson and I.A. Oltean (eds.), *Archaeology from Historical Aerial and Satellite Archives*, DOI 10.1007/978-1-4614-4505-0_11,

archaeological purposes. This is not, of course, a new or unknown topic, either in Portugal or Spain, but the reality is that insufficient attention has been paid to it and the results until now have been very limited.

The presence in Spain of archives containing aerial photographs taken for purposes other than archaeology was partly summarised in a paper given in Rome in 2006 (Sánchez-Pardo and Fumadó Ortega 2006). The main files are those from the so-called American flight, also known as Flight A (1945–1946) and Flight B (1956–1957). These are particularly useful because they were made before the extensive transformations resulting from the uncontrolled growth in the 1960s and 1970s, which contrasts with other European countries where agricultural changes started earlier, but they have been hardly studied in depth for archaeological purposes. The main photographic archives are the *Centro Nacional de Información Geográfica* (http://www.cnig.ign.es), the *Instituto Geográfico Nacional* (http://www.ign.es), and the *Centro Cartográfico y Fotográfico del Ejército del Aire* (http://www.ejercitodelaire.es). The oldest archive was that from the *Compañía Española de Trabajos Fotogramétricos Aéreos* (CEFTA) which was established in 1927. Unfortunately, the company was closed in the 1990s, and their holdings were sold off to different administrative and local offices (Fernández García 2000: 20).[1] Several companies have carried out similar works through the twentieth century, although many of them have already disappeared also.

Archives of aerial photographs available in Portugal include those from the *Instituto Geográfico do Exército* (http://www.igeoe.pt) and part of the *Esquadra 401* from *Base Aérea Nº1 de la Força Aérea Portuguesa*, located in Granja do Marquês (Sintra). Partial flights can be seen in several *Câmaras Municipaes do Concelho*, or in geography departments in some universities (Oporto, Coimbra, Lisbon), which include the Portuguese part of the American flight (1958–1960).[2] More complete and accessible coverage is located in the *Instituto Geográfico Português* (http://www.igeo.pt), where there are aerial photographs from several regional and even local flights across the entire country. The pictures were taken

[1] The photographs from the Asturias were bought by the *Departamento de Geografía de la Universidad de Oviedo*; those from Catalonia by the *Institut Geogràfic de Catalunya*; those from the Basque Country by the *Departamento de Ordenación del Territorio, Vivienda y Medio Ambiente del Gobierno Autonómico Vasco*; those from Valencia by the *Consellería de Cultura, Educació i Ciència de la Generalitat Valenciana*; those from Murcia by the *Consejería de Política Territorial Obras Públicas del Gobierno Autonómico de Murcia*; those from the Canary Islands by the *Consejería de Política Territorial del Gobierno Insular*; those from Extremadura by the *Consejería de Medio Ambiente, Urbanismo y Turismo de la Junta de* Extremadura.

[2] Iván Fumadó had the opportunity to join the *Fototeca del Departamento de Geografia da Universidade de Lisboa*, where they have the entire 'American flight'. I would like to thank the kindness of its staff. Internet site http://www.fl.ul.pt/dep_geo/.

from the very end of Second World War until the present day, with scales that vary between 1:7,500 and 1:25,000. Finally, the urban area of Lisbon has received more attention, and there are several records housed in the headquarters of the *Área Metropolitana de Lisboa* (http://www.aml.pt).

In our previous paper (Sánchez Pardo and Fumadó Ortega 2006), our goal was to underline that the usual role of aerial photographs in Spanish research projects is far removed from the complexity involved in a complete survey project. Frequently, research groups use such photographs only in an illustrative way, as no more than a collection of beautiful pictures. More recently, some scholars have made more use of the potential that this material offers, even though at a reduced level. These works try to reach a better understanding of historical processes that occur in the landscape (Ariño Gil et al. 2004), to study well-known sites in a different way (Penedo Cobo 2006; Liz Guiral and Celis Sánchez 2007) or to discover new archaeological sites (Bellón Ruiz et al. 2009). That these surveys extend across a wide chronological range from Iberian to medieval times is an extremely positive development.

The situation in Portugal is very similar. Though territorial and landscape studies have seen particular development (Alarcão 1998, 2002), use of this kind of graphic material was even lower than in Spain. The pioneer methodological works from Farinha dos Santos (1965) were praiseworthy but without much impact. With the exception of some specific studies (Gil Mantas 1985, 1996, 2003), Portuguese archaeology is still not fully aware of the significance and potential that aerial photography can provide. It is true that the management of the archaeological heritage is now helped by a new and powerful GIS framework, *Endovélico* (Bugalhão et al. 2002), created by the *Instituto Português de Arqueologia* as an instrument for the detection, preservation, and management of archaeological remains, which would be an example for Spanish institutions even though the Portuguese government does not seem to appreciate it enough (Burgalhão 2002). But it seems that aerial data have not been much explored, not even in the most recent papers and articles on archaeological cartography (Oliveira et al. 2001; Correia 2005) or new GIS applications (Strutt 2000; Langley 2006; Rua 2007). In the meantime, the *Instituto Geográfico Português*, with the largest air photography archive in Portugal, is trying to make it more accessible through its website and its helpful staff.

Our goal in this chapter is to encourage new scientific research projects with both archaeological and historical approaches that appreciate the huge source of information which can be provided by aerial photography, even when not made originally with any archaeological purpose. By way of encouragement, we have provided a few examples of its potential in Spain and Portugal, specifically on the Atlantic coast, focusing on just a few of the most attractive aspects. In the first part, we will use the American flight in a diachronic landscape study of northwest Spain. In the second, we will try to identify new archaeological sites through the systematic study of images captured in historical Portuguese flights in the 1940s and 1950s, choosing some areas of high occupational density with a wide chronological range of historical remains (Fig. 11.1).

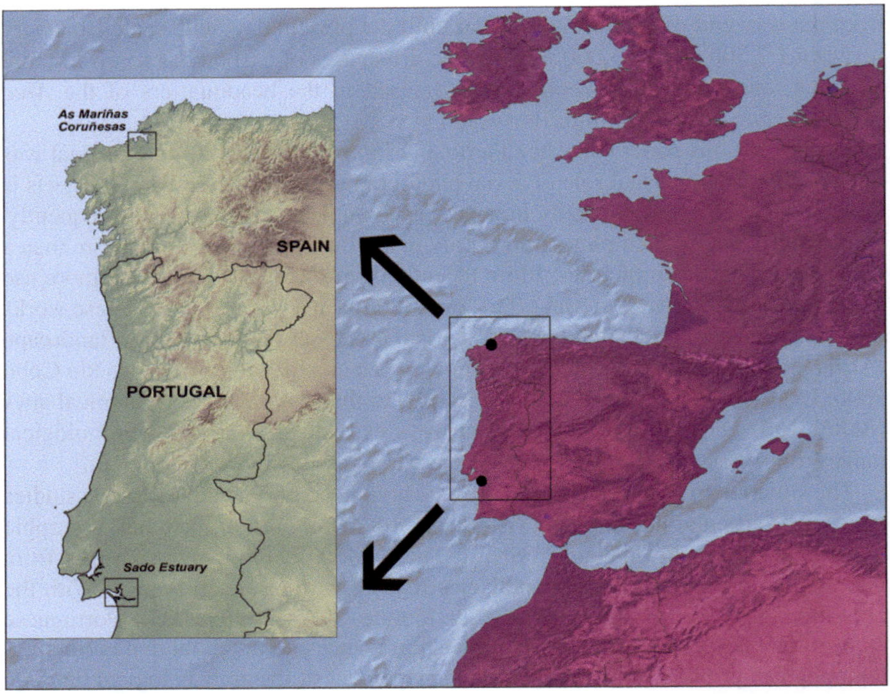

Fig. 11.1 Map of general location of the study areas in relation to south-western Europe and Iberian Peninsula

11.2 Archaeology Through the American Flight in Spain: The Case of As Mariñas Coruñesas (A Coruña, Galicia)

One of the first and most complete aerial photographic coverages of quality of Spain was the so-called American flight or Flight B, conducted between the years 1956 and 1957 by the US Air Force. Flight A from 1946 has more problems with visibility, because of altitude and cloud cover, and its application is more limited (Pellicer Corellano 1998: 7). The photographs from these flights were handed over to the Spanish Air Force as part of the negotiations between the Franco regime and the USA, and the current copyright owner is the *Centro Cartográfico y Fotográfico del Ejército del Aire* (CECAF). The collection consists of a series of black and white vertical, stereoscopic aerial photographs, with an average scale of 1:33,000 (Pellicer Corellano 1998: 7).

These photographs are exceptional and extremely valuable documents for visualising the Spanish countryside before the major changes and transformations that it has suffered over the past 50 years, offering a large amount of interesting information that would now be impossible to recover. However, despite this potential, these photographs have barely been used in an extensive or even minimally systematic way in archaeological research, with their use often limited to the analysis or the mere illustration of chosen cases. This is partly due to the limitations of these photographs for archaeological purposes, since they were taken from too high an altitude and not always under favourable conditions for the study of crop or soilmarks.

However, we believe that the photographs of the American flight still have enormous potential for archaeological research, as we will try to illustrate below. Although these photographs cannot show smaller or less pronounced marks on the ground, it is possible to see other larger archaeological sites, many of which have now disappeared. Moreover, we must remember that the value of these photographs for archaeology is not limited to the recognition of specific sites, but reflects the structure of an intact landscape that symbolises the result of a long historical development whose elements (and inter-relationships) can sometimes be traced back many centuries.

By way of example, focusing on the whole landscape over time, we would like to demonstrate the potential of the photographs of the American flight of 1956–1957 for archaeological research in Spain using a specific territorial study on the Spanish Atlantic coast: the area known as As Mariñas Coruñesas (northwest of Galicia), in the vicinity of the city of A Coruña (Fig. 11.1). This area, roughly bounded by the estuaries (rías) of A Coruña and Betanzos and defined by the inland valleys of the rivers Mero and Mandeo, has a natural and cultural coherence that makes it suitable for historical study of its territorial organisation (Sánchez Pardo 2008: 46–50). It is also, as is generally the case across all the northwest of the Iberian Peninsula, an area where industrialisation and modernisation of the countryside took place later that in the rest of Spain, so in 1957 the countryside still maintained, as it does in part even today, all its traditional characteristics.

Clearly, we cannot here conduct a comprehensive study of this area using these aerial photographs, but merely seek to show some representative cases of archaeological information regarding different historical and cultural phases that we can identify. In order to do so, we have first georeferenced all the photographs of this area in a geographical information system (GIS) that we created containing all the relevant archaeological,[3] documentary and toponymic information. Recent aerial photographs from 2003 to 2007 have also been consulted in order to compare the results. Unfortunately, given our limited resources, it has not been possible to verify most of the data in the field, so we must emphasise the need to check hypotheses and ideas by future fieldwork.

[3] Most of the archaeological information used here has been taken from the *Inventario del Servicio de Arqueología de la Dirección Xeral de Patrimonio de la Xunta de Galicia.*

11.2.1 The Hillfort Culture (Cultura Castreña)

To begin, we will focus on the 'castros', one of the most visible and recognisable archaeological remains in the Galician countryside. The 'castros' or hillforts are settlements fortified by natural or artificial defences which are characteristic of (but not exclusive to) the Iron Age and early centuries of the Roman period in the northwest of the Iberian Peninsula (González Ruibal 2007: 632–637). Despite the many differences between them, such as their different chronologies, locations and main economic activities, their elevated and fortified position and the actual abandonment of many of them after the Roman conquest make them relatively easy to identify on aerial photographs, at least in comparison with other archaeological sites. Indeed, they are one of the best recognisable elements in aerial images taken from a high altitude, such as the American flight photographs. These allow us to acquire much interesting information about the hillforts, much of which could not currently be obtained due to changes in the Galician countryside in the last 50 years. We will illustrate this with some brief case studies of the Iron Age hillforts in the area of As Mariñas Coruñesas.

Firstly, this series of significant aerial photographs allows us to obtain some details about hillforts that have either disappeared or been irreparably damaged by the extensive recent changes in the rural landscape, by house construction or by the general urbanisation of the countryside, especially in the periphery of A Coruña. Although we will probably never know the precise characteristics and chronology of these destroyed hillforts, aerial photographs from the American flight allow us to obtain at least some information about their size and shape.[4]

One example, Castro de Ans in Paleo parish (Carral muncipality), is clearly visible in the American flight aerial photography but has now disappeared as a result of the expansion of the town of Carral. The photograph shows that it was a small, oval-shaped hillfort, approximately 65 by 45 m in diameter. Similarly, Castro de Lorbé (Dexo parish, Oleiros municipality), which has now completely disappeared as a result of the construction of a residential area, is easily distinguished as an oval-shaped concentration of trees in the middle of the crops. In other cases, the hillforts have been totally or partially destroyed because of recent changes in the organisation of the fields, especially the concentration of plots, or in the management of the traditional Galician countryside because of changes in the economic activities and ways of life of its inhabitants. An example of this phenomenon could be the Castro de Gosende (Tabeaio, Carral), which is impossible to identify as a result of current changes to the plots in this area but whose shape can be distinguished in an American

[4] Dimensions provided in this chapter are always approximate, taken from the visible elements on the aerial photographs. In the case of hillforts, these elements usually correspond to the course of the walls and ramparts, their dimensions expressed in metres for the maximum diameter across each principal axis. These measurements have been provided only in the clearest cases, but we would emphasise that they are still estimates awaiting verification by fieldwork where possible.

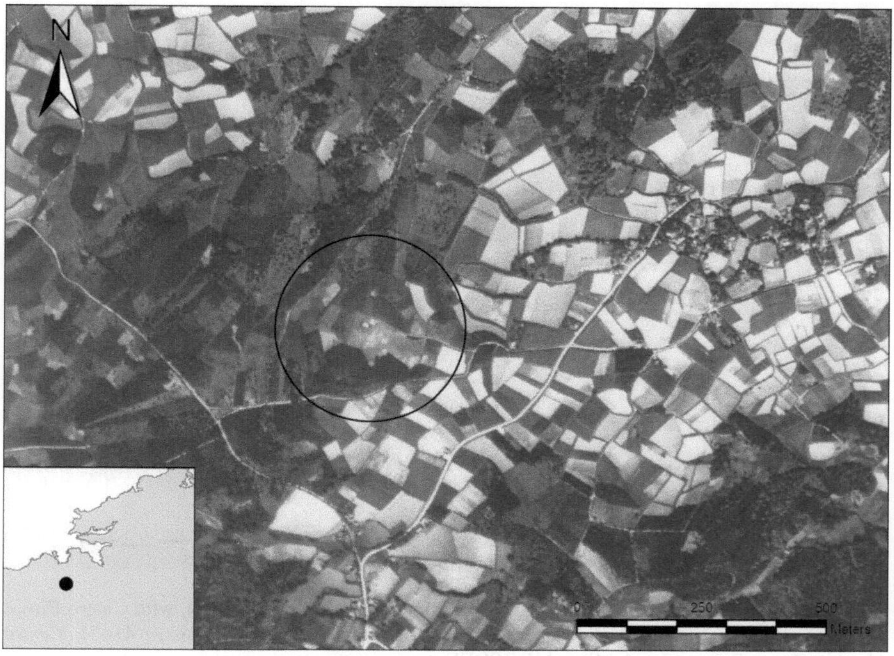

Fig. 11.2 Iron Age hillfort of Castro de Gonsende (Carral municipality). American flight photograph, 'Serie B', image 41676, *Centro Cartográfico y Fotográfico del E.A.*, MINISDEF

flight photograph (Fig. 11.2). Similarly, the Castro de Bordelle (Sarandóns, Abegondo) is now unrecognisable from the air and seems to have been damaged by the creation of forest roads, but the photographs of 1957 allow us to identify its perimeter, which indicates an oval shape approximately 115 by 83 m.

Secondly, the photographs allow us to study evidence of any hillforts that, although not destroyed, are now unrecognisable because of the significant increase in forest cover. Indeed, one of the major changes in the Galician countryside in the past 50 years has not been so much its destruction as its desertion, with consequent growth of vegetation and forest. Thus, the hillfort of Castro de Reboredo (Ouces, Bergondo) is now covered by trees and other vegetation that make it impossible to identify. However, in the American flight photograph, it can be seen very clearly, especially the course of its walls and different ramparts, indicating that it is one of the largest hillforts in this area (Fig. 11.3). Another example is the Castro de Abegondo (Abegondo, Abegondo), which is easily recognisable in the photographs as an oval shape of approximately 125 by 85 m. The same applies to the hillforts of Carnoedo (Carnoedo, Sada), with dimensions of 114 by 89 m, Armental (Pravío, Cambre), Callobre (Cuíña, Oza dos Rios), Bergondo (Bergondo, Bergondo) and Castro da Torre (Quembre, Carral), whose name and location could also indicate that it was a medieval or modern tower.

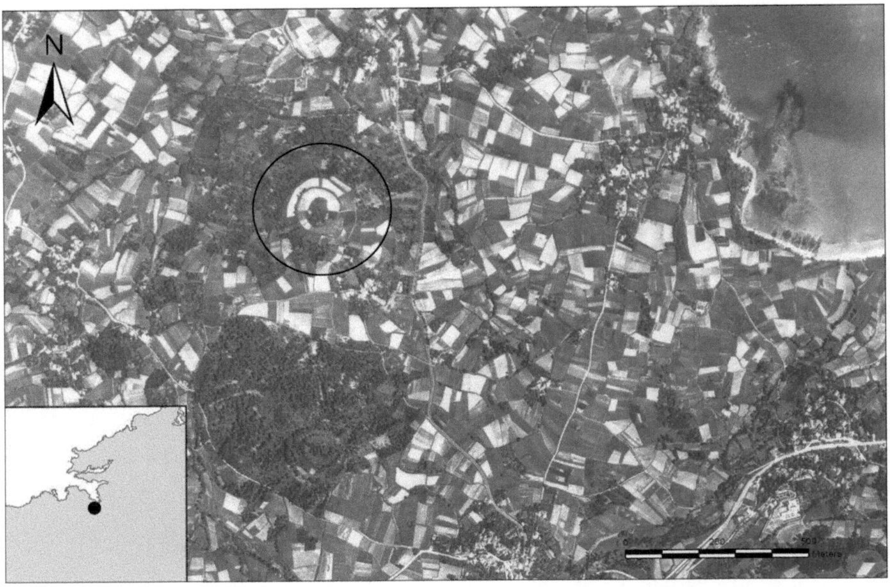

Fig. 11.3 Iron Age hillfort of Castro de Reboredo (Bergondo municipality), with Castro Eiroas visible in woodland to the south. American flight photograph, 'Serie B', image 41631, *Centro Cartográfico y Fotográfico del E.A.*, MINISDEF

Thirdly, in some cases, the photographs also help to support previous hypotheses about the possible existence of a hillfort at a particular place, based on an indicative name, but otherwise archaeologically unproven. At Castromán (Dexo, Oleiros), for example, there is no known physical evidence of the existence of a hillfort other than its name, and neither current nor recent aerial images are helpful. However, an oval enclosure can be seen in this area in an American flight photograph that could well be that hillfort. Also, the aerial photograph of 1957 shows some significant marks that may help us to demonstrate the existence of a hillfort nearby at O Castro (Dorneda, Oleiros). But such indications can be confirmed or rejected only by future archaeological research.

Finally, the photographs might allow us in some cases to locate evidence of potential hillforts which are currently unknown or unpublished, although of course it would be necessary to confirm them through archaeological fieldwork. There are usually traces in areas that have suffered major changes in recent decades. One example might be at Castros in the southwest part of Vilarraso parish (Aranga municipality), where the significant place name Rego dos castros is also found. As far as we are aware, no archaeological site has ever been identified or proposed here, and the current aerial photographs do not show any significant evidence. However, a photograph from the American flight shows an oval shape approximately 115 by 85 m in diameter. It is not possible to undertake archaeological verification as this evidence has been destroyed by the construction of a small reservoir. Nearby at A Gulpilleira (Curtis, Curtis), an oval-shaped enclosure measuring 81 by 71 m was

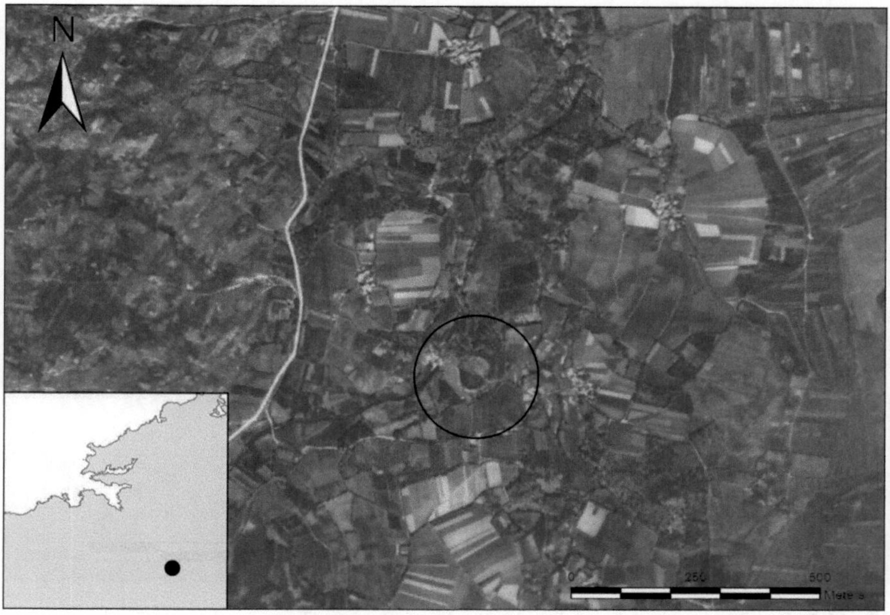

Fig. 11.4 A possible Iron Age hillfort at A Gulpilleira (Curtis municipality). American flight, 'Serie B', image 41824, *Centro Cartográfico y Fotográfico del E.A.*, MINISDEF

also evident in an American flight image. From its appearance and location on a small hill by a river, it could possibly be a hillfort not previously catalogued (Fig. 11.4). Little is visible today because of vegetation growth in the area. Another possible example, located in the southwest of Cañás parish (Carral), is oval-shaped and about 94 by 70 m in diameter. It is situated in an area (Monte Lourido) where Roman mining activities have been verified archaeologically but is no longer recognisable because of the growth of vegetation. This may be a hillfort which was exploiting local mineral resources, although its identification requires verification by archaeological fieldwork.

11.2.2 Medieval and Modern Fortresses

The American flight images can also be of great interest in the study of towers, castles and any kind of military fortifications characteristic of the medieval and modern centuries. We find a considerable density of place names in the study area referring to fortifications (tower, castle, etc.), for many of which there is no archaeological verification. Specifically, we have identified 62 place names of this type, of which only 20 correspond to known archaeological sites, leaving 42 possible cases of interest to be pursued. There is also the potential to study the problem of the relationship between medieval fortifications and protohistoric hillforts in Galicia, as

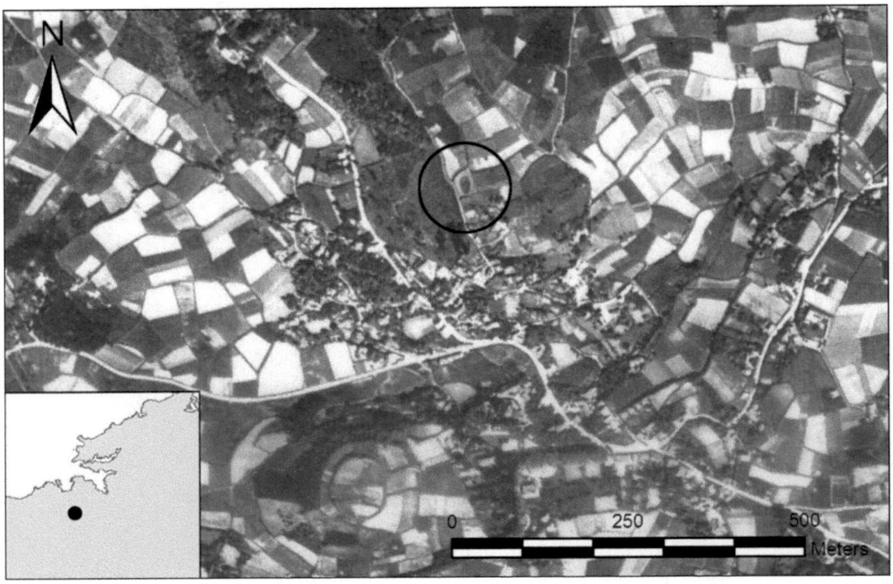

Fig. 11.5 Possible remains of a medieval or modern fortress in 'Torre das Arcas' (Cambre municipality). American flight, 'Serie B', image 41634, *Centro Cartográfico y fotográfico del E.A.*, MINISDEF

well as whether there is any archaeological difference between the place names 'Castillo' and 'Castro'. Accordingly, we offer below some suggestions about possible medieval- and modern-period fortifications in As Mariñas Coruñesas, combining the study of place names with evidence from the American flight photographs.

A first example is O Castelo in the northern part of Rodeiro parish (Oza dos Rios), which is probably the same place as Castrelo named in a document dating to the year AD 1118 from the monastery of Sobrado de los Monjes (Loscertales de García de Valdeavellano 1976: 172–175). This would indicate that this hypothetical fortress is probably earlier in date than the twelfth century. There is nothing of significance evident on current aerial photographs, but in those of 1957, we can see a well-defined, oval-shaped enclosure approximately 61 by 49 m in diameter that could be the remains of a fortress. Unfortunately, these remains have recently been destroyed to build a small reservoir.

Another example might be O Castelo, in the southeast of Paleo parish (Carral). In the American flight photographs, it is possible to distinguish an oval feature which is still visible on recent aerial photographs about 83 by 58 m in diameter. This feature, partly defined by a row of trees, could possibly represent the remains of a medieval or modern fortification.

Finally, Torre das Arcas, near to the town of Cambre (Cambre, Cambre), has seen major urban expansion in recent decades, which prevents the observation of any significant remains. However, the American flight clearly shows a dark, oval outline, possibly a cropmark, of an enclosure about 27 by 15 m in diameter (Fig. 11.5).

This may be the remains of a fortress, but fieldwork is required to confirm this, although this may be difficult to carry out because significant areas of land have been taken for the construction of houses since the taking of the photograph.

11.2.3 The Structure of the Traditional Landscape

To conclude this short review of the value of the American flight for archaeological purposes in any territorial study, we would like to mention briefly a different, but important, potential use of these images. In addition to their use for finding ancient archaeological remains, we must not forget their wider importance as a record of the contemporary traditional rural landscape, such as the structure of traditional villages, the organisation of the crops and the fields, and the network of roads linking the different villages. These elements of the landscape can provide valuable historical information as they can often be traced back several centuries, but at the time of the photography had not yet undergone the many changes arising from countryside industrialisation and urbanisation. In the case of the American flight in Galicia, this phenomenon is particularly significant. Several authors have emphasised the considerable historical continuity in the basic structure of the Galician traditional rural landscape since at least the Middle Ages (Bouhier 2001: 1219–1224; Criado Boado and Ballesteros Arias 2002: 463). At the time of the American flight in 1957, this structure remained virtually intact, providing clues to the history of our study area.

The village of Meangos (Meangos, Abegondo) is a good example of a typical traditional rural village in Galicia which is mentioned in a document of the year AD 942 from the Monastery of Sobrado de los Monjes (Loscertales de García de Valdeavellano 1976: 160–163) (Fig. 11.6). The photograph from the American flight reflects the village structure before the transformations which it has undergone in recent decades. It reveals several small groups of houses located in a little valley at the foot of a hill, distributed in clusters around several roads and surrounded by field plots. The village church is located away from the houses further up the hill, but it functions as the main element of unity and cohesion for the community, while on the top of the hill, there is a protohistoric hillfort which may be considered the forerunner of the current village. Thus, we have here all the key elements that define the historical development of the Galician countryside in the past two millennia: the hillfort, the open village, the roads and the church. The study of their spatial relationships, considered in detail elsewhere (Bouhier 2001; Sánchez Pardo 2008; Criado Boado and Ballesteros Arias 2002; Fernández Mier 1999), can give us some clues about their formation and evolution: the gradual replacement of hillforts by settlements in lower areas for cultural reasons, such as the emulation of Roman models, for economic reasons, such as proximity to farmlands, or for social reasons, such as the progressive disappearance of Iron Age social organisation as a result of Romanisation; the introduction of the church in a place with previous sacred connotations during the medieval period; and the gradual dispersion of settlements in order to better exploit the fertile soils. Although this village still retains

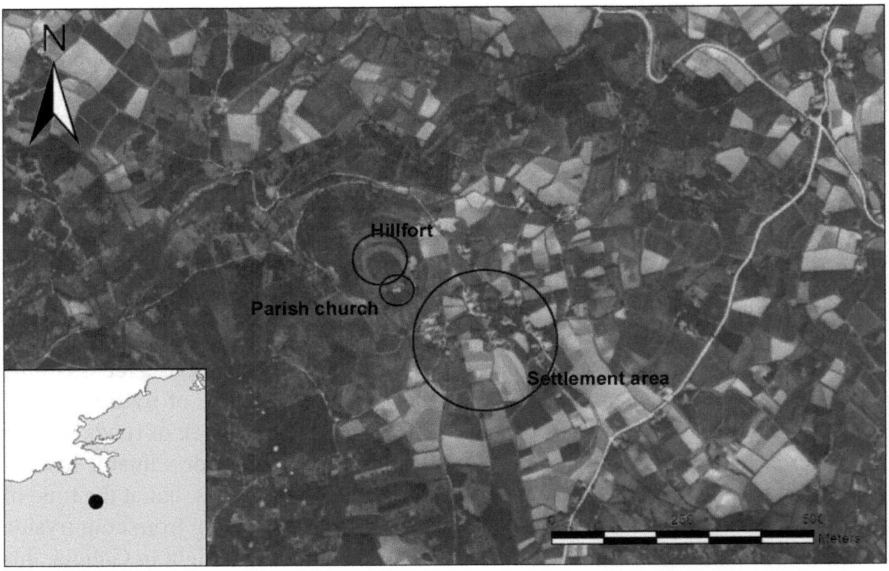

Fig. 11.6 Structure of a Galician traditional rural village at Meangos (Abegondo municipality). American flight photograph, 'Serie B', image 41816, *Centro Cartográfico y fotográfico del E.A.*, MINISDEF

much of its traditional structure, there are several elements that have changed or disappeared as a result of transformations in the organisation of crops and of land, the construction of new houses, the desertion of farmlands and the modification and asphalting of roads. Thus, the aerial photographs of the American flight should be seen as a key element in the archaeological study of the landscape.

11.3 Aerial Photographs from Yesterday, Archaeology of Tomorrow

11.3.1 Geographical and Historical Context

The second part of this chapter focuses on a probable new site that is not currently recorded in the Portuguese *Endovélico* GIS database (see above). In our opinion, the photographs illustrated here justify both a ground survey and a thoughtful revision of the *Instituto Geográfico Português* aerial files. The area of interest is located in the Portuguese Middle-South Atlantic coast, in the region of the Sado Estuary between Setúbal and the Baixo Alentejo region. The site focus is a 51 m high mound (Fig. 11.7), immediately to the south of the Ribeira de São Martinho, in the last western foothills of the gentle chain named the Serrinha (187 m high). This mountain range is well known because of the exploitation of its rich mineral resources since the Late Bronze and Iron Ages.

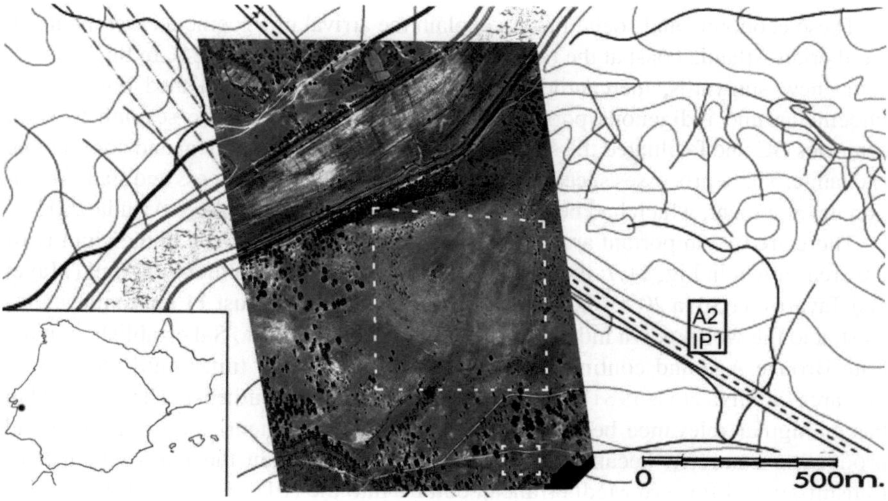

Fig. 11.7 Location of sites south of Ribeira de São Martinho (Sado Estuary). Part of a vertical photograph taken by the *Força Aérea Portuguesa* superimposed on the military map of the *Instituto Geográfico do Exército*

The Ribeira de São Martinho is an ancient branch of the River Sado from the Cenozoic Era. Like others in the vicinity, it is now an area of alluvial sediment filled with detritus material, mostly sand and clay, partly used for agriculture. These first hills form a geological frontier between two kinds of soils: the alluvial ones, saturated but deep; and the podozoic ones, acid and shallow, with a large amount of added organic material that has slowly mineralised, and covered by degraded forest. The area photographed has not seen any agricultural work in recent decades and preserves a typical Mediterranean low forest landscape (Dos Santos Castanheira 2005).

In terms of potential human presence, the area was hugely rich in resources for hunting, fishing and seafood gathering. Thus, in the whole of the Baixo Alentejo region, human groups were located near the coastline without occupying the interior. The great biological productivity that the Sado Estuary offers resulted in a significant delay in the transformation of economic systems, and the adoption of Neolithic techniques of farming and cattle raising were delayed here. Hence, sedentism and the construction of architectural structures large enough to leave significant archaeological evidence did not take place until the beginning of the third millennium BC in the interior and a little later in the Estuary zone. Nevertheless, Alentejo has remarkable megalithic structures from the fifth millennium BC. During the nineteenth to thirteenth centuries BC, the region saw a large demographic increase with new settlements, some of them fortified, of rectangular huts and associated necropolises containing monumental tombs and tumuli with several burials in each. Over succeeding centuries, the social hierarchy process increased its rhythm in parallel with intensive control over the territory. This social development during the Bronze Age was stronger on the Alentejo coast and in the Sado Estuary because of its abundant mineral resources (Tavares da Silva and Soares 2006).

These economic and social factors explain the arrival of Phoenician merchants on the Iberian Atlantic coast at the beginning of the first millennium BC. In this context, some new shipyards, mercantile centres and settlements, established both by the Phoenician and indigenous population, were founded during the seventh to sixth centuries BC and facilitated the stimulation of cultural, technological and commercial exchange. These processes speeded up the social stratification process and state formation development, which had been embryonic since the Bronze Age (Arruda 2002).

There are two important archaeological sites from the Iron Age in the vicinity of the area shown in Fig. 11.7: the Phoenician establishment-sanctuary of Abul (Mayet and Tavares da Silva 2000; Tavares da Silva 2005),[5] located just 11 km to the southwest, and the well-known indigenous settlement at Alcácer do Sal established in the Late Bronze Age and continuously occupied from Roman times until the present (Tavares da Silva et al. 1981), located some 13 km to the southeast. The site at Abul has a singular relevance because, after its first establishment as an exchange and production centre, it became the main sanctuary and then the major channel of transmission of Eastern Mediterranean culture into the Atlantic coast of the Iberian Peninsula. In Roman times, lacking any cult function, Abul was transformed into an important amphora production site (first to third centuries AD), thanks to the high quality clays of the Sado Estuary.

Unlike other regions in Roman Lusitania, rural landscapes in Baixo Alentejo did not see a spread of small settlements (of 2 ha or less). The population was gathered around bigger villas, which were able to control land holdings of between 200 and 400 ha. Middle-sized urban settlements, such as *vici* or *civitates*, were usual as well. Thus, small settlements in this area are usually explained as pre-Roman habitats potentially surviving into the Early Middle Ages (Alarcão 1998). However, in the Late Roman period, Baixo Alentejo experienced further concentration of property and, in the Islamic and Middle Ages, occupation of the entire area declined. During the repopulation after the *Reconquista*, powerful entities such as military orders, lords and the monarchy controlled large areas of land in order to structure the territory and, once again, that process favoured the concentration of settlers (Beirante 1993).

Only in the last two centuries did Baixo Alentejo and the Sado Estuary again show demographic growth, but obviously with different settlement patterns, mainly concentrated in the urban centres such as Alcácer do Sal and Setúbal. In recent decades, this expansion has stopped and an internal migration to Lisbon and its metropolitan area is taking place. In parallel, it is important to stress that only over recent decades has the Portuguese landscape begun to suffer a strong transformation process, as is apparent from our photographs, with the construction of the new motorway. This suburban and infrastructural development adds further value to the aerial photographs that were taken just before such landscape alteration processes began.

---

[5] Though the archaeologists did use aerial photographs in this research project, they focused on a very limited area around their sites, excluding the study of the surrounding region (Mayet and Tavares da Silva 2000: 245–259 and pls. 32–37).

11.3.2 Air Photo Interpretation

Turning to the specific examples in more detail, in Fig. 11.7, we can see on the military map no. 467 of the *Instituto Geográfico do Exército* a branch of the Sado Estuary at a scale 1:25,000, onto which we have rectified[6] and superimposed part of a vertical photograph taken at an approximate scale of 1:8,000 by the *Força Aérea Portuguesa* in 1949 in a flight made for cartographic purposes. The area involved is today crossed by the A2-IP1 motorway (marked on the map) that links Lisbon with the southern coast of the Algarve. In the middle of the photograph, we can see the low mound referred to in the previous section and towards the top a stretch of the Ribeira de São Martinho. Running down the centre of the first enlargement (Fig. 11.8), we can see a stream bed that separates two slight mounds. The more westerly, on the left, has various remains on its top, both earthworks and surface structures, the latter indicating some walls with parallel and perpendicular alignments. The site is elliptical in shape, some 55 m long and 20 m wide. Some linear cropmarks extend away from the hilltop in three directions, probably ancient links to the site from northeast, southwest and southeast. The latter two would fit with the main routes towards Abul and Alcácer do Sal, respectively. In the lower part of Fig. 11.8, two further cropmarks are visible. The first defines a perfect circle approximately 45 m in diameter; the second indicates the presence of a small rectangular structure with dimensions of approximately 3 by 5 m. The second enlargement (Fig. 11.9) shows a large, compact group of irregular rounded cropmarks, each with a diameter of a metre or two, as far as it is possible to calculate on the basis of the scale of the image.

The obvious interpretation of the large circular structure is as a tumulus. However, most tumuli excavated in the south of the Iberian Peninsula have smaller dimensions, little more than 20 m in diameter, as in Setefilla (Aubet Semmler 1982), Carmona (Fernández Cantos and Amores Carredano 2000), Cástulo, Cerrillo Blanco (Ruíz 2008: 779–782), or even smaller. However, there are also large tumuli on the north Atlantic coast of Morocco, in Arcila (Gozalbes Cavrioto 2006), and in other necropolises in south Portugal, such as Mealha Nova (Arruda 2001: 244, Fig. 9), which seem to have similar large circular structures of some 40 or 50 m in diameter. It is usual in Iberian Late Bronze or Iron Age contexts to find several simple tombs associated with tumular burials. Without further fieldwork, it is impossible to confirm the identification of the features recorded in Figs. 11.7, 11.8, and 11.9, but the probable tumulus and the group of small cropmarks nearby suggest a typical funerary landscape next to the settlement on the hilltop.

The above features are not easily explained as modern human activity, nor do they fit with the demographic dynamics of the Sado Estuary and Baixo Alentejo either in medieval times or in the Roman period (Alarcão 2002). However, this

[6] Rectification of photograph 402 from the roll 49.14 from the *Instituto Geográfico Português* was undertaken using AirPhoto3.

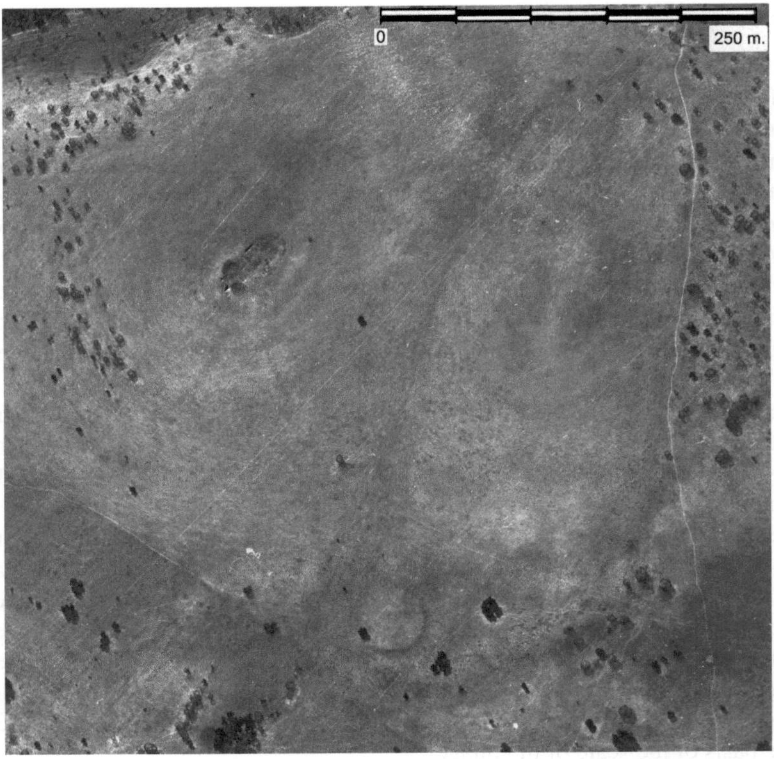

Fig. 11.8 Detail of the larger area south of Ribeira de São Martinho (Sado Estuary) indicated on Fig. 11.7

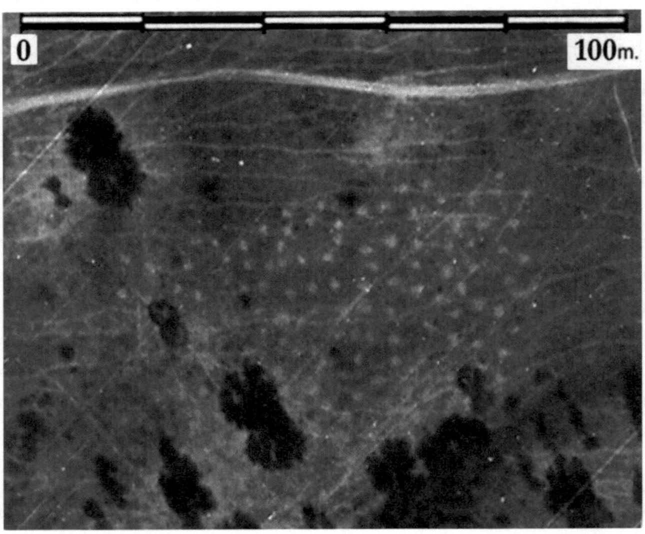

Fig. 11.9 Detail of the smaller area south of Ribeira de São Martinho (Sado Estuary) indicated on Fig. 11.7

territory shows strong demographic and economic growth during the Late Bronze II (ninth to seventh centuries BC) and the Orientalising period (seventh to fifth centuries BC) with a multiplication of small groups of habitats in the landscape of mid- and southern Portugal (Schubart 1975; Pellicer Catalán 2000; Tavares da Silva 2005). Thus, the remains might reasonably be interpreted as a modest-sized settlement complex, perhaps of pre-Roman origin.

11.4 Conclusions

In conclusion, these two examples, one from Spain and one from Portugal, illustrate just a small part of the huge number of ways in which the analysis of aerial photography from historical archives can be useful for major archaeological projects. These documents offer diachronic information of great value about a landscape that has suffered heavy transformations in recent decades. They can be very useful not only for the detection of unrecorded remains, but also support the study of the historical structure of rural landscapes. With our contribution, we hope to encourage the directors of research projects in the Iberian Peninsula to take an interest in these aerial photographic archives that have not yet been sufficiently explored.

Acknowledgments We would like to thank Brais Xosé Currás Refojos for his valuable help in relation to Roman mining activities and Jorge Pardo Vicente and Javier Martínez Jimenez for all their valuable help in the translation of this chapter.

Bibliography

Alarcão, J. (1998). A paisagem rural romana e alto-medieval em Portugal. *Conimbriga, 37*, 89–119.

Alarcão, J. (2002). *O dominio romano em Portugal*. Sintra: Europa-America.

Ariño Gil, E., Gurt Esparraguera, J. M., & Palet Martínez, J. M. (2004). *El pasado presente: arqueología de los paisajes en la Hispania romana*. Salamanca: Universidad de Salamanca.

Arruda, A. M. (2001). A Idade do Ferro pós-orietalizante no Baixo Alentejo. *Revista Portuguesa de Arqueologia, 4*(2), 207–291.

Arruda, A. M. (2002). *Los fenicios en Portugal. Fenicios y mundo indígena en el centro y sur de Portugal (siglos VIII-VI a.C.)*. Cuadernos de Arqueología Mediterránea 5–6. Barcelona: Carrera Edició.

Aubet Semmler, M. E. (1982). Los enterramientos bajo túmulo de Setefilla (Sevilla). *Huelva Arqueológica, 6*, 49–70.

Beirante, M. Â. (1993). A 'Reconquista' cristã. In J. Serrão & A. H. de Oliveira Marques (Eds.), *Nova História de Portugal, Tomo II – Portugal, das invasões germânicas à 'Reconquista'* (pp. 253–363). Lisbon: Presença.

Bellón Ruiz, J. P., Gómez, F., Sánchez, A., Wiña, L., Ruiz, A., Molinos, M., Gutiérrez, L., Mozas, F., Rueda, C., García, M. A., Martínez, A. L., &Lozano, G. (2009). Bæcula. Análisis arqueológico del escenario de una batalla de la Segunda Guerra Púnica. In Á. Morillo, N. Hanel, & E. Martín (Eds.), *Limes XX. XX Congreso Internacional De Estudios Sobre La Frontera Romana – XXth International Congress of Roman Frontier Studies León (España), Septiembre, 2006* (pp. 253–265). Madrid: Ediciones Polifemo.

Bouhier, A. (2001). *Galicia. Ensaio xeográfico de análisis e interpretación de un vello complexo agrario*. Santiago de Compostela: Consellería de Agricultura, Gandería e Política Agroalimentaria.
Bugalhão, J., Lucena, A., Bragança, F., Neto, F., Sousa, M. J., Gomes, S., Pinto da Costa, J., Caldeira, N., Viralhadas, P., & Fraga, T. (2002). Endovélico. Sistema de Gestão e Informação Arqueológica. *Revista portuguesa de Arqueologia, 5*, 277–283.
Burgalhão, J. (2002). Instituto Português de Arqueologia em processo de extinção. *Al-Madam, 11*, 45–48.
Correia, M. (2005). Novos dados para a Carta Arqueológica do Concelho do Alcochete. *Al-Madam, 13*, 130–132.
Criado Boado, F., & Ballesteros Arias, P. (2002). La Arqueología rural: contribución al estudio de la génesis y evolución del paisaje tradicional. In *I Congreso de Ingenieria Civil, Territorio y Medio Ambiente. Volumen I* (pp. 461–479). Madrid: Colegio de Ingenieros de Caminos, Canales y Puertos.
Dos Santos Castanheira, A. (Ed.). (2005). *Atlas de Portugal*. Lisbon: Instituto Geográfico Português.
Farinha Dos Santos, M. (1965). Aplicação da Fotografia Aérea no levantamento de Cartas Arqueológicas. *Arquivo de Beja, 22*, 5–9.
Fernández Cantos, A., & Amores Carredano, F. (2000). La necrópolis de la Cruz del Negro (Carmona, Sevilla). In C. Aranegui Gascó (Ed.), *Argantonio: rey de Tartessos* (pp. 157–164). Sevilla: El Monte.
Fernández García, F. (2000). *Introducción a la fotointerpretación*. Barcelona: Ariel.
Fernández Mier, M. (1999). *Génesis del territorio en la Edad Media. Arqueología del paisaje y evolución histórica en la montaña asturiana*. Oviedo: Universidad de Oviedo.
Gil Mantas, V. (1985). Arqueologia Urbana e fotografia Aérea: Contributo para o Estudo do Urbanismo Antigo de Santarém, Évora e Faro. I Encontro Nacional de Arqueologia Urbana. *Trabalhos de Arqueologia, 3*, 13–26.
Gil Mantas, V. (1996). Teledetecção, cidade e territorio: Pax Iulia. *Arquivo de Beja, 3*(1), 5–30.
Gil Mantas, V. (2003). Indícios de um Campo Romano na Cava do Viriato? *Al-Madam, 12*, 40–42.
González Ruibal, A. (2007). *Galaicos. Poder y comunidad en el Noroeste de la Península Ibérica (1200 a. C.-50 d. C.)*. A Coruña: Museo arqueolóxico e histórico provincial de A Coruña.
Gozalbes Cavrioto, E. (2006). El monumento protohistórico de Mezora (Arcila, Marruecos). *Archivo de Prehistoria Levantina, 26*, 323–348.
Langley, M. M. (2006). Est in agris: a spatial análisis of Roman *villae* in the region of Monforte, Alto Alentejo, Portugal. *Revista portuguesa de Arqueologia, 9*, 317–328.
Liz Guiral, J., & Celis Sánchez, J. (2007). Topografía antigua de la ciudad de Lancia (Villasabariego, León, España). *Zephyrus, 60*, 241–263.
Loscertales de García de Valdeavellano, P. (1976). *Tumbos del monasterio de Sobrado de los Monjes*. Madrid: Dirección General del Patrimonio Artístico y Cultural.
Mayet, P., & Tavares Da Silva, C. (2000). *L'établissement phénicien d'Abul (Portugal). Comptoir et sanctuaire*. Paris: de Boccard.
Oliveira, A. C., Silva, A. R., Estêveão, F., & Deus, M. M. (2001). *Carta Arqueológica do municipio de Loures*. Loures: Câmara Municipal de Loures.
Pellicer Catalán, M. (2000). El proceso orientalizante en el occidente ibérico. *Huelva arqueológica, 16*, 89–134.
Pellicer Corellano, F. (1998). *Introducción a la fotografía aérea*. Zaragoza: Azara Editores.
Penedo Cobo, E. (2006). El yacimiento visigodo de Buzanca 2. *Zona arqueológica, 8*(2), 594–602.
Rua, H. (2007). Os sistemas de informação geográfica na pesquisa arqueológica: um modelo preditivo na detecção de *uillae* em meio rural. *Revista portuguesa de Arqueologia, 10*, 259–274.
Ruíz, A. (2008). Iberos. In F. Gracia Alonso (Ed.), *De Iberia a Hispania* (pp. 733–844). Barcelona: Ariel.

Sánchez Pardo, J. C. (2008). *Territorio y poblamiento en Galicia entre la Antigüedad y la Plena Edad Media.* Unpublished PhD thesis, Universidad de Santiago de Compostela.

Sánchez Pardo, J. C., & Fumadó Ortega, I. (2006). Aerial Archaeology in Spain: Historiography and expectations. In S. Campana & M. Forte (Eds.), *From space to place: 2nd International Conference on Remote Sensing in Archaeology. Proceedings of the 2nd International Workshop, CNR, Rome, Italy, December 4–7, 2006* (pp. 65–72). British Archaeological Reports S1568. Oxford: Archaeopress.

Schubart, H. (1975). *Die Kultur der Bronzezeit im Südwesten der Iberischen Halbinsel.* Madrider Forschungen 9. Berlin: W. de Gruyter.

Strutt, K. (2000). Use of a GIS for regional archaeological analysis: Application of computer-based techniques to Iron Age and Roman settlement distribution in North-West Portugal. In G. Fincham, G. Harrison, R. Holland, & L. Revell (Eds.), *TRAC 1999, Ninth proceedings of the theoretical roman archaeology conference* (pp. 118–141). Oxford: Oxbow.

Tavares Da Silva, C. (2005). A presença fenícia e o processo de orientalização nos estuarios do Tejo e Sado. In F. J. Jiménez Ávila & S. Celestino Pérez (Eds.), *III Simposio Internacional de Arqueología de Mérida: Protohistoria del Mediterráneo Occidental* (pp. 749–765). Mérida: Instituto de Arqueología de Mérida.

Tavares Da Silva, C., & Soares, J. (2006). *Setúbal e Alentejo Litoral*. Tomar: Arkeos.

Tavares Da Silva, C., Soares, J., Beirão, C. M. M., Ferrer Dias, A., & Coelho Soares, A. (1981). Excavações arqueológicas no Castelo do Alcácer do Sal (campanha de 1979). *Setúbal Arqueológica, 6–7*(1980–1981), 149–218.

Chapter 12
Soviet Period Air Photography and Archaeology of the Bronze Age in the Southern Urals of Russia

Natal'ya S. Batanina and Bryan K. Hanks

Abstract This chapter examines the use of air photography during the Soviet Period and the utilisation of this imagery for archaeological research. A detailed case study is provided on the Southern Ural Mountains region of Russia where archaeologists focusing on the Middle to Late Bronze Age (2100–1500 BC) have utilised black-and-white air photography to identify numerous archaeological sites ranging from the Bronze Age to the medieval period. In recent years, the integration of air photography, geophysical prospection and stratigraphic excavation has produced important insights into the spatial characteristics and diachronic phasing of prehistoric settlement and cemetery patterning. These successful research programmes provide a valuable model for similar field programmes being conducted throughout the territories of the Russian Federation and other regions of the world.

12.1 Introduction

The twentieth century was an extremely dynamic period in Russia's history as it included not only the creation and collapse of the Soviet Union and the First and Second World Wars, but also the attainment of immense tracts of new territory in the

N.S. Batanina
Historical-Cultural Reserve Arkaim, Chelyabinsk, Russia
e-mail: someone2005@pisem.net

B.K. Hanks (✉)
Department of Anthropology, University of Pittsburgh, Pittsburgh, PA, USA
e-mail: bkh5@pitt.edu

W.S. Hanson and I.A. Oltean (eds.), *Archaeology from Historical Aerial and Satellite Archives*, DOI 10.1007/978-1-4614-4505-0_12,

northern Eurasian region. This period of expansion and colonisation included the extension of settlement into new environmental zones, a substantial increase in livestock grazing, mineral exploitation, construction of hydroelectric dams and an emphasis on intensified agricultural production. Such activities impacted on the environment of these new territories and in some cases substantially influenced prehistoric archaeological sites. Fortunately, during every decade in the second half of the twentieth century, the Soviet Union carried out systematic black-and-white aerial photography within its territorial boundaries. The photographs produced during this time have become a valuable historical resource for archaeological research as they recorded sites either no longer extant or less well preserved as a result of developments during the second half of the twentieth century. Today, many scholars in regions of the former Soviet Union utilise this archival aerial photography within their archaeological research programmes.

12.2 Early Soviet Period Remote Sensing

The first aerial photographs in Russia were made in 1918 near the city of Tver for the preparation of topographic maps. Starting in 1925, during the early years of the USSR, further data was collected through the special aerial photography department of the Geographical Society. In the first 10–12 years, the data obtained were used primarily for topographic mapping as aerial photography provided an efficient means of surveying. In 1929, the Leningrad Academy of Sciences formed the Research Institute of Geodesy and Cartography; in 1930, this institute was reorganised into the Research Institute for Aerial Photography (Komarov 1969). During the 1930s–1950s, the reorganisation of this institute was repeated numerous times and ultimately transferred to Moscow. By the mid-twentieth century, the scope of aerial photography was greatly increased and various new techniques were developed for deciphering aerial data. In addition to mapping, aerial photography was used to inventory forests, to provide for land management and reclamation, to obtain important data for hydrological projects, for mineral prospection and in designing roads and surveying for pipeline construction. As a result, aerial reconnaissance and photography played a significant role in the new economic policies of the USSR.

Aerial photography was first used for archaeological research in the Soviet Union in the 1930s–1940s in the Chorasmia oasis region of central Asia. In this period, S.P. Tolstov led a highly successful archaeological-ethnographic programme of research along the Syr Darya and Amu Darya rivers, and large-scale aerial photography was utilised. Subsequent aerial surveys, especially those from 1946 to 1973 that were undertaken specifically for archaeological purposes, produced excellent results, and over 200 archaeological sites were identified. These included settlements, necropolises and ancient irrigation systems. N.I. Igonin and V.V. Andrianov made additional refinements to the methods of aerial photography during this time, such as in the use of scale, the time of year and day that the photographs were taken, the best use of lighting, etc. (Igonin 1969; Andrianov 1965). Experience gained from these experts was used, with appropriate amendments, for the landscape zone in the Tatar Autonomous Soviet Socialist Republic (e.g. Bilyarsk settlement) (Khalikov and Igonin 1974). Starting in 1960 in Ukraine, K.V. Shishkin contributed importantly to the development of air photography and produced detailed studies of archaeological monuments in the lower Dnieper River region and in the area around the ancient Greek city of Olbia (Shishkin 1966). As a result, a number of new archaeological sites were identified, and ancient roadways and systems of land use were documented.

While it is clear that the use of aerial photographic data played a significant role in the development of archaeological research during the Soviet Period, access to the vast archives of historical imagery from this era is still largely restricted by government agencies and not widely available for public or academic use. Nevertheless, in recent years, several projects, and their subsequent publications, have utilised and disseminated this important imagery and highlighted its utility for the detection, identification and interpretation of archaeological sites. In some cases, completely new categories of monuments have been discovered, such as the Middle Bronze Age 'Country of Towns' (*strana gorodov*) in the steppe zone of the south-eastern Ural Mountains (Zdanovich and Batanina 2002, 2007). A total of 22 nucleated, fortified settlements linked to this development were discovered through the analysis of Soviet Period air photography coupled with archaeological reconnaissance and excavation (Fig. 12.1). This regional settlement pattern has been linked to the emergence of the Sintashta culture and has received increasing international interest since the 1990s (Gening et al. 1992). As many scholars have suggested, the appearance of this unique settlement form in the central steppe region of Eurasia may be linked to more intensive copper mining, bronze metal production and regional trade (Anthony 2007; Hanks 2009; Hanks and Linduff 2009; Zdanovich and Zdanovich 2002). In addition, the excavation of Sintashta cemeteries has produced the earliest-dated spoke-wheeled chariot technology in the world (Anthony and Vinogradov 1995; Anthony 2007; Hanks et al. 2007). Chariot technology, coupled with clear evidence of settlement fortification, strongly suggests that the role of warfare was a crucial element driving socio-technological developments during the Middle to Late Bronze Age in this region (2100–1500 BC).

The remainder of this chapter is devoted to a discussion of Soviet Period aerial photographs utilised in this region that were applied in collaboration with archaeological reconnaissance and excavation. Although initially conducted as part of a programme of rescue archaeology in response to plans for dam construction, several research programmes were continued and have contributed importantly to the development of archaeological remote sensing in the Russian Federation and have led to improved understanding of late prehistoric archaeology.

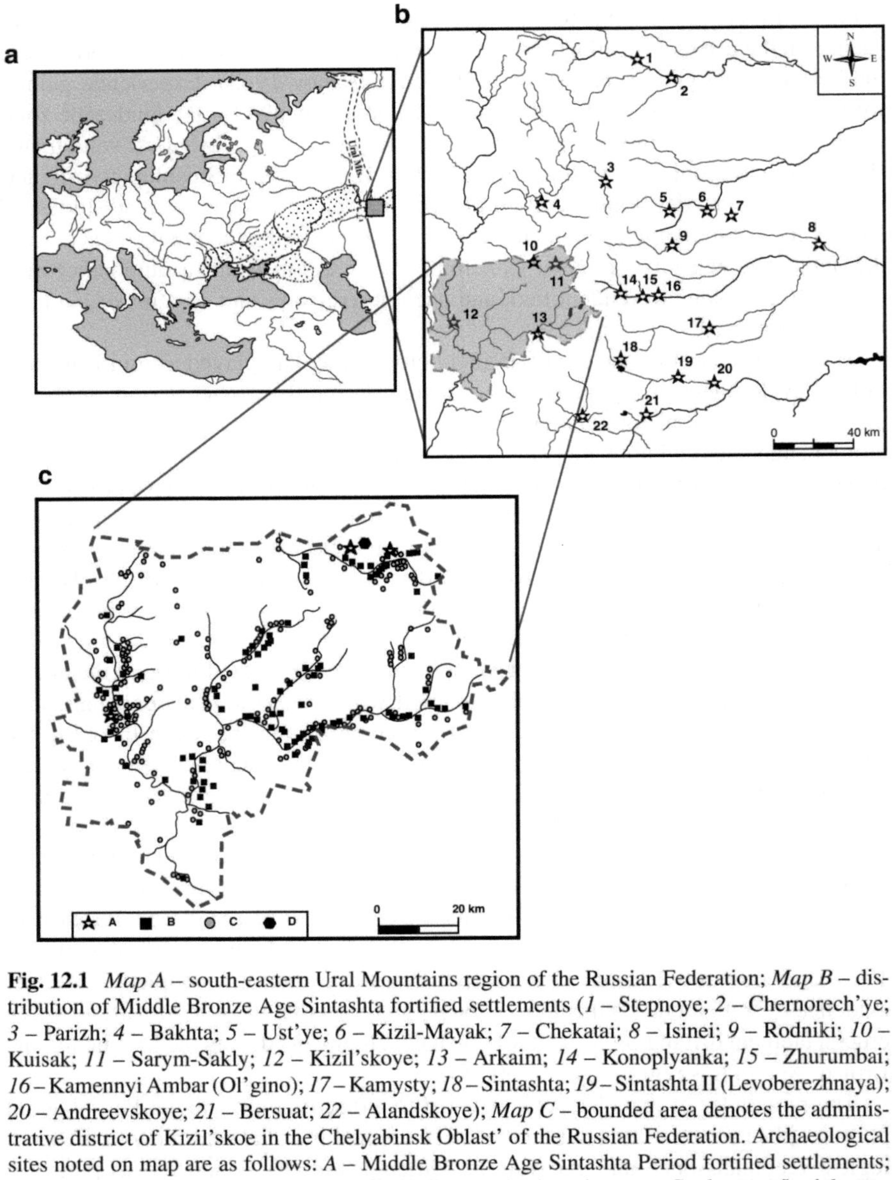

Fig. 12.1 *Map A* – south-eastern Ural Mountains region of the Russian Federation; *Map B* – distribution of Middle Bronze Age Sintashta fortified settlements (*1* – Stepnoye; *2* – Chernorech'ye; *3* – Parizh; *4* – Bakhta; *5* – Ust'ye; *6* – Kizil-Mayak; *7* – Chekatai; *8* – Isinei; *9* – Rodniki; *10* – Kuisak; *11* – Sarym-Sakly; *12* – Kizil'skoye; *13* – Arkaim; *14* – Konoplyanka; *15* – Zhurumbai; *16* – Kamennyi Ambar (Ol'gino); *17* – Kamysty; *18* – Sintashta; *19* – Sintashta II (Levoberezhnaya); *20* – Andreevskoye; *21* – Bersuat; *22* – Alandskoye); *Map C* – bounded area denotes the administrative district of Kizil'skoe in the Chelyabinsk Oblast' of the Russian Federation. Archaeological sites noted on map are as follows: *A* – Middle Bronze Age Sintashta Period fortified settlements; *B* – unfortified Late Bronze (possible Middle Bronze Age) settlements; *C* – kurgans/burial complexes, *D* – Bronze Age copper mine Vorovskaya Yama (Map C adapted and redrawn from Zdanovich et al. 2003)

12.3 Late Soviet Period Remote Sensing and Archaeological Research

The steppe and forest-steppe ecological zones of the Southern Ural Mountains region of Russia were extensively photographed from 1954 to 1999. Air photographs were taken primarily during the months of April through November when the earth's surface was free of snow cover (Fig. 12.5). The scale of the photographs ranged from 1:10,000 to 1:50,000, which provided a wide range of imagery that has been used for a number of different objectives. For example, aerial photographs were intensively used for geological prospecting during the last half of the twentieth century, and the Soviet government formed a national geological survey organisation that worked systematically with aerial photographic data for this purpose.

A pioneering scholar involved in early remote sensing research in the Southern Urals was Iya M. Batanina (1935–2011). She was a member of the Geological Faculty at Ural State University (located in Yekaterinburg, Russia) and was involved in focused geological surveys in the Urals region. Over the years, she worked with several geological institutions and specialised in the development of remote sensing methods for geological surveying and the prospection of mineral resources. By 1990, and at the age of 55, Iya Batanina brought her extensive knowledge into the field of archaeology, which gave her the opportunity to assist regional archaeologists with the identification and interpretation of numerous anthropogenic features and sites discernable through aerial photography.

A strong interest in the use of aerial photography for archaeological research in the Southern Urals had developed already by 1986. This was stimulated by regional administrative plans to construct a dam on the Bol'shaya Karaganka River, which is situated in the southern zone of the Chelyabinsk Oblast' (administrative region). Archaeological research that had commenced in 1969 in this area, under the direction of V.F. Gening, had uncovered a series of archaeological sites dating from the Mesolithic through the medieval period along the Sintashta River and its tributaries. Continuing research during the 1970s and early 1980s brought to light the eponymous settlement-cemetery complex of Sintashta and a completely new category of fortified Bronze Age settlement with associated cemeteries (more recently labelled 'The Country of Towns') (Fig. 12.2). In 1986, plans for the construction of a dam in the region resulted in intensive emergency archaeological surveys within the catchment of the proposed reservoir. This included collaboration with specialists such as Iya Batanina and the incorporation of detailed aerial photographic analysis. These efforts led to the discovery and identification of a second fortified Sintashta period settlement known today as Arkaim. Increasing debate over the destruction of these important heritage sites attracted worldwide attention, and plans for the construction of the dam were eventually overturned in 1991 (see overviews in Shnirelman 1999; Zdanovich 1999). Shortly thereafter, a commission was enacted to protect the heritage of Arkaim, its environs and the other known fortified Sintashta period settlements (Arkaim Museum and Reserve: http://www.arkaim-center.ru/).

Successful collaboration between geologists and archaeologists in the microregion around Arkaim continued from 1990 to 1995, and a total of 71 archaeological

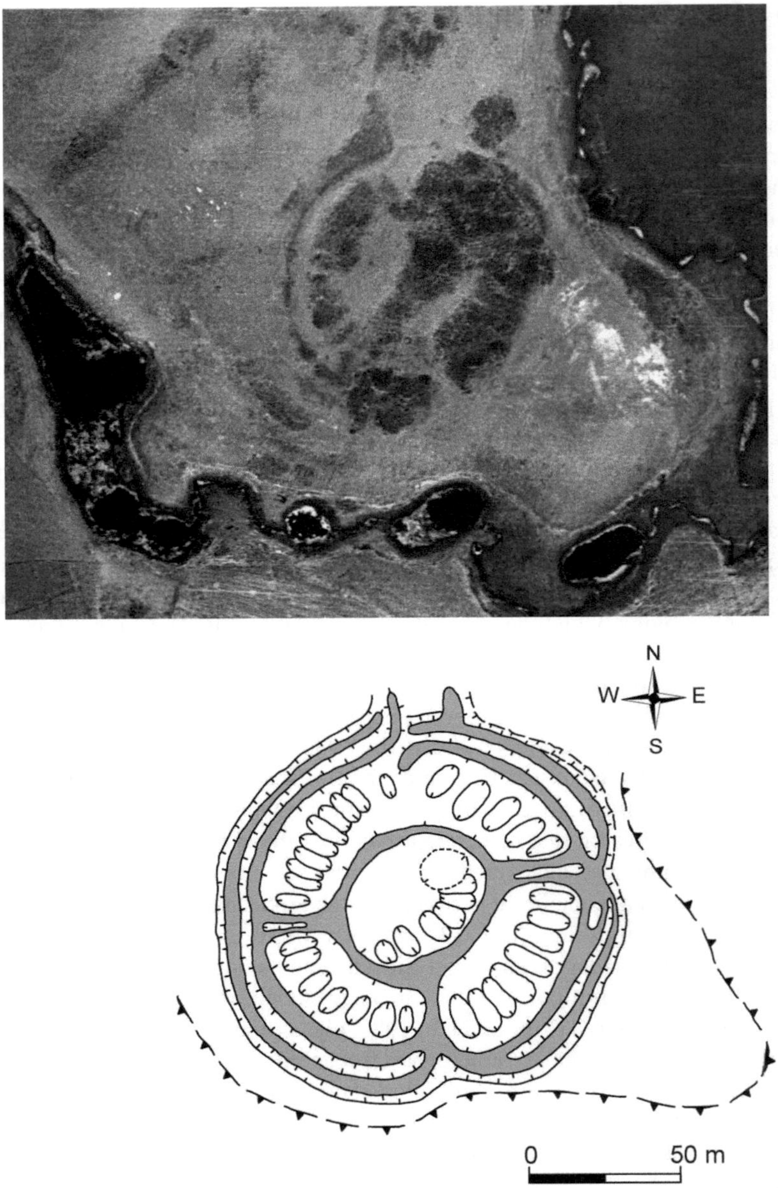

Fig. 12.2 *Upper* – portion of an August 17th 1974 Soviet Period aerial photograph of the Middle Bronze Age fortified settlement of Sarym-Sakly (scale: 1 cm–22 m); *lower* – aerial photographic interpretation by Iya Batanina (Redrawn by B. Hanks)

sites from different periods were identified. This research became a further proving ground for the development of a methodology for archaeological reconnaissance that was guided by the interpretation of aerial photography. Numerous ground surveys and stratigraphic excavations were employed to ground truth anthropogenic features identified through Soviet Period aerial imagery.

These important discoveries in the last two decades of the twentieth century brought to light the first known fortified settlements in the Southern Urals and a host of additional prehistoric and historic archaeological sites (Fig. 12.1). The successful remote sensing methods that were developed during the 1990s in connection with Arkaim were applied to other districts within the Chelyabinsk Oblast', and in 2004, the first archaeological atlas of identified sites was published. This atlas focused on the territory within the contemporary administrative district of Kizil'skoe (Russian Federation) and details the spatial distribution, classification and description of 789 archaeological sites (including settlements, cemeteries and singular ritual and/or mortuary monuments) (Figs. 12.1, Map C, 12.3, 12.6, and 12.7). Two additional atlas publications are in preparation for the Nagaybakskiy and Kartalinsky districts, both of which are situated in the Chelyabinsk Oblast'.

12.4 Key Methods: Integrating Air Photography and Archaeological Interpretation

In general, the archaeological interpretation of aerial photographs in the Southern Urals region has pursued two main objectives: (1) the search for new sites, the mapping of these sites and subsequent ground survey and exploration and (2) to provide a detailed study of the monuments through stereoscopic analysis and the use of computer programmes for geographic information systems processing and archiving. In the Chelyabinsk Oblast', archaeologists have available to them aerial archives from the years 1954, 1956, 1960, 1970, 1983–1988 and 1999. Not all of this imagery is useful for purposes related to the remote sensing of archaeological sites. As experience has shown, the most successful images from the steppe and forest-steppe zones of the Southern Urals are those taken in the spring (during the early growing season) and autumn (after harvest) and at times that are early or late in the day. Long shadows, which are formed from the oblique angle of the sun (shadow sites), stress relief irregularities and create the maximum stereoscopic effect. The optimal scale for aerial imagery is typically in the range of 1:10,000–1:35,000.

Finer scale imagery does not help in the detection of additional archaeological monuments, but does help in the study of the larger identified settlements (Figs. 12.2, 12.3, and 12.6). The finer scale imagery is also useful for the analysis of landscape zones and the possible effects of anthropogenic activities. Crucial factors that also affect the usefulness of older imagery are the weather at the time the photographs were captured and the print quality of the photographs themselves – as even a slight haze disguises many of the finer details of the earth's surface and the topographical details of prehistoric monuments.

All aerial photographic imagery produced during the Soviet Period is classified as 'confidential'. Since the early 1990s, and the decline and subsequent collapse of the Soviet Union, the degree of secrecy of the imagery has changed and it is now classified as 'For Official Use Only'. In order to purchase such imagery, a special permit is required. Nevertheless, it is possible in most cases to publish portions of air photographs produced during this time, and today more regional archaeologists are making use of these opportunities.

As noted above, the steppe and forest-steppe landscapes during the second half of the twentieth century underwent a substantial transformation because of Soviet policies for economic development. These activities strongly influenced the preservation of archaeological monuments. Numerous sites, especially late prehistoric settlements and mortuary features, which were located along major rivers, their tributaries and in flat arable land, were affected by agricultural activities (ploughing and cultivation) and the construction of dams. General knowledge about the existence of these monuments would not be possible without the use of archived aerial imagery – especially those produced during the 1950s before the later surge of Soviet Period economic development. Furthermore, the aerial record produced during the 1950s has proven to be the best suited for archaeological reconnaissance, and the majority of research being done in the Southern Urals today relies on this older imagery. With such imagery, it is possible to establish not only the presence of archaeological sites but also their general spatial characterisation (the Russian terminology for this is planigraphy), which is very important for Bronze Age studies. The ideal situation for aerial photographic study is when several sets of photographs are available from different years. Depending on the angle of light captured in the photographs, the imagery can vividly highlight various elements of the microrelief, especially the multiphase settlement constructions linked to the Sintashta culture development. In Fig. 12.3, for example, three different photographs, taken in 1956, 1967 and 1969, show substantial variation in what they reveal in the rectangular fortification site of Stepnoye. Analysis of the internal structure of the monument for each of the photographs has produced quite different results that allowed for a more complete characterisation of the monument and its possible phasing.

12.4.1 Aerial Photographic Analysis of Bronze Age Settlements and Cemeteries

The geographical territory where the Sintashta culture has been identified, and the fortified settlement pattern distinguished, is bounded on the west by the relatively low relief of the Ural Mountains range. It is bounded in the east by the West Siberian lowlands, to the north by the steppe and forest-steppe ecological boundary and in the south by the Orenburg steppes (approximately 450 km × 120 km). Twenty-two[1]

[1] The 22nd fortified settlement named Kizil-Mayak was identified very recently (Batanina and Batanina 2009).

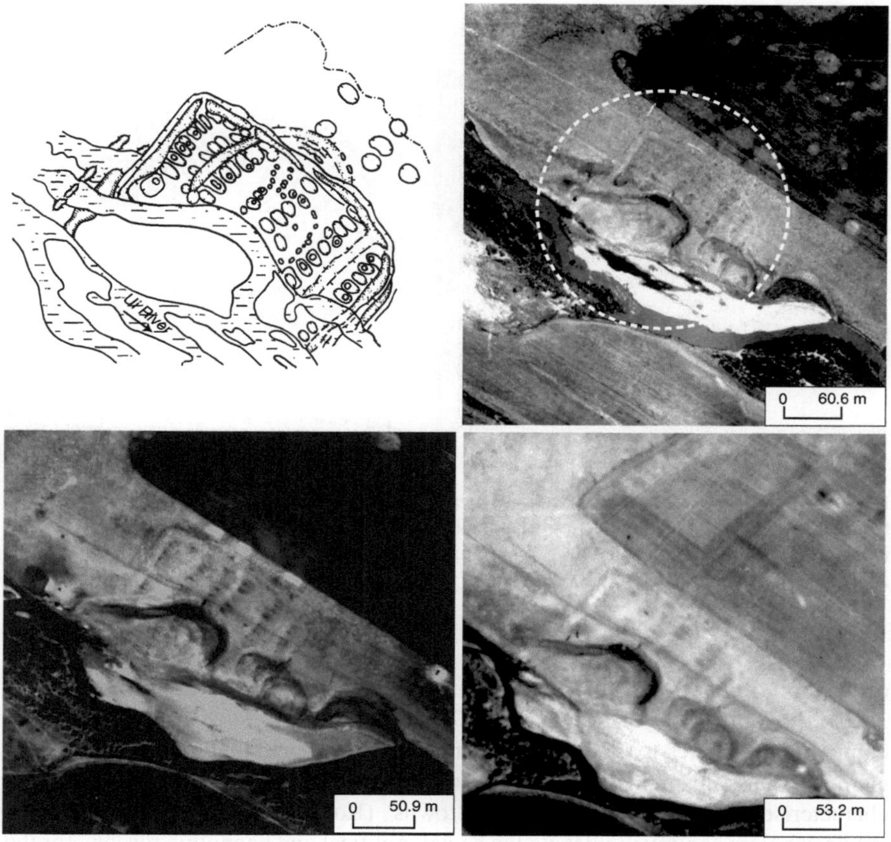

Fig. 12.3 *Upper left* – interpretation of aerial photograph of the Middle Bronze Age Stepnoye settlement by Iya Batanina (After Zdanovich and Batanina 2007: 160); *upper right* – portion of aerial photograph of Stepnoye taken on June 17th 1956 (Area of aerial photographic interpretation denoted by white-dotted line); *lower left* – portion of aerial photograph of Stepnoye taken on September 25th 1967; *lower right* – portion of aerial photograph of Stepnoye taken on September 25th 1969

identified settlements dating to the Sintashta Middle Bronze Age (2100–1700 BC) have been tentatively labelled the 'Country of Towns' and all are found within the Sintashta cultural region (Zdanovich and Zdanovich 2002). They have been characterised as 'towns' because of the systematic planning and organisation of the enclosed zones of habitation. Some scholars have considered this a clear 'proto-urban' development, while others have debated what the settlements actually reflect in terms of demographic aggregation and other 'urban' trends (see various viewpoints in Jones-Bley and Zdanovich 2002; Hanks and Linduff 2009). Some of the settlements have upwards of 60 rectangular house units (Fig. 12.4) contained within the fortified areas, and population estimates have varied widely from a few hundred

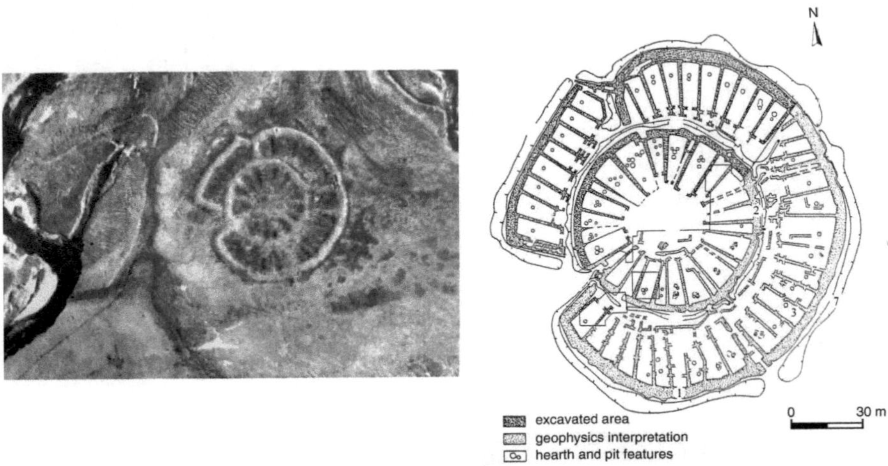

Fig. 12.4 *Left* – a portion of a 1978 Soviet Period aerial photograph of the Middle Bronze Age fortified settlement of Arkaim (After Zdanovich and Batanina 2007: 26); *right* – plan of Arkaim indicating area of excavation and unexcavated areas interpreted through geophysics data (Adapted from Zdanovich and Zdanovich 2002: 256)

occupants to over 1,000 for each of the settlements (Koryakova and Epimakhov 2007: 72; Kohl 2007: 14; Epimakhov 1996). Some scholars, such as Phil Kohl, have compared these sites to late prehistoric nucleated settlements in the Near East and elements of urbanism there and suggested that Sintashta fortified sites may be better characterised as 'villages' rather than 'towns' (Kohl 2007: 122, see also12–14). While scholars continue to debate the social, political and economic reasons for the appearance of these sites, aerial photographic analysis has provided the basic foundation for their identification in the landscape, their regional distribution and the nature of construction phasing.

I.M. Batanina and N.V. Levit can be credited with the identification of 19 of the 22 settlements through their detailed analysis of archived aerial photography produced during the 1950s and 1960s. The settlements of Zhurumbai, Sintashta 2, Kamysty and Konoplyanka were only identifiable through air photography, as ground survey alone failed to recognise associated topographical relief connected with fortification features and domestic structure depressions. Sintashta settlements are generally located on the first terrace above the floodplain of the rivers and seem strategically placed in the landscape to take advantage of natural barriers for defence, such as small hills and the rivers. Air photography primarily highlights the fortification features whereby ramparts and ditches have a clear structure and appear as bright white and dark grey in the imagery (Figs. 12.2, 12.3, and 12.4). The internal space in the monuments is quite patterned and can often be discerned through air photographs. Sintashta settlements that are round or oval in overall shape typically have dwellings that are arranged radially and adjoin on one end of the defensive ramparts. In most cases, the central part of the settlement was undeveloped and probably acted as a central courtyard (Fig. 12.4).

Most of the fortified settlements are multilayered stratigraphically and when excavated reveal construction and habitation features connected with the Middle to Late Bronze Age (2100–1500 BC). Detailed aerial photographic analysis of these multilayered fortified settlements, such as Isiney, Stepnoye, Ust'ye and Rodniki, makes it possible to discern the diachronic nature of their construction phases. The earliest phases are circular or oval with later phases becoming more rectangular or square. By the end of the Late Bronze Age (approximately 1200 BC), occupation at these sites, and others within the region, had ceased to emphasise fortification. Several of the Sintashta settlements also appear to have periods of inactivity followed by later phases of reoccupation. These subsequent occupations frequently disturbed the earlier stratigraphic levels, but in most cases conformed to the use of the internal space of the earlier fortified site (Fig. 12.8).

One of the important methods developed by I.M. Batanina and N.V. Levit through their joint research on Sintashta settlements was the utilisation of different-dated aerial photographs to track the changing nature and destruction of the sites (Batanina 1995; Batanina and Ivanova 1995). For example, the topographical height of the defensive walls of the Arkaim fortified settlement was estimated to be 1.2 m in 1954. By 1988, however, the relief of the walls was only approximately 0.7 m. The erosion of the fortification features was likely accelerated by the grazing of livestock over the site during the height of the Soviet Period, which led to the drastic deflation of the monument relief (Zdanovich and Batanina 2007: 45).

Unfortified settlements of the Late and Final Bronze Age (1700–1200 BC) also show evidence of diachronic occupation and multilayered stratigraphy (Figs. 12.5 and 12.6). The more specific dating of these monuments, and their archaeological culture affiliation, can only be determined after excavation (Zdanovich and Lyubchanskii 2004). Unfortified Late and Final Bronze Age settlements can be identified in aerial photographs by dark spots related to house depressions and stereoscopically by their negative microrelief. Identified unfortified settlements typically occupy a niche in the landscape and are located on the first terrace of the rivers at a height of 2.5–5.0 m above the water's edge. The disposition of these settlements depends largely on the size and shape of the terraced areas, which are often quite narrow. These settlements are usually constructed in one, two or three rows of house structures and can stretch along the river from 100 to 200 m or more (Fig. 12.5). Settlements situated on less spatially restricted terraces comprise house depressions that are located more unevenly and form compact groups, such as the site of Peschanka-5 (Fig. 12.6).

It is also important to note that Bronze Age fortified and unfortified settlements that have been affected by agricultural cultivation often show signs of ashy soils that are visible as bright white spots within the darker background of arable soils around the sites (Fig. 12.3). These features are connected with midden deposits that were formed in most cases near the external perimeter of the enclosed settlements during their occupations. A good example of this is the Sintashta period settlement of Stepnoye, which clearly reveals three discrete areas of ashy soil and midden formation

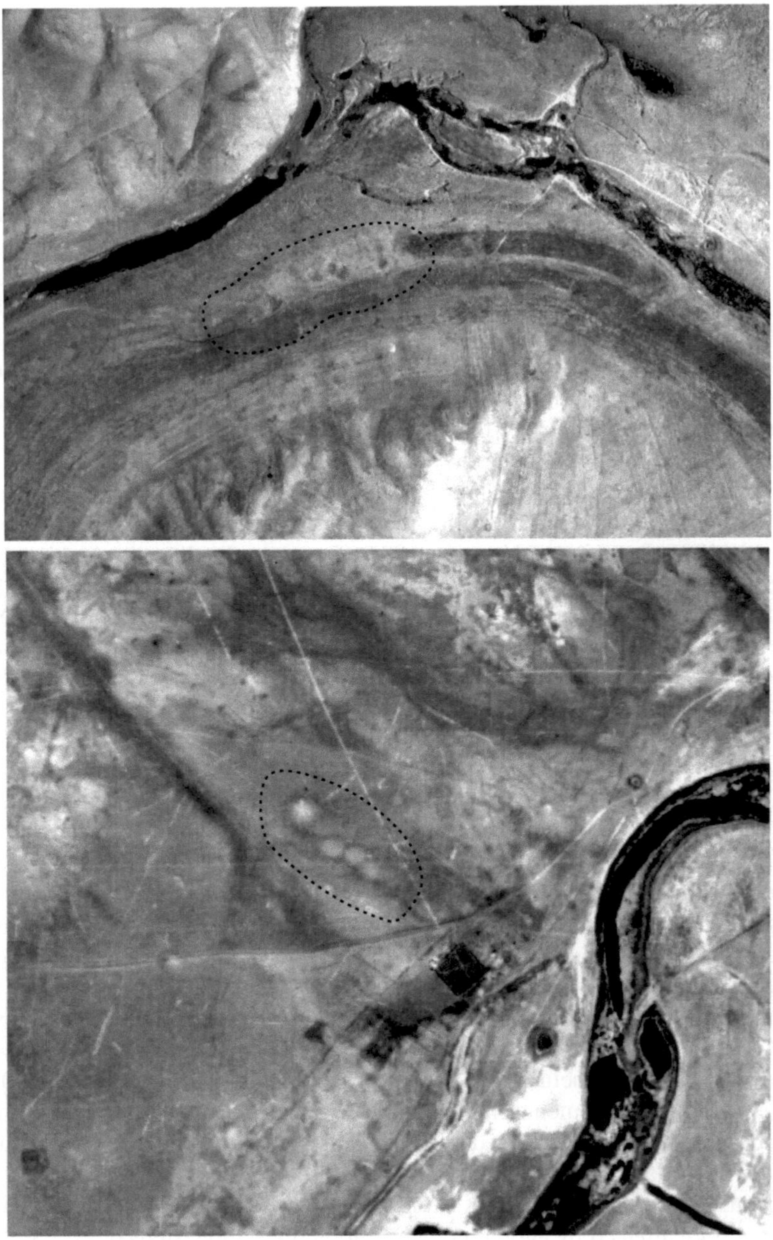

Fig. 12.5 *Upper* – portion of a July 3rd 1956 Soviet Period aerial photograph of the unfortified settlement of Cherkasy II (scale: 1 cm–102 m). Several Bronze Age house feature depressions are discernable within the *dotted area*. *Lower* – portion of a June 14th 1974 Soviet Period aerial photograph showing a series of Bronze Age cemetery barrows (kurgans) within the *dotted line* from site No. 717 in the Chelyabinskaya Oblast' (scale: 1 cm–29 m)

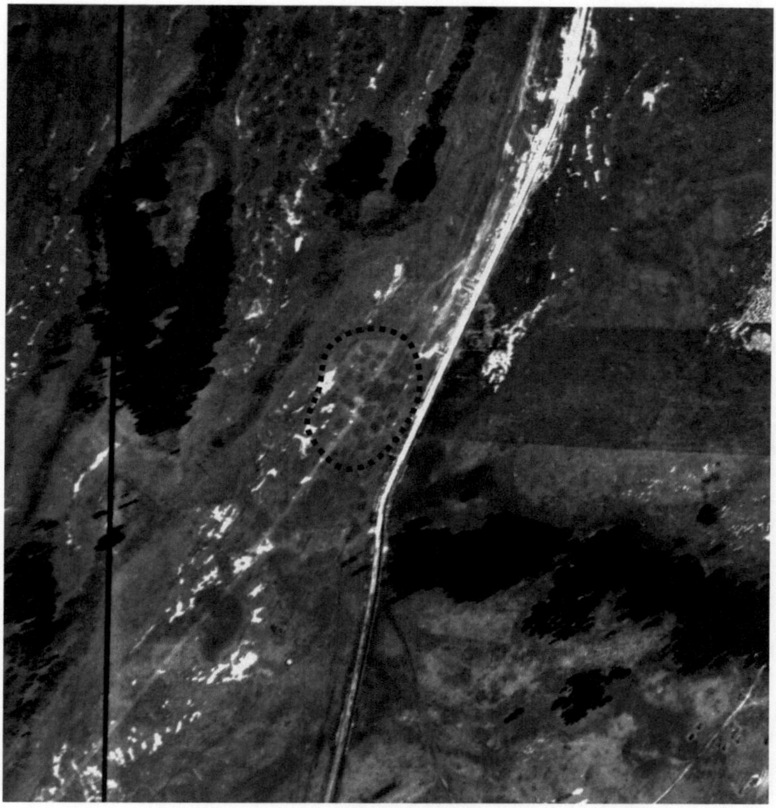

Fig. 12.6 A portion of a September 5th 1956 Soviet Period aerial photograph of the unfortified Bronze Age settlement of Peschanka-5. House feature depressions of the settlement are visible within the *black-dotted circle* (Scale: 1 cm–80 m)

near the north perimeter of this enclosure. Recent excavations at this site have revealed that these midden deposits were formed at a depth of up to 1 m and are primarily composed of ashy soil with mixed artefacts such as faunal remains, metallurgical slag, lithics, ceramic sherds and other materials connected with Bronze Age occupation of the settlement.

The analysis of Soviet Period air photographs has also proven extremely effective for the identification of single and grouped Bronze Age *kurgans* (funerary barrows). Bronze Age burial mounds are usually located along the river terraces and on the gentle slopes of hills at a further distance from the river courses. Barrows from both the Bronze and Iron Ages typically form a chain or a compact group of features, and Bronze Age barrows are frequently located in association with settlements (Figs. 12.3, 12.5, and 12.7). The main identifying features of the barrows are direct indicators such as geometric shape, microrelief of the mounds, colour differentiation of the surface soils and the overall spatial nature of grouped barrows or

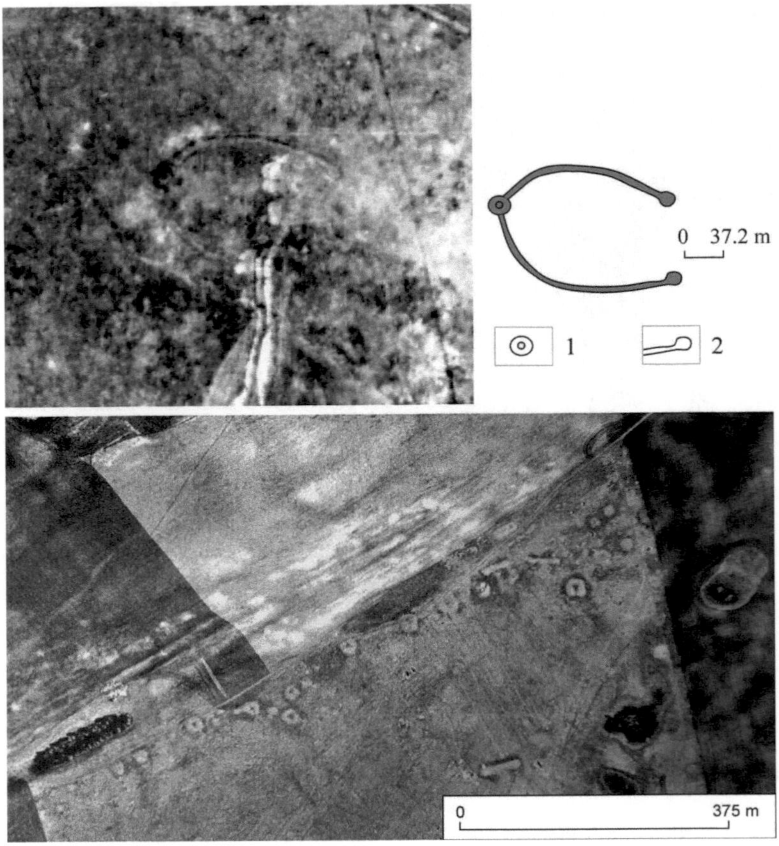

Fig. 12.7 *Upper left* – portion of a July 2nd 1956 Soviet Period aerial photograph of a Kurgan 's usami' (scale 1:25,000) with illustration interpretation on the right (1 – central mound feature; 2 – curvilinear 'whiskers' with terminal features); *lower* – portion of a May 30th 1976 Soviet Period aerial photograph of the Solenyi Dol cemetery showing 'dumb-bell'-shaped (gantelevidnymi) barrow features (Bredinskiy region, scale 1:12,500, after Zdanovich et al. 2003: 43)

cemeteries. Indirect signs of these monuments include variation in vegetation and surface shadows cast by the barrows that allow for the topographical relief of the monuments to be seen through stereoscopic analysis. Bronze Age barrows are rounded or slightly elongated in shape, and their sizes range from 8 to 10 m in diameter for the smaller constructions and 20–30 m for larger barrows. Barrows that are located in non-arable locations are identified as mounds through stereoscopic analysis. Barrows that have been strongly affected by ploughing in arable locations are identified by strong contrasts in soils whereby the mounds appear as bright spots surrounded by darker soils that are connected with contemporaneous ditch constructions or modern arable soils.

12.4.2 Recent Archaeological Research Programmes

Archaeological research programmes focusing on Middle Bronze Age settlements and cemeteries in the Chelyabinsk Oblast' have continued since the earlier phases of research that began in the 1980s and 1990s. International projects including excavation and remote sensing have been underway at the Sintashta fortified sites of Stepnoye and Kamennyi Ambar since 2007 and Ust'ye since 2011. In addition, a radiocarbon dating project focused specifically on the dating of Middle, Late and Final Bronze Age sites within the larger south-eastern Urals region was undertaken from 2005 to 2007 and was subsequently published in both Russian and English (Epimakhov et al. 2005; Hanks et al. 2007). This programme of study produced a total of 40 AMS calibrated dates from 16 Bronze Age settlements and cemeteries spanning a time period of 2140–880 cal. BC and has provided an important absolute chronology for key transitions in settlement and mortuary patterning.

The use of geophysical prospection has also been employed with increasing frequency within the Southern Urals region. Both international and Russian teams have undertaken this research using a variety of techniques that include proton magnetometry, fluxgate gradiometry, electrical resistivity, caesium magnetometry and ground penetrating radar. To date, however, very little of this research has been published. This is important, as the combination of aerial photographic analysis with geophysical prospection offers an extremely effective approach to the identification and interpretation of settlement and mortuary feature patterning. Additional publications will likely be forthcoming in the near future given the state of international research being undertaken in the Sintashta settlement region.

The publication of geophysical surveys conducted at the Sintashta period settlements of Kamennyi Ambar and Stepnoye, in 2005 and in 2008, respectively, has made it possible to compare the results with previous aerial photographic interpretations of the sites (Merrony et al. 2009). In 2005, a fluxgate gradiometer survey (Geoscan FM 18) was employed over the enclosed area of the Kamennyi Ambar fortified settlement. A sampling interval of 1.0 m was chosen as a result of time and budgetary constraints with the objective of providing a complete survey of the entire enclosed area. As a result of this rather coarse sampling procedure (with only 400 data points for each 400 m^2 block), the visualisation of the subsurface features was not highly detailed. Nevertheless, a total area of 48,400 m^2 was completed in a few days and easily defined key characteristics of the site (Fig. 12.8). These included fortification ditches, probable construction phases and several highly magnetic anomalies interpreted as high-temperature features, such as hearths, kilns and/or metallurgical furnaces. Several linear features were also present that probably relate to domestic architecture and the division of internal space within the enclosed area.

The geophysical survey at Kamennyi Ambar provided interesting results when compared with earlier interpretations of the settlement through aerial photographic analysis (Fig. 12.8). Occupation of the Kamennyi Ambar settlement occurred in both the Middle Bronze Age and Late Bronze Age phases – with the ditch phasing visible in the gradiometry survey probably connected with the earlier phase(s) of

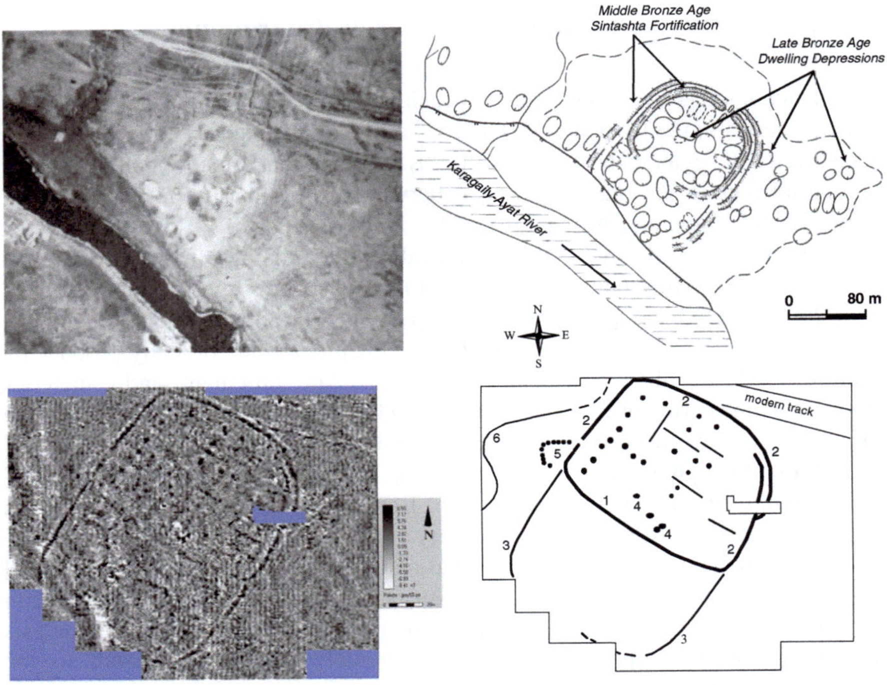

Fig. 12.8 *Upper left* – portion of a June 2nd 1954 Soviet Period aerial photograph of the Kamennyi Ambar (Ol'gino) fortified settlement (Zdanovich and Batanina 2007, 98); *upper right* – adaptation of aerial photographic interpretation by Iya Batanina, with detail of Middle and Late Bronze Age phases (Zdanovich and Batanina 2007: 99); *lower left* – greyscale plot of fluxgate gradiometer survey (Merrony et al. 2009: 425); *lower right* – interpretation of phases of Stepnoye settlement based on geophysical data: *1*, *2* – primary ditch phases, *3* – secondary ditch phase extension, *4* – strong magnetic anomalies, possible kilns or metallurgical furnaces, *5* – features possibly connected with entrance, *6* – geological anomaly, possible drainage ditch

construction at the site. The Soviet Period aerial photography and aerial photographic analysis of the site (Fig. 12.8) also provide evidence for Late Bronze Age occupation wherein several large, oval-shaped house depressions are visible across the surface of the site. These features were not discernable through the geophysical survey, and it is likely that these later occupation features disturbed earlier, and stratigraphically lower, phases of occupation. In combination, the aerial photographic analysis and geophysical prospection have provided an important foundation of spatial information for future excavation programmes at the site. A large-scale Russian-German collaborative project has been operating at this site since 2007, and information stemming from the excavations there will provide an excellent opportunity to compare aerial photographic, geophysical and excavation data as publications from the project are forthcoming.

In 2008, both a fluxgate gradiometry survey (Geoscan FM256) of 19,000 m^2 and an electrical resistivity survey (Geoscan RM12) of 26,400 m^2 were undertaken over

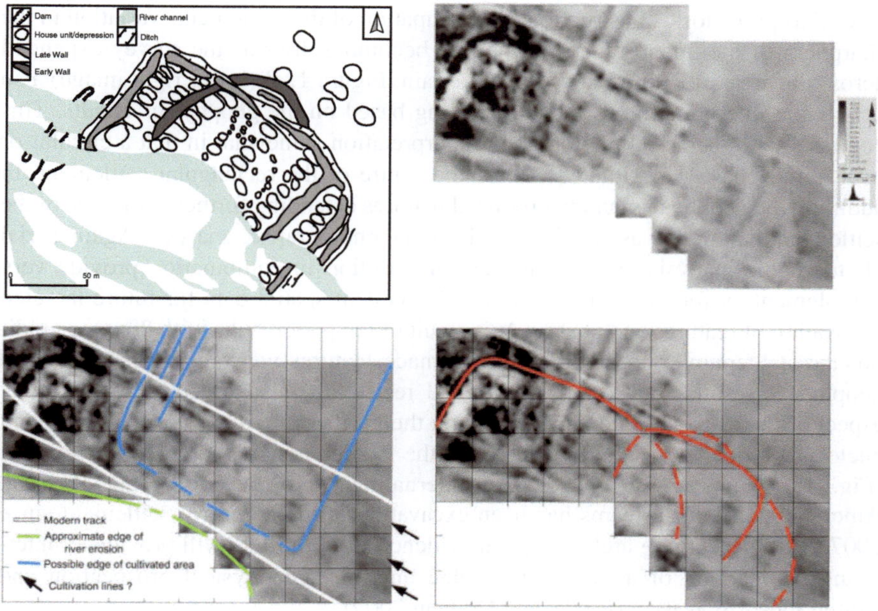

Fig. 12.9 *Upper left* – redrawn and adapted aerial photographic interpretation of Stepnoye fortified settlement from Zdanovich and Batanina 2007: 160 denoting key construction characteristics (Prepared by D. Pitman); *upper right* – greyscale plot of electrical resistivity survey of Stepnoye settlement (Prepared by C. Merrony); *lower left* – interpretation of modern-day features from geophysical plot (Prepared by C. Merrony); *lower right* – interpretation of phases of Stepnoye settlement based on geophysical data (Prepared by C. Merrony)

the Sintashta settlement of Stepnoye. Figures 12.3 and 12.9 provide an important comparison between the aerial photographic analysis of this site and the geophysical results from the electrical resistivity survey. Interestingly, the detailed fluxgate gradiometry survey (64 points taken per 1.0 m) provided very poor results, and it was impossible to discern any magnetic anomalies that appeared related to either the aerial photographic interpretation or prehistoric features clearly visible on the surface, including fortification banks, ditches and house depressions (see Merrony et al. 2009: 426–427 for extended discussion and associated imagery). The soils at Stepnoye vary markedly from those encountered at the site of Kamennyi Ambar and consist of coarser sandy material that has very low magnetic susceptibility. In response to this, a second geophysical survey was conducted over the site utilising electrical resistivity. The greyscale plot from this survey, detailed in Fig. 12.9, compares very well to the aerial photographic analysis of the site and interpretations regarding the chronological phasing there. Several key features are discernable in the greyscale plot and have been interpreted as relating to modern agricultural activities as well as prehistoric construction and occupation of the settlement. Variation in the sharpness of the image relates to the soil becoming drier as the survey expanded across the area after a period of intensive rain. Figure 12.9 (lower right image) also details the interpretation of the site phasing based on the geophysics results. This contrasts with the aerial photographic interpretation somewhat in that a circular or oval initial phase was situated east of the centre of the rectangular settlement. In addition, the geophysics survey produced features extending further to the east of the settlement than what was identified in the aerial photographic analysis. Again, aerial photographic analysis and geophysical prospection in combination provide very complementary forms of information and excellent spatial data for future targeted excavation of features and phases. As a result of the poor results with fluxgate gradiometry at Stepnoye, no hearth and/or furnace features were distinguishable. The geophysics results (both gradiometry and resistivity) also failed to identify any aspect of the ashy soil/midden deposits on the northern perimeter of the settlement enclosure, which are clearly visible in the Soviet aerial photographic imagery (Fig. 12.3, upper-right image). An international project comprising Russian, American and British teams has been excavating at the Stepnoye settlement since 2007, and forthcoming archaeological evidence from this site will provide an interesting comparison of aerial photographic analysis, geophysical prospection and stratigraphic excavation (Hanks and Doonan 2009).

12.5 Conclusion and Future Trends in Remote Sensing

The1990s in the Southern Urals witnessed tremendous progress in the decipherment of aerial images for archaeological research. This led to an improved understanding of the spatial and diachronic nature of human occupation in the region (approximately 3500 BC to AD 1000) and shifting patterns of settlement and ritual monument construction. Unfortunately, sites of the Palaeolithic to Neolithic are largely hidden from aerial photography, as they lie buried under substantial horizons of sediment. It is also impossible to identify many of the megalithic monuments (primarily dating to the late Iron Age and medieval period) within the region because of their diminutive size. Nevertheless, aerial photography has proven itself indispensable for the detection of such unique monuments as the complex stone barrows known as kurgan 's usami' (barrows with a 'moustache') and burial mounds with a 'dumb-bell' shape. Such features are visible through aerial photography because they are predominantly comprised of surface constructions with stone mounds and linear and curvilinear stone arrays (Fig. 12.7).

Currently, archaeological mapping based on the interpretation of aerial photographs has covered nearly 10,000 km^2 in the steppe and forest-steppe ecological zones of the Southern Urals, and more than 1,000 monuments have been identified. Nearly all the fortified settlements connected with the Sintashta 'Country of Towns' development discussed above were located through aerial photography. In addition, with the exception of a few dozen previously known kurgan 's usami' monuments, 'dumb-bell-shaped' mounds and unfortified settlements of the Late

Bronze Age, the identification of archaeological monuments in the Southern Urals was made by I.M. Batanina and N.V. Levit.

The initial efforts made by Batanina, Levit and regional archaeologists in the 1980s and 1990s led to the development of important aerial photographic methods and the construction of a large database of identified archaeological sites. The use of aerial remote sensing and ground-based geophysical prospection, as reviewed for the Southern Ural Mountains region of Russia, is proving to be a powerful multidisciplinary approach to the further identification and interpretation of late prehistoric monuments.

The use of historical aerial photography reviewed in this chapter has provided an important foundation for contemporary and future archaeological research. However, the use of newly available satellite imagery combined with historical air photography is rapidly developing among scholars in regions of the former Soviet Union. The use of satellite imagery for archaeological purposes has, until very recently, been hampered by a lack of good-resolution imagery. As the availability of higher spatial resolution satellite data improves, it is sure to become more widely used by archaeologists. Several scholars, such as G.E. Afanas'ev, M.R. Korobov and G.P. Garbuzov, have contributed importantly to the integration of these methods in the past decade (Garbuzov 2003; Afanas'ev 2004), and multispectral satellite data is also being drawn upon (Antimonov 2010). Experimentation in the use of satellite imagery was done initially to compare with the results of aerial photography analysis. However, the emphasis is now shifting increasingly to the use of both modern and archival satellite imagery – including data produced in Russia, such as the declassified military KFA-1000 imagery from the 'Resurs-F1' satellite deployed in the 1970s, and commercially available foreign satellite imagery (Landsat, IRS, SPOT, etc.). In Russia, one of the current leaders in this field is the Archaeology Institute of the Russian Academy of Sciences, which established in 2002 a focus on archaeological-geological information systems. One of the important achievements from this has been the organisation of an annual round-table conference (since 2003) on 'Archaeology and Geoinformatics', which is dedicated specifically to the development and refinements of remote sensing methods in archaeology. The annual meetings produce a number of important presentations and articles with additional data made available on CD-ROM media. These meetings provide an important opportunity for archaeologists to come together with specialists in remote sensing and geographic information systems (GIS) to discuss the refinement of methods and interpretation of remote sensing and GIS data.

These recent trends in the use of available satellite data within archaeological research reflect the important historical role that remote sensing has had in Russia and in the territories of the former Soviet Union. Nearly a century has passed since the first air photographs were taken in 1918 in Tver, Russia for topographical mapping. The further development of remote sensing methods and their integration within archaeological research are sure to remain a dynamic field in the years ahead. As more archaeological projects begin to make use of these important methods, additional discoveries are sure to be made that will add to our existing knowledge about prehistoric life ways and long-term diachronic social change in this important region of Eurasia.

Acknowledgments The authors would like to thank the editors of this volume for their sincere patience in the preparation of our paper. We also owe a debt of gratitude to our numerous field colleagues in Russia, United States, China, England and Slovenia. A special word of thanks is owed to our colleagues Dmitri Zdanovich, Elena Kupriyanova, Andrei Epimakhov, Roger Doonan, Derek Pitman and Colin Merrony for their assistance with the geophysical surveys at Stepnoye and Kamennyi Ambar. The geophysical prospection data reviewed for the settlements of Stepnoye and Kamennyi Ambar was achieved through field projects funded by the National Science Foundation (BCS #0726279) and the University of Pittsburgh and is gratefully acknowledged. We are also grateful to Chelyabinsk State University for permission to reproduce portions of the Soviet Period air photography that appear in this chapter. All opinions and errors in this chapter are the sole responsibility of the authors.

Bibliography

Afanas'ev, G. E. (2004). Osnovnyye napravleniya primeneniya GIS – DZ-tekhnoligiy v arkheologii. *Arkheologiya i geoinformatika.* Vyp. 2 (Electronic resourse). Moscow: CD-ROM.

Andrianov, B. V. (1965). Deshifrirovaniye aerfotosnimkov pri izuchenii drevnikh orositel'nikh sistem. In B. A. Kolchin (Ed.), *Arkheologiya i estestvennye nauki* (pp. 261–266). Moscow: Nauka.

Anthony, D. (2007). *The Horse, the wheel and language. How Bronze Age riders from the Eurasian Steppes shaped the modern world.* Princeton: Princeton University Press.

Anthony, D., & Vinogradov, N. (1995). Birth of the Chariot. *Archaeology, 48*, 36–41.

Antimonov, N. P. (2010). Obnaruzheniye arkheologicheskikh pogrebennykh ob'yektov na mul'tispektral'nykh kosmicheskikh snimkakh s pomoshch'yu spetsializerovannoi programmy "Image Media Center 5.0". *Geomatics, 3*, 67–71.

Batanina, I. M. (1995). *Otchot. Arkheologicheskaya karta zapovednika"Arkaim".* Chelyabinsk [unpublished report, on file at the Historical-Cultural Reserve Arkaim, Chelyabinsk, Russia].

Batanina, I. M., & Batanina, N. S. (2009). Kizil-Mayak – novoe ukreplennoe poselenie epokhi bronzi v Uzhnom Zaural'ye. In U. B. Ashimov, K. M. Baypakov, A. A. Tukachev, M. I. Netiaga, A. A. Pleshakov, & Z. S. Samashev (Eds.), *Materiali mezhdunarodnoy nauchnoy konferensii Margulanovskie chtenia* (pp. 18–22). Petropavlovsk: North-Kazakhstan State University.

Batanina, I. M., & Ivanova, N. O. (1995). Arkheologicheskaya karta zapovednika Arkaimi. Istoriya izucheniya arkheologicheskikh pamyatnikov. In G. B. Zdanovich & N. O. Ivanova (Eds.), *Arkaim: issledovaniya, poiski, otkrytiya* (pp. 129–195). Chelyabinsk: Kamenniy Poias.

Epimakhov, A. V. (1996). Demograficheskiye aspekty sotsiologicheskikh rekonstruktsiye (po materialam sintashtinsko-petrovskikh pamyatnikov). In V. Ivanov (Ed.), *XIII Urailskoe arheologicheskoye soveshchaniye* (pp. 58–60). Ufa: Vostochnyi Universitet.

Epimakhov, A., Hanks, B., & Renfrew, A. C. (2005). Radiouglerodnaya Kronologiya Pamyatnikov Bronzovogo Veka Zaural'ya. *Rossiiskaya Arkheologiya, 4*, 92–102.

Garbuzov, G. P. (2003). Arkheologicheskiye issledovaniya i distantsionnoye zondirovaniye zemli iz kosmosa. *Rossiiskaya Arkheologiya, 2*, 45–55.

Gening, V. F., Zdanovich, G., & Gening, V. (1992). *Sintashta* (Vol. 1). Chelyabinsk: Uzhno-Uralskoe knizhnoe izdatelstvo.

Hanks, B. (2009). Modeling early metallurgical production and societal organization in the Bronze Age of North Central Eurasia. In B. Hanks & K. Linduff (Eds.), *Social complexity in prehistoric Eurasia: Monuments, metals and mobility* (pp. 146–167). Cambridge: Cambridge University Press.

Hanks, B., & Doonan, R. (2009). From scale to practice: A new agenda for the study of early metallurgy on the Eurasian Steppe. *Journal of World Prehistory, 22*, 329–356.

Hanks, B., & Linduff, K. (Eds.). (2009). *Social complexity in prehistoric Eurasia: Monuments, metals and mobility*. Cambridge: Cambridge University Press.

Hanks, B., Epimakhov, A., & Renfrew, A. C. (2007). Towards a refined chronology of the Bronze Age of the Southern Urals, Russia. *Antiquity, 81*, 333–367.

Igonin, N. I. (1969). Issledovaniye arkheologicheskikh pamyatnikov po materialam krupnomasshtabnoi aerofotos'yomki. In S. P. Tolstov & A. V. Vinogradov (Eds.), *Istoriya, arkheologiya i entografiya Srednei Azi* (pp. 257–267). Moskva: Nauka.

Jones-Bley, K., & Zdanovich, D. B. (2002). *Complex societies of Central Eurasia from the 3rd to the 1st Millennium BC: Regional specifics in light of global models* (Journal of Indo-European Studies Monograph series 45). Washington, DC: Institute for the Study of Man.

Khalikov, A. Kh., & Igonin, N. I. (1974). Aerofotos'emka krupnykh arkheologicheskikh ob'yektov. *Vestnik Rossiysko Akademii Nauk, 7*, 67–72.

Kohl, P. (2007). *The making of Bronze Age Eurasia*. Cambridge: Cambridge University Press.

Komarov, V. V. (1969). Itogi i perspektivy rasvitiya aerometdov v SSSR. In G. G. Samoylovich (Ed.), *Doklady komissii aeros'emki i fotogrammetrii* (pp. 3–17). Leningrad: Geograficheskoye Obshchestvo SSR.

Koryakova, L. N., & Epimakhov, A. V. (2007). *The Urals and Western Siberia in the Bronze and Iron Ages*. Cambridge: Cambridge University Press.

Merrony, C., Hanks, B., & Doonan, R. (2009). Seeking the process: The application of geophysical survey on some early mining and metalworking sites. In T. K. Kienlin & B. W. Roberts (Eds.), *Metals and societies. Studies in honour of Barbara S. Ottaway* (Universitätsforschungen zur prähistorischen Archäologie, pp. 421–430). Bonn: Habelt.

Shishkin, K. V. (1966). Primeneniye aerofotos'emki dlya issledovaniya arkheologicheskikh pamyatnikov. *Sovetskaya Arkheologiya, 3*, 116–121.

Shnirelman, V. (1999). Passions about Arkaim: Russian nationalism, the Aryans, and the politics of archaeology. *Inner Asia, 1*, 267–282.

Zdanovich, G. (1999). Arkaim Archaeological park: An cultural-ecological reserve in Russia. In G. Panel-Philippe & G. Stone-Peter (Eds.), *The constructed past: Experimental archeology, education and the public* (pp. 283–291). London: Routledge.

Zdanovich, G. B., & Batanina, I. M. (2002). Planography of the fortified centers of the Middle Bronze Age in the Southern Trans-Urals according to Aerial Photography Data. In K. Jones-Bley & D. G. Zdanovich (Eds.), *Complex Societies of Central Eurasia from the Third to the First Millennia BC: Regional specifics in the light of global models* (Chelyabinsk State University, Chelyabinsk, Russian Federation/Journal of Indo–European Studies Monograph series, pp. 121–147). Washington, DC: Institute for the Study of Man.

Zdanovich, G. B., & Batanina, I. M. (2007). *Arkaim – Strana Gorodov*. Chelyabinsk: Krokus.

Zdanovich, S. Ya., & Lyubchanskii, I. E. (2004). *Istoricheskii ocherk Karaganskoi doliny: drevnost', srednevokov'ye. Arkaim. po Stranitsam Drevnei istorii Yuzhnogo Urala.* Chelyabinsk.

Zdanovich, G. B., & Zdanovich, D. G. (2002). The 'Country of Towns' of Southern Trans-Urals and some aspects of Steppe Assimilation in the Bronze Age. In K. Boyle, A. C. Renfrew, & M. Levine (Eds.), *Ancient interactions: East and West in Eurasia* (pp. 249–263). Cambridge: McDonald Institute Monographs.

Zdanovich, G., Batanina, I., Levit, N., & Batanin, S. (2003). *Arkheologicheskii Atlas Chelyabinskoi Oblasti. Vyp. 1. Step'-Lesostep'. Kizil'skii Raion*. Chelyabinsk: Uralskoye knishnoye isdatelstvo.

Chapter 13
Historical Aerial Imagery in Jordan and the Wider Middle East

Robert Bewley and David Kennedy

Abstract Hundreds of thousands of aerial photographs of Middle Eastern countries have been taken since the beginning of the First World War. The majority has been destroyed, but tens of thousands survive in archives in several countries. Identification of and research on these collections has grown rapidly in recent years. Although the potential value of these 'historical' photographs has long been known, the rapid and dramatic development in Middle Eastern countries, affecting archaeological sites and landscapes, has accelerated the need for such records. This chapter surveys the present state of our knowledge of historical archive aerial photography, illustrates its use through specific examples and sets out a programme for increasing access to the 'archive' and opening it up to researchers through a web-based collection.

Abbreviations and Web Addresses

Aerial Photos of Israel 1917–1919, University of Haifa Library, Digital Media Center http://digitool.haifa.ac.il/R/PJYB6P3BY5JNLXPARVQK44EA4A2VRGIV5XBCE6EAN6Y2NYGEKE-02684?func=collections-result&collection_id=1280

APAAME Aerial Photographic Archive for Archaeology in the Middle East http://www.flickr.com/photos/APAAME/collections/

HAS Hunting Aerial Survey of Jordan, 1953

R. Bewley (✉)
Honorary Visiting Professor at the Institute of Archaeology, UCL, London SW1W 8NR, UK
e-mail: bob.bewley@btinternet.com

D. Kennedy
Classics and Ancient History, The University of Western Australia,
35 Stirling Highway, Crawley, WA 6009, Australia
e-mail: david.kennedy@uwa.edu.au

W.S. Hanson and I.A. Oltean (eds.), *Archaeology from Historical Aerial and Satellite Archives*, DOI 10.1007/978-1-4614-4505-0_13,

JADIS	The Jordanian Archaeological Data Information System (Palumbo 1994).
MEGA-J	Middle Eastern Geodatabase for Antiquities: http://www.megajordan.org/
TNA	The National Archives, Kew, UK: http://www.nationalarchives.gov.uk/

13.1 Introduction

We know from work in European countries (Bewley 1997; Bewley and Rączkowksi 2002; Cowley et al. 2010) that access to, and interpretation of, historical aerial photographs is fundamentally important for undertaking research into landscape archaeology. If such archives can be made accessible, there is a further requirement to build up the necessary interpretative skills; finding an archive is only the beginning, and one component, of what may be a long but fruitful journey in understanding our past.

Historical aerial imagery exists for probably every country in the Middle East. Most commonly it is held in the country in question, but for many regions of the developing world, including most of the Middle East, the situation is more complex: much early imagery was taken and is still held by the former colonial power. Moreover, not only is such historical imagery split between at least two countries, but it has often become scattered and lost from sight and held in both public and private hands. The fundamental problems for would-be researchers are determining what is archived in which country; gaining access; and finding out what was taken, when, and for what purpose. Only then can one turn to the question: 'In what ways and to what extent are these old photographs of research value to archaeologists'?

The purpose of this chapter is to explore the impact and value of the historical imagery available for Jordan in particular, but also, as we expand our horizons, to some extent for the wider Middle East (Fig. 13.1). Details of what exists for Jordan have been accumulating and more is continually coming to light (cf. Kennedy 1982, 1985, 1997, 2002a, 2004; Kennedy and Bewley 2010). The situation more widely in the region is highly varied and far less developed (below), with the sole exception of Israel. In the wider Middle East, despite impressive pioneering work between the World Wars (below), even access to current or recent aerial photographs is often highly restricted or forbidden. Unsurprisingly, therefore, in an environment where neither indigenous nor foreign archaeologists give much thought to recent imagery, the potential and significance of historical imagery has yet to be fully realised. This is especially true for those who are responsible for the management of archaeological resources in the region. Historical aerial photographs of archaeological sites have been obtained in the past by departments of antiquities in various Middle Eastern countries, but they seem often unknown to researchers and seldom to be utilised except as 'interesting' illustrations in reports with little to explain their contribution. This is a missed opportunity. There is an increasing number of researchers whose study and understanding of the region's past will be enhanced by access to historical aerial imagery. Indeed, as the entire region has undergone urban, as well as agricultural development, with breath-taking rapidity in the last half century, old photographs have become yet more important as records of places and

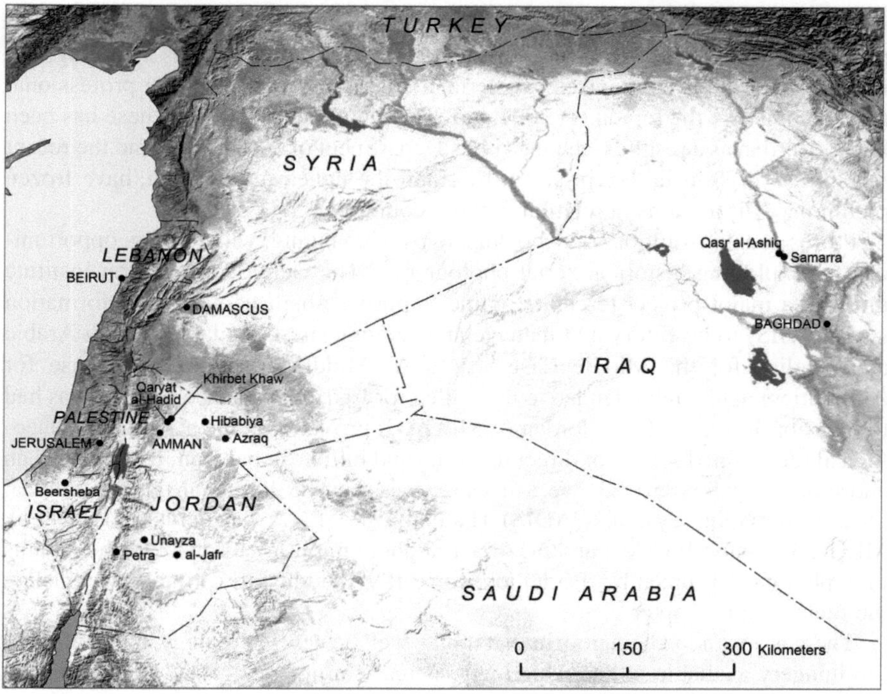

Fig. 13.1 Map showing places mentioned in the text (Drawn by Mat Dalton)

landscapes which are now utterly transformed. Old aerial photographs are especially valuable in a way even the best ground view can seldom be. In the case of the oldest aerial photographs, from the First World War, they are almost a century old and are themselves now artefacts.

There are hurdles to overcome, the first being a lack of knowledge about, and also access to, archives and the imagery. The Internet now permits a partial solution by providing a simple platform on which historical images can be displayed, existing online archives linked for direct access and references provided; it can be a dynamic resource, constantly revised and updated. Two simple examples illustrate the point. First, to add to known collections (Kennedy and Bewley 2010: 193–5), recent research by Kennedy on historical imagery of the Middle East held in Britain uncovered several hundred prints at The National Archives (TNA) in Kew taken by German and British aerial photographers over Palestine, Transjordan and Iraq in 1917–1918 and during the 1920s. The ready support of TNA has allowed them all to be digitised and put on display on the Aerial Photographic Archive for Archaeology in the Middle East (APAAME) website under 'Historic Aerial Photography'. On the same website may be seen the second example, an expanding collection of digitised aerial photographs published in books and articles, long out of copyright and often forgotten (cf. Kennedy 2012).

A second hurdle is a lack of training in landscape archaeology (including air photo interpretation) in the local universities in the region. This is changing, but very slowly. We have organised two aerial archaeology workshops for professional archaeologists in the region (in 2007 and 2009), but the impetus for these has been lost for Jordan as a result of changes in its Department of Antiquities, and the recent revolutions in Tunisia, Egypt and Libya, and the upheavals in Syria, have frozen preliminary discussions in relation to those countries.

Professional institutions are showing signs of changing, which offers opportunities for exploiting historical aerial photographs. The Getty Conservation Institute initiated a major project to create online 'a purpose-built geographic information system (GIS) to inventory and manage archaeology sites' available in both Arabic and English for the Middle East – MEGA (Middle Eastern Geodatabase for Antiquities). Although intended to be trialled for Iraq, the initial component has had to focus on Jordan. MEGA-Jordan (= MEGA-J) provides online access to archaeological records and scope for direct updating and infinite expansion. It represents an enormous step forward relative to its predecessor, The Jordanian Archaeological Data Information System (JADIS) (Palumbo 1994). A desideratum is to link MEGA-J to our APAAME archive of aerial photographs of Jordan, which contains both photographs taken by the authors since 1997 and historical imagery, which is the focus of this chapter.

There is a paradox here, too, in that it may well be new technology and new satellite imagery available online which will increase an interest in and opportunity to use historical imagery. The quality of Google Earth and Bing Map aerial imagery allows for much easier access to what was once a specialist domain, and fieldworkers who increasingly turn to Google Earth are drawn to thinking, too, about conventional aerial photographs. A recent article (Kennedy and Bishop 2011; cf. Kennedy 2011a) on discoveries in Saudi Arabia generated huge media interest not just because of the archaeological subject matter (a variety of prehistoric sites known as 'The Works of Old Men') but also because it involved remote sensing, using satellite imagery, available on the Internet and readily accessible to anyone and everyone, professional and amateur, anywhere in the world. That research using Google Earth revealed something else: as we see routinely in Jordan, development and road building in even desert areas is destroying archaeological remains and making historical photographs a precious record.

13.2 The Resource

An example of the way in which existing imagery has been used to understand past landscapes is *Rome's Desert Frontier from the Air* (Kennedy and Riley 1990; c.f. Millar 1991). At that time there was no aerial reconnaissance for archaeology in the region and the research for the book was undertaken purely on the basis of available aerial photographs. There were three main historical sources: photographs taken and published by Poidebard (1934; cf. Mouterde and Poidebard 1945); the RAF imagery taken at various intervals from 1918 onwards and mainly held at the Institute of Archaeology in University College London; and photographs taken by the commercial aerial survey company, Hunting. The book presents over 100 photographs which are from historical sources, 79% from the three mentioned above and the rest from a variety of international sources from Australia, Britain, France, Germany, Lebanon, Israel, and the USA. Although many had been published before, books such as that of Poidebard were rare book collectors' items and the photographs were seldom familiar to scholars. In this instance the aerial view of scores of Roman military sites, of all kinds, over a wide area could be brought together, reinterpreted and presented in a fresh, evocative way to revive and advance scholarship.

Poidebard was principally interested in Roman military structures and his archive consists overwhelmingly of low oblique photographs of sites that have been extensively preserved. However, he recorded other site types including a superb example of a Kite – a prehistoric animal trap (called a Kite because of the shape of the enclosures and the approaching walls). Unlike Roman forts and towns, preservation of Kites has seemed less important in Syria and elsewhere. They are numerous (at least 2,500 are known to exist), large (up to 2–3 km overall in some instances) and consist of slight, crudely built walls (perhaps only 50–100 cm high as they survive today). What that means in practice can be seen by comparing one Poidebard photograph of an especially well-preserved example with the same site on Google Earth on 1 January 2009 (Fig. 13.2).

It is worth noting that the publications by Poidebard (three books and a score of articles), and not just the underlying collections, represent historical archives in their own right. Further afield such collections are rarer. Stein's coverage of Iraq and Transjordan is far more slender (Gregory and Kennedy 1985: ch. 9), and for Iran, we have Schmidt (1940), supplemented by Gerster's work in the late 1970s (Gerster 2008). There are not even single books for most other Middle East countries – Iraq, Saudi Arabia, Yemen and Oman. For Israel, however, we have the evocative selection of old and new in Kedar (1999) and the superb online archive of imagery of all kinds, including aerial photographs (Aerial Photos of Israel 1917–1919).

Apart from the work of Poidebard, the most important historical imagery for improving our knowledge of the archaeology of the region, but especially of Jordan, has been the Hunting Aerial Survey (HAS) of 1953, which covered the western part of Jordan (Fig. 13.3); the HAS is made up of c. 4,000 large vertical, black and white photographs. Diapositive copies of the negatives of the entire survey were provided to Kennedy in the early 1980s. Subsequent systematic interpretation of the imagery, and the maps created, provided the estimate that there were c. 25,000 potential archaeological sites visible on these photographs alone (Kennedy 1997, 2001b, 2002a, b). By contrast, at that time, the official database of Jordan's archaeology (JADIS) recorded only 8,680 sites for the entire country, but 'projected' that the total for Jordan as a whole would be between 100,000 and 500,000 (Palumbo 1994: 1.9). The most recent version of MEGA-J (accessed in December 2012) has increased the official database tally to 10,900 (with 45,373 elements associated with these sites). Aerial reconnaissance by the authors since 1997, covering all the landscape zones in Jordan, has underscored how out of date the current databases are. For example,

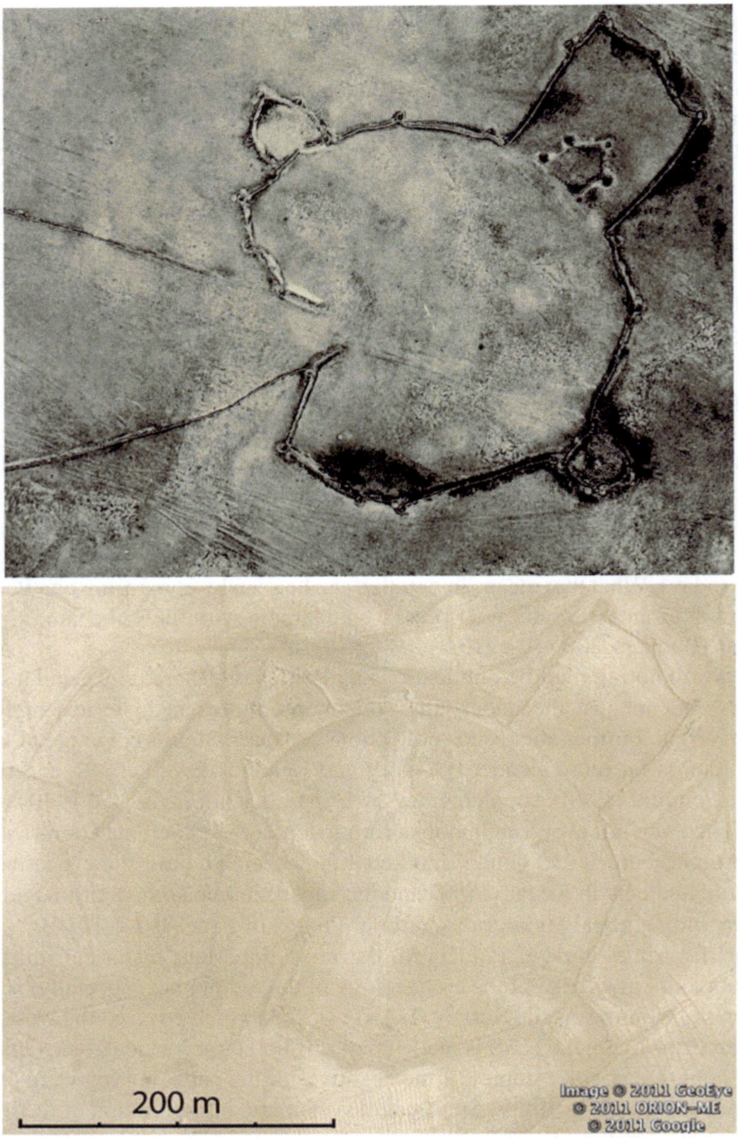

Fig. 13.2 Khan Abu ash Shamat – 3357.II Kite 1 (**a**) as recorded by Poidebard (1934: Pl. XIV); (**b**) as seen on Google Earth on 1st June 2009 (© 2011 Google Earth; © 2011 Geoeye; © 2011 ORION-ME. Compiled by Rebecca Banks)

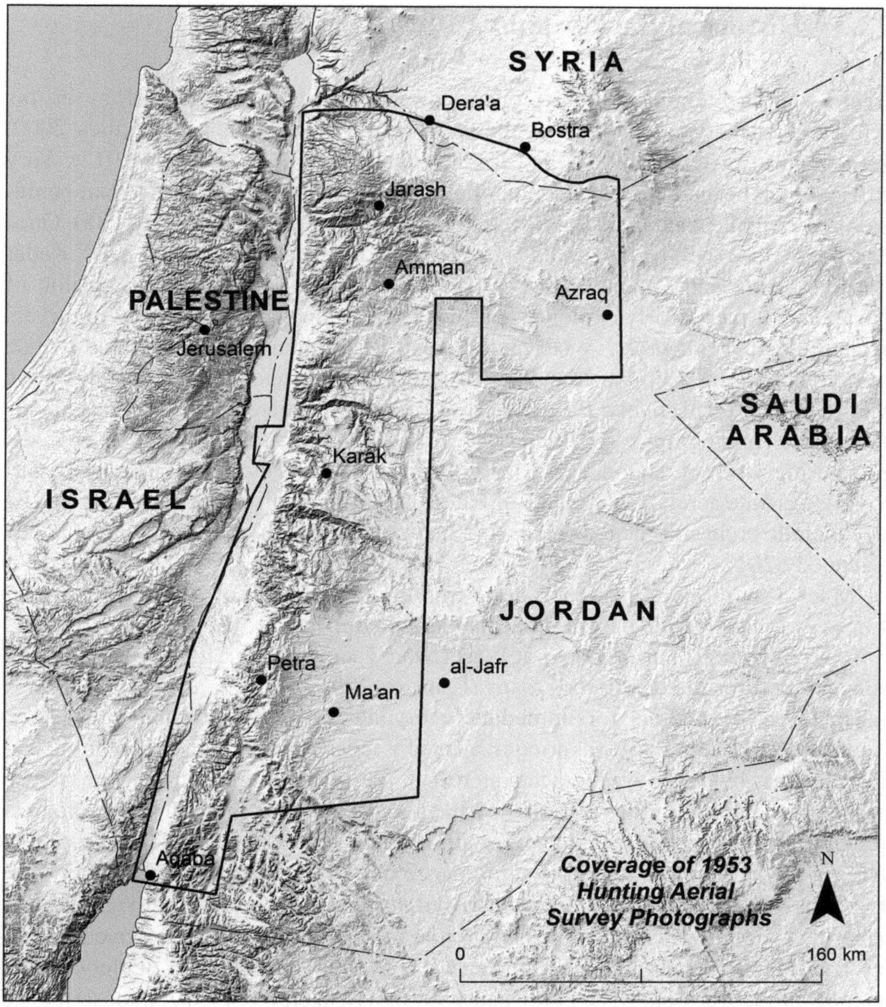

Fig. 13.3 Area covered by the Hunting Survey 1953 (Drawn by Julie Kennedy and Mat Dalton)

MEGA-J, following JADIS, records 332 Kites where our aerial reconnaissance and interpretation of HAS imagery of 1953 and current Google Earth imagery has found a total of 1,236.

There is every reason to suppose that historical imagery of these and similar types of sites exist for all of the countries of the Middle East; and their study could be equally rewarding (cf. below and Fig. 13.9). With the exception of Israel, however, there has been no systematic attempt to catalogue, much less draw together, all the surviving photographs.

13.3 Historical Background

There is little literature on the pioneers of aerial survey for archaeology in this region; most of what there is concerns Syria (cf. Nordiguian and Salles 2000; Denise and Nordiguian 2004) and Jordan (Crawford 1954; Kennedy 2012). Very little can yet be said about Mesopotamia (Iraq, see below), and the Arabian peninsula is a virtual closed book to this day (cf. Kennedy and Bishop 2011: 1285). Once again Israel is much better served (cf. Gavish and Biger 1985; Gavish 1987; Kedar 1999). During and after the First World War (1914–1918), the entire region of 'Arabia', till then poorly explored and mapped, became known to aerial observers from the main protagonists – German, British, Australian and Turkish, and British and French in the post-war Mandates of Iraq, Palestine and Transjordan and Syria and Lebanon, respectively. Remarkably, during the war itself, the Germans undertook aerial photography specifically of archaeological sites in Palestine with the pioneering work of Theodor Wiegand (1920) in the Negev. Thousands more surviving wartime aerial photographs of Palestine, taken by all the combatants, explicitly include archaeological remains and landscapes, which are now much changed (Kennedy 2012).

Much less research has been done on the reconnaissance activities of the air forces in Mesopotamia despite the considerable impetus provided by equally large military operations during the First World War. There, too, aerial photographs were fundamental for the creation of maps (Lambert 2003: *passim*), and they were also taken in large numbers for immediate everyday use on the various fronts (e.g. Tennant 1920: 200). There was no equivalent of Wiegand's Denkmalschutzkommando in Palestine, but there was a clear interest in the visible archaeology. One person who saw the opportunities afforded by the aerial view over the rich archaeological remains of this part of the 'cradle of civilisation' was Lt. Col. Beazeley, an engineer and surveyor. He understood the significance of aerial reconnaissance and air photographs, not just for their military use, but also their archaeological significance. Although shot down on one of his aerial missions in May 1918, he survived captivity and delivered two notable lectures to The Royal Geographical Society within months of the war's end. The published versions (Beazeley 1919, 1920) describe the mapping of the ancient city of Samarra, on the Tigris 125 km north of Baghdad (cf. Northedge 1985, 2005). His aerial surveys provided the evidence for the recognition and interpretation of the size, significance and population of this immense Early Islamic city. Excavation had begun before the war on parts of its extensive (57 km^2) and well-preserved ruins. The German Air Force in Mesopotamia took many photographs of the site as early as October 1917 for the use of the excavators, but none was published till 1923 (Leisten 2003: 27). Beazeley's is perhaps the first significant use of air photographs, using photogrammetry to create properly scaled maps, which he then used to develop an interpretation which led to historical analysis of the significance of this (early urban) landscape (Crawford and Keiller 1928: 4; cf. Stein 1919: 200). Beazeley published few aerial photos, and overall Crawford found the results disappointing: '…he reproduced only one vertical air photograph [a mosaic],

Fig. 13.4 Samarra (**a**) Beazeley's photograph of c.1918 (Beazeley 1920); (**b**) similar view of c.1918 (RAF 1918 II Pl. 36); (**c**) Google Earth view dated 16th June 2004 (© 2011 Google Earth; © 2011 Digital Globe. Compiled by Rebecca Banks)

and it cannot be regarded as a success' (Crawford 1929: 498). This might seem surprising as Crawford, more than most, understood the need to create photomosaics and interpretative maps. However, Fig. 13.4 is a useful comparison of both Beazeley's published photograph of the early modern town of Samarra surrounded by Early Islamic ruins it has partially obliterated and the very similar but far superior view probably taken in the same year (1918). Nevertheless, these early photographs, many of which survive in published form or in archives, British and German, remain a precious record of Samarra in 1916–1918. Together with those taken in the 1920s onwards, they were used to produce a marvellous plan of the entire site photogrammetrically (Northedge 1985, 2005: 27–8).

Northedge has itemised (2005: 27) the aerial photographs available to him for the detailed plan of the entire site. It is worth quoting in full as it seems likely that even lesser sites may have been covered by one or more of these or other aerial surveys:

'1. Two mosaics taken by the RAF in 1917, and used for creating the 6' to the mile map of Samarra (Royal Geographical Society, London)'.
2. Three rolls totalling 155 photographs taken by the RAF in October 1928, during training missions from Habbaniyya airbase, and deposited in the Institute of Archaeology, University College London. The scale is about 1:8000.
3. A series of vertical photographs taken by the Iraqi Air force in 1937. Prints in the Department of Antiquities, Samarra.
4. Three runs oriented north-south, totalling 55 stereoscopic photographs, taken by Hunting Aerosurveys in 1953. The scale is about 1:24,000.
5. A series of runs east-west probably taken by KLM Aerocarto. Prints in the Department of Antiquities, Samarra. Not available to this study'.

Nor is this a complete catalogue. To note only the largest of several additional collections known to us, there are 29 photographs taken between January and May 1918 by the German Air Force. These are now held in Berlin at the Deutsches Archäeologisches Institut (DAI), along with almost 1,000 more from various parts of the Ottoman fronts, from the Macedonian through Dardanelles to Caucasus and Mesopotamia between 1915 and 1918, including 'Palestine'.

13.4 Pioneers

Given these early examples, it is no surprise that the significance and importance of the technique of aerial photography were swiftly recognised in this region. The totality was modest, but the financial constraints of the post-war years imposed tight limits. The scale of the region meant that large parts had been inaccessible to all but a few hardy pioneers. The wonder is not that so little was done but that there was a steady take-up and usage of the technique. A few examples illustrate the point and bring out, too, the usefulness of the archives surviving from the 1920s to 1950s.

Foremost were the regular surveys and publications by Antoine Poidebard in Syria beginning in May 1925 and continuing for a generation, but with his great book of 1934 often regarded as the principal outcome. The flights proved that the region's varied landscapes were well suited for the technique of aerial photography to be successful. In addition, much of the work of other pioneer air photographers elsewhere in the Middle East was reported in the early volumes of *Antiquity*; no coincidence in that the editor and founder was none other than the so-called father of aerial photography for archaeology, O.G.S Crawford.

The British maintained an interest in the region throughout the early twentieth century. Crawford was flown in Iraq in 1928 (1929) and a decade later the RAF supported two seasons (1938–1939) of flying by Sir Aurel Stein in search of the Roman frontier. The former resulted in a substantial collection of glass negatives still held in University of London; the photographs taken by Stein have been more scattered but have largely survived (Kennedy and Bewley 2009: 69–70; Gregory and Kennedy 1985: ch. 9).

In Transjordan, Sir Alexander Kennedy used aerial photography in his surveys of Petra (Kennedy 1925: vi–vii, cf. 1924: 278–9) in attempting to understand the landscape setting, as well as the distribution and range of sites. Petra is notoriously difficult to capture from the air because of its scale and variety of structures (cf. Kennedy and Bewley 2004: 126–8), but Kennedy was able to secure the services of the RAF squadron at Amman in 1923[1] and again in 1924 to carry out a vertical survey of a wide area around the city centre as well as a number of oblique views which he also used in his book. The mosaic represents a very considerable achievement – it covers c. 85 km^2 and consists of several dozen photographs. For an audience familiar with ground views, but unable to visualise the context and landscape, this would have been revelatory. But it was more than a novel view or general photomap; Kennedy evidently examined closely those frames that covered the central area of the city:

'The air-plane view shows also very distinctly the lines of the Roman streets and the outlines of some of the principal buildings or places, and many other points quite unrecognizable on the ground' (Kennedy 1924: 278–9).

Kennedy may have had a predecessor in 1922 in Jordan (and possibly Iraq). A passing reference in Crawford and Keiller (1928: 4) draws attention to the abstract of a lecture given in 1923 by the enigmatic Robert Alexander MacLean of the University of Rochester (NY), who reports casually (1923: 68): 'This last summer I went by aeroplane from Amman in Transjordania to visit some Roman ruins at "Kasr Azraq" in the Syrian desert'. His lecture was illustrated by at least two aerial photographs. Just how he managed to arrange such a flight may be hidden in a subsequent report that he 'served as staff captain under General Maude [the British commanding general in Mesopotamia], being engaged particularly in intelligence and political work'.

Systematic work like that done for Kennedy formed the background to the much more opportunistic photography that followed. The RAF was responsible for patrolling the Mandate of Transjordan, but also creating the Cairo to Baghdad Air Mail Route (Hill 1929). In the course of those tasks, many individual sites were photographed. In 1928, Crawford arrived from Iraq and subsequently carried off several hundred of the aerial photographic negatives which are now housed in London (Kennedy 2012).

[1]TNA AIR5/1239 'Middle East Monthly Summaries 1921–1924' reports 14 Squadron was involved in taking photos of archaeological sites for the Palestine Government during January and February 1923.

13.5 Recent Discoveries

A recent discovery is a small archive of British aerial photographs of the region examined (by Kennedy) at The National Archive (TNA CB5/2 Nos: 357–460). Studying these, as well as the wartime German photographs (above), has added areas and targets which can be photographed today to gain a better understanding of the sites and the changes and threats to them (see 'Historic Aerial Photography' under http://www.flickr.com/photos/APAAME/collections/).

Wartime flying against the Vichy forces in French Syria probably lies behind at least some of the aerial surveys still held by the RAF in the UK, covering the border area. Hunting Surveys were commissioned to undertake an aerial survey of the western part of Jordan in 1953 (HAS) – the same year as they undertook extensive survey work in Iraq (above). Recently, some 300 of the oblique photographs, taken by the Hunting pilots and crew whilst transiting from Cairo and during their stay in Jordan, have come to light and require further detailed examination (they are in English Heritage's archive, the National Monuments Record, in Swindon) (Kennedy and Bewley 2010: 195).

The photographs for the HAS aerial survey were a real catalyst in the creation of an aerial reconnaissance campaign in Jordan. Funding would not have been forthcoming without some evidence that a campaign would be successful; obtaining support within Jordan was strengthened by recent research and publication of the HAS material (e.g. Kennedy 1997, 2001b). Indeed, the early years of the current programme, pre-GPS and Google Earth, was guided by the interpretation of the HAS imagery, prints from which were employed in-flight to locate sites.

13.6 Impact and Use of Historical Imagery

Since 1997 the Aerial Archaeology in Jordan project has flown 287 hours (on 71 flights), with several thousand target sites photographed. However, what has been the impact of the work so far, using both historical and recent imagery? There are two significant contributions resulting from studying the historical imagery, often now combined with the current reconnaissance programme: the first is the number of new sites discovered and additional information gained about known sites; the second is understanding the scale of destruction and pressure on the landscape, even in countries like Jordan and Syria (Kennedy and Bewley 2010; cf. Gerster and Wartke 2003: *passim*).

The experience of most aerial archaeologists working in European countries is of landscapes which have been cultivated. Archaeological sites are made visible as cropmarks, soilmarks, parchmarks or very slight remnants only visible in particular lighting and soil conditions. In Jordan, however, the vast majority of archaeological sites are visible as stoneworks or earthworks; the main requirement to photograph them is suitable lighting conditions. To highlight the value and impact of combining active aerial reconnaissance with historical imagery, we have selected a few examples and have given references to other articles which have reported new discoveries from historical imagery.

13.6.1 Samarra

Samarra is an immense site (above) and lies in a densely settled part of the Tigris valley; arable land is at a premium. Much of the ancient site has been protected, but the spread of modern Samarra itself has long ago destroyed structures under the early modern walled town and now beyond its walls (Fig. 13.4). Comparing an aerial photograph of an area on the west bank of the Tigris (away from the centre of the city) with the same area as recorded in a high-resolution satellite view on Google Earth in 2005 (Fig. 13.5a, b) shows that not only has much of the First World War trenching been obliterated, but the clear traces of buried structures in the centre have been lost.

13.6.2 First World War Aerial Photographs

German aerial photos from the First World War have proven especially useful for the study of Roman sites. In 1995, Fabian (1995) discovered a small fort on a German photograph of Beersheva for an area now totally obliterated by subsequent development. The photograph is the sole record of what may be the garrison place cited in a Late Roman document (cf. Kennedy 2002a: 35).

More recently, on a contemporary German photograph (2 April 1918) of a stretch of the Hedjaz Railway and a bridge south of Zarqa in northern Jordan, it was observed that the frame included a large part of a major Roman site at Qaryat al-Hadid, Zarqa. That in turn enabled more recent RAF aerial photographs of 1951 and HAS photographs of 1953 to be examined and collectively exploited to give shape and character to a small Roman town and fort noted by nineteenth-century Western travellers (Kennedy 2002c). These more recent photographs also reveal the site in the process of being destroyed and by 2000 it had been entirely overwhelmed by the spread of Zarqa.

A further German photograph has recently been examined to shed light on a more recent past and to guide a flight in Jordan at Qala'at Unayza and West Unayza. Part of the Palestine Front in 1917–1918 was in what is now southern Jordan. Both British and German aircraft operated over the triangle between Mudawarra Aqaba and Tafila, where the forces of the Arab Revolt faced the Turkish Fourth Army. The RFC/RAF certainly took many photographs (Kennedy 2012), but these have not yet been traced. Some German ones have been found, including a revealing low oblique view (Fig. 13.6).

Unayza is best known for the small Haj Fort on the Pilgrim Road and the large reservoir nearby that sustained garrison and pilgrims alike (there is even a hint that the Haj Fort overlies an earlier, possibly Roman fort). During the First World War the importance of the place lay in the railway station, water supply and a recently built branch running off to the west to tap timber supplies around Shaubak. The photograph (Fig. 13.6) shows the railway lines, station and ancient square reservoir in the foreground and the Haj Fort near the centre. The Turkish forces had dug extensive trenches to defend the station and water supply; much of this is no longer visible. Equally interesting is the German annotation: the white arrow is a misplaced north sign (it is

Fig. 13.5 Samarra area – Qasr el-Ashiq on the west bank of the Tigris opposite Samarra, Iraq in (**a**) 1918 (?). (**b**) The same area on Google Earth 1st June 2005 (© 2011 Google Earth; © 2011 Digital Globe. Compiled by Mat Dalton)

in fact pointing west), but it was thought by one author to be drawing attention to a feature. Coincidentally it is – our flight in October 2011 recorded the extensive traces of trenches and artillery emplacements on the hill beneath the arrow (Fig. 13.7) which we had not seen on earlier flights in this area and had not suspected.

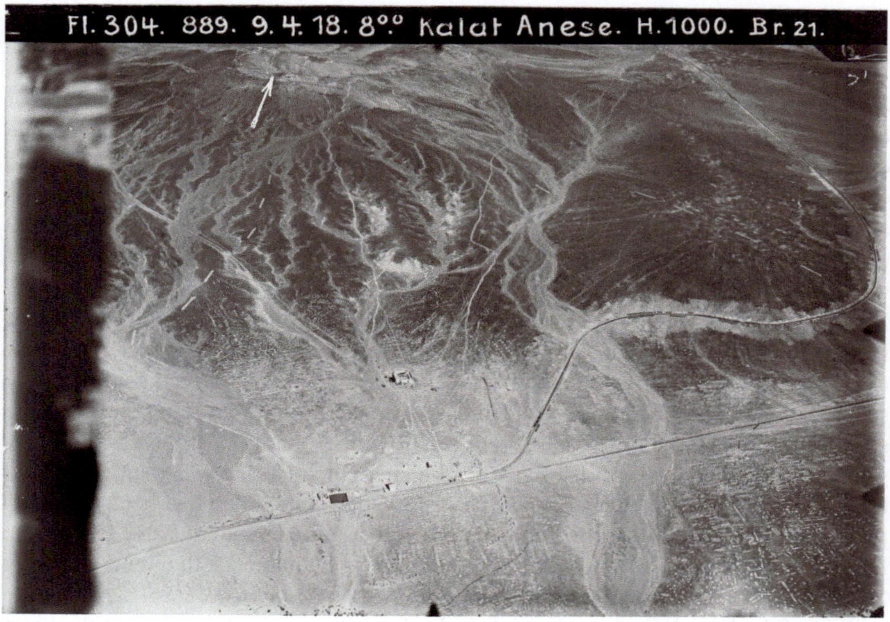

Fig. 13.6 Unayza on 9 April 1918 in a German aerial view. The *arrow* annotated by the Germans points to the site in Fig. 13.7 (Bayerisches Kriegsarchiv 1234_1)

Fig. 13.7 Unayza Turkish fortifications (APAAME_20111010_RHB-0198) (Photographed by Bob Bewley)

13.6.3 Early Mandate Aerial Photographs

Agricultural development has led to numerous examples of destruction as some of the earliest photographs of the Mandate period record. For example, the double destructive power of an olive plantation and quarrying has removed most of the evidence for the 'lost' Early Islamic village at Hibabiya, Jordan. Air photographs are now the only source of its existence; our own reconnaissance in 1998 and 1999 could not easily locate the site using the Hunting images. Even with the correct location in our GPS, it took some time to recognise this transformed landscape. Despite having been visited by a number of archaeologists, the date and function of the site were puzzling. The aerial photograph published in 1929 characterised it as a prehistoric 'fishing village' on the edge of a seasonal lake. Artefacts found nearby seemed to support the date, but the 'village' seemed unlikely for that period. As the site is totally destroyed, we had to rely on the reinterpretation of the aerial photograph and comparison with parallels for the site. Happily, this was then confirmed by the rediscovery of pottery collected over 80 years ago which had been misinterpreted at the time. We can now recognise the site as a Late Roman/Early Islamic (c. late sixth/early seventh to mid-/late eighth century AD) nomad settlement (Kennedy 2011b), the structures being interpreted as houses and gardens.

One of the least-known settlements in Jordan is that around the oasis at the heart of the Al-Jafr Basin, c. 50 km east of the Pilgrim Road and the nearest Hajj fort at Unayza (above, 13.6.2). Although the oasis has revealed evidence from the Roman period, including Latin graffiti, it is a much more recent construction that concerns us here. Following the First World War and the creation of the British Mandate of Transjordan, one of the best-known leaders of the Great Arab Revolt, Auda Abu Tayi, settled at the Al-Jafr oasis and used Turkish prisoners of war to build a 'palace'. It was incomplete at the time of his death in 1924, but this RAF aerial photograph of 1927 captures it in a rare image (Fig. 13.8a). By 2011, the situation had changed dramatically with the outer enclosure largely destroyed to ground level and the inner residence more than half collapsed (Fig. 13.8b).

13.6.4 The Hunting Aerial Survey of Jordan (1953)

The principal developments in the landscape of Jordan have occurred from the 1970s onwards. Aerial photographs of 1953 are images of a world soon to be transformed.

One of the RAF pilots who first flew over the volcanic lavafields of the northeast (Harret al-Shaam) in the 1920s noticed large and numerous sites which he referred to as 'Kites' (Rees 1929: 395), one of several categories of site known collectively to the Bedouin as the 'Works of the Old Men'. Ever since then, archaeologists have been trying to understand the size, scale, date and distribution of these sites. What is clear is that Kites are prehistoric, perhaps as early as the seventh millennium BC (the Pre-Pottery Neolithic). To date (December 2011), our project has recorded 1,236 Kite sites in the Jordanian part of this *harrat* and more will certainly be found.

Fig. 13.8 Al-Jafr Castle (**a**) in 1927 (The National Archives). (APAAME_19260902_TNA_RAF14_AIR5-1157 (2)-6); (**b**) in 2011 (APAAME_20111013_DLK-0239) (Photographed by David Kennedy)

Many, however, have already been damaged and some have been totally destroyed and would not have been known to us had we been reliant solely on the present aerial reconnaissance. However, we have extensive coverage at 1:25,000 scale in the HAS photographs of 1953 for precisely the area west of the Azraq Oasis in which the most widespread damage and loss has occurred. The detailed interpretation of the HAS photos for the Jimal area (Kennedy 1998) recorded 59 Kites even in an

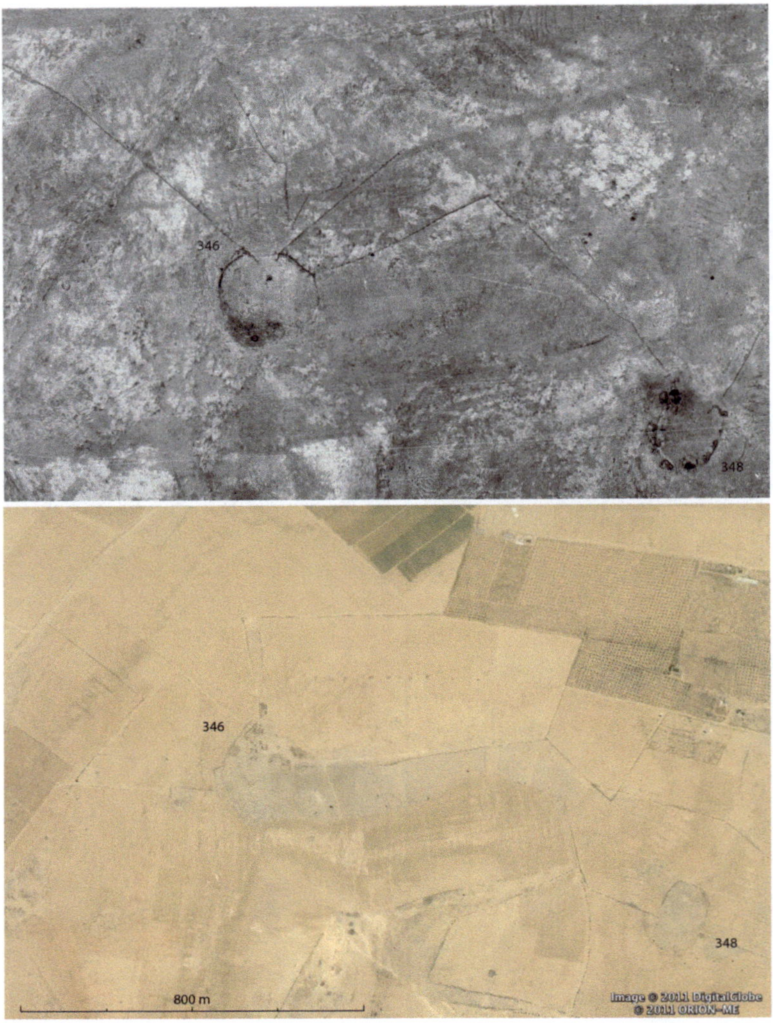

Fig. 13.9 (**a**) Two Kites recorded on a HAS survey photograph of 1953 (APAAME_19531031_HAS_58-025) and (**b**) the same area recently on a high-resolution Google Earth image of 1st June 2009 (© 2011 Google Earth; © 2011 Digital Globe; © 2011 ORION-ME. Compiled by Rebecca Banks)

area already beginning to be extensively cultivated in 1953. There has been a huge extension of cultivation as deep water sources have been tapped in the last 25 years. Now, almost all traces of these Kites have been lost or survive in part as sections of field wall but are recognisable only as such *because* we have the 1953 images to guide us. Figure 13.9 shows one example of a Kite visible in 1953 and the same area today on a high-resolution Google Earth image. Without the historical imagery, the overall distribution pattern of these sites would be significantly distorted.

Fig. 13.10 Khirbet Khaw – fort, caravanserai, houses and tombs (APAAME_19980506_DLK-0115) (Photographed by David Kennedy)

The archaeological remains at Khirbet Khaw near Zarqa were known to a succession of travellers in northern Jordan between 1876 and 1930. Although one party reported a 'town', others were dismissive, one of the more influential offering the unflattering view that it was 'an uninteresting ruin on a hilltop' (cf. Kennedy 2001a: 174). During the 1930s, the site was enclosed within a military compound and 'lost' for further study until rediscovered in the course of systematic interpretation of the HAS aerial photographs. At a stroke, the true scale and character of the remains were revealed: an impressive small town including a caravanserai and previously unknown Roman fort (Kennedy 2001a). Once again, the historical image guided our present flight planning and has allowed us to photograph the site as it is today (Fig. 13.10).

13.7 Conclusions

The cases above could be multiplied easily both by reference to published and new examples emerging regularly in the course of research. The Aerial Photographic Archive for Archaeology in the Middle East (APAAME) consists today overwhelmingly of photographs taken since 1997. It began, however, with the HAS photographs of 1953. In the last few years, those have been complemented by thousands more, even older, photographs and emerging information of yet more collections that can be exploited in the coming years. The more historical imagery discovered the better for research: many sites visible even in 1953 are now damaged or utterly destroyed. The ideal approach to any research should be through interpretation of a range of photographs from those of the First World War to the most recent satellite imagery. At the moment, Jordan is the only country which allows aerial reconnaissance for archaeology, so that we can combine it with the study of historical imagery; even so, there are yet significant gaps. The extensive archive of Poidebard's photographs for Syria has long been available, but there are few photographs dating from after 1945 and no current flying. In Iraq, the situation is worse, but recent archival research gives grounds for optimism about what may be traced in the future (and we know that imagery from the Gulf Wars exists but has not been examined systematically for its archaeological content). A recent development is the improved access now available for Iraq (and elsewhere) of CORONA satellite imagery from the 1960s (http://corona.cast.uark.edu/index.html; see Chap. 4 by Fowler, this volume).

The significance of the historical imagery is not just that it exists but that it is made known and accessible, otherwise the boxes may as well remain closed. Web-based archives can help to meet this need. The APAAME Flickr site already includes many historical photographs. Many others have already been acquired, but not yet displayed while copyright issues are resolved, and there are more to follow. Finding them should be a priority.

The parts of the region for which historical imagery seems likely to be most abundant are Syria, Iraq, Lebanon, Jordan, Palestine and Israel. Historical imagery will certainly exist for Iran, Turkey, Saudi Arabia, Yemen and Oman, both in the government archives of those countries and in the records of oil and geological exploration companies, but they have generally not been open to researchers. High-resolution satellite imagery of these countries has firmly outflanked and negated any possible military significance of traditional aerial photography of 25–60 years ago. Obtaining access to these older photographs would help enormously to transform our understanding of the archaeology and history of those lands.

As archaeologists, we aim to increase our understanding of the changes to the landscape brought about by human intervention and thus gain a better understanding of human behaviour. Aerial survey is a very efficient and useful tool, but it is only one technique for discovering who lived where and when. As yet, it is an underused method (on a global scale); the best starting point for anyone studying ancient landscapes is to examine the existing imagery, however recent or historical.

Bibliography

Beazeley, G. A. (1919). Air photography in archaeology. *The Geographical Journal, 53*(5), 330–335.

Beazeley, G. A. (1920). Surveys in Mesopotamia during the war. *The Geographical Journal, 54*(2), 109–127.

Bewley, R. H. (1997). From military to civilian: A brief history of the early development of aerial photography for archaeology. In J. Oexle (Ed.), *Aus der Luft – Bilder unserer Geschichte: Luftbildarchäologie in Zentraleuropa* (pp. 10–21). Dresden: Landesamt für Archäologie mit Landesmuseum für Vorgeschichte.

Bewley, R. H., & Rączkowski, W. (Eds.). (2002). *Aerial archaeology. Developing future practice* (NATO Science Series – Series 1: Life and Behavioural Sciences, Vol. 337). Amsterdam: IOS Press.

Cowley, D. C., Standring, R. A., & Abicht, M. J. (Eds.). (2010). *Landscapes through the Lens. Aerial photographs and the historic environment*. Oxford: Oxbow.

Crawford, O. G. S. (1929). Air photographs of the Middle East. *The Geographical Journal, 73*, 497–512.

Crawford, O. G. S. (1954). A century of air-photography. *Antiquity, 28*, 206–210.

Crawford, O. G. S., & Keiller, A. (1928). *Wessex from the air*. Oxford: University Press.

Denise, F., & Nordiguian, L. (2004). *Une aventure archéologique: Antoine Poidebard, photographe et aviateur*. Marseille/Arles/Beirut: Presse Université St. Joseph.

Fabian, P. (1995). The Late-Roman military camp at Beer Sheba: A new discovery. In J. H. Humphrey (Ed.), *The Roman and Byzantine near east* (pp. 235–240). Ann Arbor. *Journal of Roman Archaeology*, Supplementary Volume 14.

Gavish, D. (1987). An account of an unrealized aerial cadastral survey in British Mandatorial Palestine, 1918–1948. *The Geographical Journal, 153*, 93–98.

Gavish, D., & Biger, G. (1985). Innovative cartography in Palestine 1917–18. *Cartographical Journal, 22*(1), 38–44.

Gerster, G. (2008). *Paradise lost, Persia from above*. London/New York: Phaidon.

Gerster, G., & Wartke, R.-B. (2003). *Flugbilder aus Syrien von der Antike bis zur Moderne*. Mainz-am-Rhein: Von Zabern.

Gregory, S., & Kennedy, D. L. (1985). *Sir Aurel Stein's Limes report. (The full text of M.A. Stein's unpublished Limes Report (his aerial and ground reconnaissances in Iraq and Transjordan in 1938-39)* (BAR, International Series 272). Oxford: BAR.

Hill, R. (1929). *The Baghdad air mail*. London: Edward Arnold.

Kedar, B. (1999). *The Changing Land between the Jordan and the Sea: Aerial photographs from 1917 to the Present*. Yad Ben-Zvi and Israel Ministry of Defense: Jerusalem and Tel Aviv.

Kennedy, A. B. W. (1924). The rocks and monuments of petra. *The Geographical Journal, 63*(4), 273–295.

Kennedy, A. B. W. (1925). *Petra. Its history and monuments*. London: Country Life.

Kennedy, D. L. (1982). *Archaeological explorations on the Roman frontier in north east Jordan. The Roman and Byzantine military installations and road network on the ground and from the air*. (Including unpublished work by Sir Aurel Stein and with a contribution by D.N. Riley) (BAR, International Series 132). Oxford: BAR.

Kennedy, D. L. (1985, September). Ancient settlement in Syria. *Popular Archaeology*, 42–44.

Kennedy, D. L. (1997). Aerial archaeology in Jordan: Air photography and the Jordanian Hauran. In G. Bisheh (Ed.), *Studies in the history and archaeology of Jordan, VI* (pp. 77–86). Amman: The Department of Antiquities of Jordan.

Kennedy, D. L. (1998). The area of Umm el-Jemal: Maps, air photographs and surface survey. In B. de Vries (Ed.), *Umm el-Jimal, I* (pp. 39–90). Portsmouth. *Journal of Roman Archaeology*, Supplementary Volume 26.

Kennedy, D. L. (2001a). Khirbet Khaw. A Roman town and fort in northern Jordan. In N. J. Higham (Ed.), *Archaeology of the Roman Empire: A tribute to the life and works of Professor Barri Jones* (BAR, International Series 940, pp. 173–188). Oxford: BAR.

Kennedy, D. L. (2001b). History in depth: Surface survey and aerial archaeology. In K. Amr (Ed.), *Studies in the history and archaeology of Jordan, VII* (pp. 39–48). Amman: The Department of Antiquities of Jordan.

Kennedy, D. L. (2002a). Aerial archaeology in the Middle East: The role of the military – Past, present ... and future? In Bewley, R. H., & Rączkowski, W. (Eds.). *Aerial archaeology. Developing future practice* (pp. 33–48, 346–347). (NATO Science Series – Series 1: Life and Behavioural Sciences, Vol. 337). Amsterdam: IOS Press.

Kennedy, D. L. (2002b). Two Nabataean and Roman sites in southern Jordan: Khirbet el-Qirana and Khirbet el-Khalde. In V. Gorman & E. Robinson (Eds.), *Oikistes* (pp. 361–386). Leiden: Brill.

Kennedy, D. L. (2002c). Qaryat el-Hadid: A 'lost' Roman military site in northern Jordan. *Levant, 34*, 99–110.
Kennedy, D. L. (2004). *The Roman army in Jordan* (2nd ed.). London: Council for British Research in the Levant.
Kennedy, D. L. (2011a). 'The works of old men' in Arabia: Remote sensing in interior Arabia. *Journal of Archaeological Science, 38*, 3185–3203.
Kennedy, D. L. (2011b). Recovering the past from above. Hibabiya – An Early Islamic village in the Jordanian desert? *Arabian Archaeology and Epigraphy, 22*, 253–260.
Kennedy, D.L. (2012). Pioneers above Jordan. Revealing a prehistoric landscape. *Antiquity, 86*, 474–491.
Kennedy, D. L. (Forthcoming). 'Big circles': A new type of prehistoric site in Jordan and Syria, *Zeitschrift für Orient-Archäologie*.
Kennedy, D., & Bewley, R. (2004). *Ancient Jordan from the air*. London: Council for British Research in the Levant.
Kennedy, D., & Bewley, R. (2009). Aerial archaeology in Jordan. *Antiquity, 83*, 69–81.
Kennedy, D., & Bewley, R. (2010). Archives and aerial imagery in Jordan. Rescuing the archaeology of Greater Amman from rapid urban sprawl. In Cowley, D. C., Standring, R. A., & Abicht, M. J. (Eds.). *Landscapes through the Lens. Aerial photographs and the historic environment*. (pp. 193–206). Oxford: Oxbow.
Kennedy, D. L., & Bishop, M. C. (2011). Google Earth and the archaeology of Saudi Arabia. A case study from the Jeddah area. *Journal of Archaeological Science, 38*, 1284–1293.
Kennedy, D. L., & Riley, D. N. (1990). *Rome's desert frontier*. London: Batsford.
Lambert, P. J. (2003). *The forgotten air war: Airpower in the Mesopotamian Campaign*. Unpublished MA, U.S. Army Command and General Staff College, Fort Leavenworth, Kansas.
Leisten, T. (2003). *Excavation of Samarra: Volume 1: Architecture – Final Report of the First Campaign 1910–1912*. Mainz-am-Rhein: Von Zabern.
MacLean, R. A. (1923). The aeroplane and archaeology. *American Journal of Archaeology, 37*(1), 68–69.
Millar, F. G. B. (1991). Review of Kennedy and Riley 1990. *The Classical Review, 41*(1), 189–191.
Mouterde, R., & Poidebard, A. (1945). *Le Limes de Chalcis*. Paris: Geuthner.
Nordiguian, L., & Salles, J.-F. (2000). *Aux origines de l'archéologie aérienne. A. Poidebard (1878-1955)*. Beirut: Presses de l'Université Saint-Joseph.
Northedge, A. (1985). Planning Samarra: A report for 1983–4. *Iraq, 47*, 109–128.
Northedge, A. (2005). *The historical topography of Samarra, Samarra Studies I*. London: British School of Archaeology in Iraq.
Palumbo, G. (1994). *JADIS. The Jordanian archaeological data information system*. Amman: Department of Antiquities and American Center for Oriental Research.
Poidebard, A. (1934). *La trace de Rome dans le désert de Syrie. Le Limes de Trajan à la Conquête Arabe. Recherches Aériennes (1925–1932)*. Paris: Geuthner.
RAF. (1918). *Notes on aerial photography. The interpretation of aeroplane photographs in Mesopotamia, II*. Mesopotamia: RAF.
Rees, L. W. B. (1929). The Transjordan desert. *Antiquity, 3*, 389–406.
Schmidt, E. (1940). *Flights over the ancient cities of Iran*. Chicago: Oriental Institute.
Stein, A. (1919). Air photography of ancient sites. *The Geographical Journal, 54*(3), 200.
Tennant, J. E. (1920). *In the clouds above Baghdad*. London: Palmer.
Wiegand, T. (1920). *Wissenschaftliche Veröffentlichungen des Deutsch-Türkischen Denkmalschutz-Kommandos. I. Sinai*. Berlin/Leipzig: De Gruyter.

Chapter 14
"Down Under in the Marshes": Investigating Settlement Patterns of the Early Formative Mound-Building Cultures of South-Eastern Uruguay Through Historic Aerial Photography

José Iriarte

Abstract This chapter presents a review of the use of aerial photography for reconnaissance and archaeological survey in the investigation of the Early Formative cultures of south-eastern Uruguay. Historic aerial photographs have been crucial for both locating and obtaining a complete inventory of mound sites associated with these archaeological cultures, known as the 'Constructores de Cerritos', since the landscape of south-eastern Uruguay has been dramatically transformed by the drainage of wetlands for rice cultivation in the last decades. This chapter describes the distinctive features on the aerial photographs that archaeologists have used to identify mound sites, including their dimensions, shape, distinct vegetation cover, surrounding circular borrow pits, associated cattle trails and shadow cast. It shows how this information has allowed archaeologists to investigate settlement patterns in the wetlands of India Muerta and briefly discusses the implications for the rise of Early Formative societies in south-eastern South America.

14.1 Introduction

Uruguay is one of the few countries in South America to have a complete aerial photographic survey of the whole country since the mid-1960s. Because of major modern developments which have altered the landscape significantly, this has now also become a historic archive. In this regard, aerial photographs are now allowing archaeologists, among other social and natural scientists, to recover lost information from areas of the country that have been dramatically transformed by development.

J. Iriarte (✉)
Department of Archaeology, University of Exeter, Laver Building, North Park Road, Exeter, EX4 4QE, UK
e-mail: J.Iriarte@exeter.ac.uk

W.S. Hanson and I.A. Oltean (eds.), *Archaeology from Historical Aerial and Satellite Archives*, DOI 10.1007/978-1-4614-4505-0_14,

This is the case for the vast expanse of wetlands in south-eastern Uruguay that were drained for rice cultivation during the 1970s. As a result, the hydrological regime and the topography of the region has been completely altered and many archaeological sites have been destroyed (PROBIDES 2000). In fact, historic aerial photographs from this region are the only source which will allow us to obtain detailed data on topography, hydrology, and vegetation before the 1970s and get a complete inventory of archaeological sites. This chapter presents a review of the use of aerial photography for reconnaissance and archaeological survey in the investigation of the Early Formative cultures of south-eastern Uruguay. It starts by describing the aerial photography services available in Uruguay. Then, it presents a brief summary of the archaeology of the Early Formative cultures of south-eastern Uruguay, followed by a description of the distinctive features of the aerial photographs that were used to identify mound sites during archaeological reconnaissance of new regions. It then goes on to describe the use of aerial photography in order to identify sites with mounded architecture in the wetlands of India Muerta (Fig. 14.1b) and to reconstruct settlement patterns in our study area. Finally, the concluding section summarizes the implications of using aerial photography for archaeological reconnaissance in the region.

14.2 The Servicio Geográfico Militar Aerial Photographic Database in Uruguay

In Uruguay, the Servicio Geográfico Militar (SGM) is in charge of the preparation, updating, conservation, distribution and evaluation of all cartographic material in accordance with the mission that was assigned to the Uruguayan Army by Law 15,688 issued on 1984. This law gives the SGM the role of supporting the planning of security and national development activities in the country. The SGM is also in charge of establishing, preserving and extending the national network of triangulation, levelling, gravimetry and Earth's magnetism. More recently, the SGM was given the duty of maintaining and operating a Geographic Information System (GIS) to support management and decision making.

The SGM has a database of aerial photographs consisting of partial historic flights at different scales taken in 1922; a single flight in 1943 using the Trigometron System at 1:40,000 scale; and a set of aerial photographs covering the entire country taken in 1966–1967 at scales of 1:40,000 and 1:20,000. Combining aerial photographs with field reconnaissance, the SGM has produced 1:50,000 topographic maps of the entire country with contour levels every 10 m. The maps also depict hydrographic and hypsographic information, vegetation, roads, industrial details, cities, towns and other modern cultural elements.[1]

[1] SGM resources can be accessed through its web page at http://www.ejercito.mil.uy/cal/sgm/

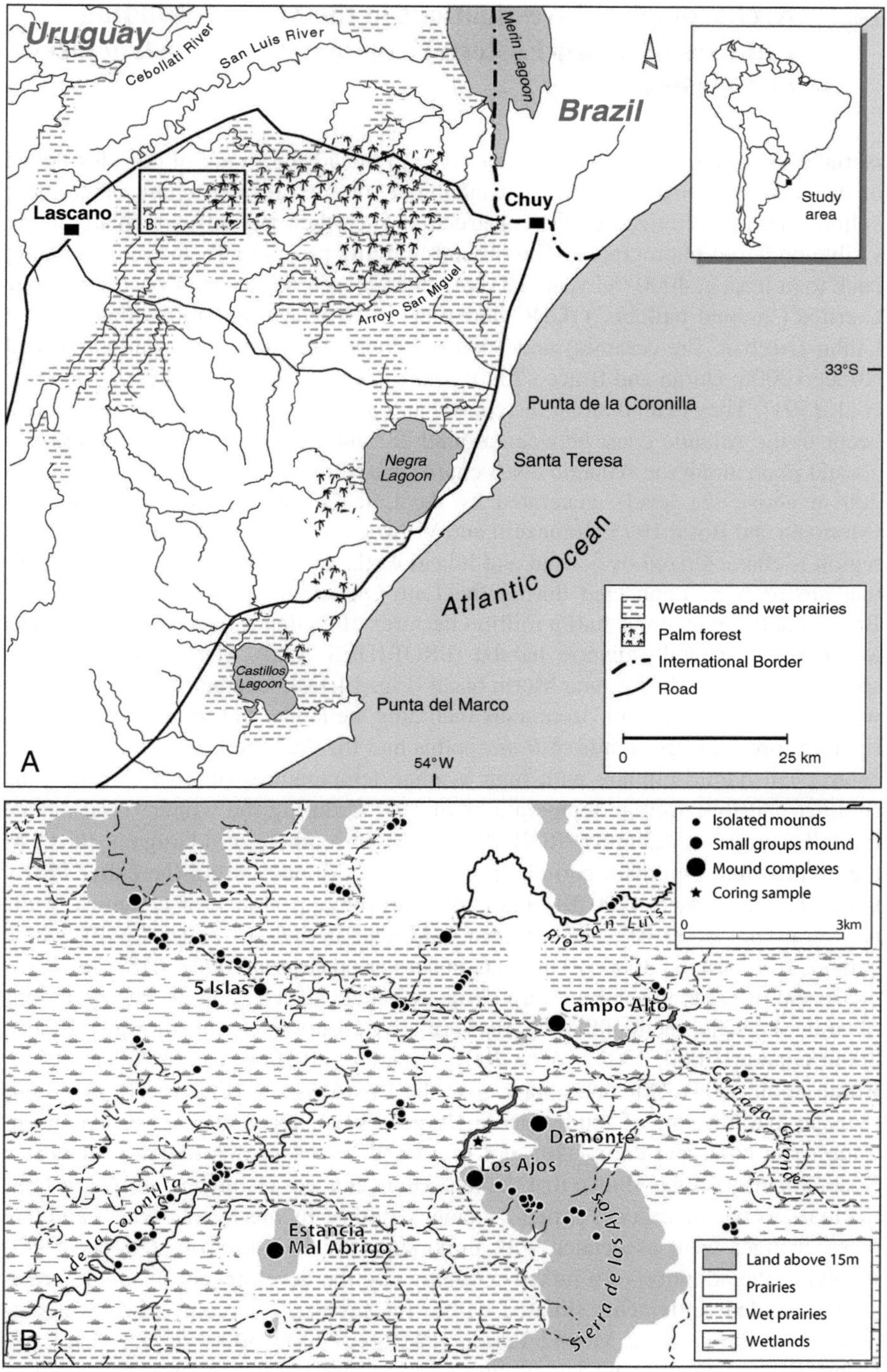

Fig. 14.1 (**a**) Map of south-eastern Uruguay showing the location of the India Muerta wetlands. (**b**) Distribution of mound sites in the India Muerta wetlands in the southern sector of the Laguna Merín basin. Map based on 1:50,000 topographic maps (Lascano D23, Averias D22, and Cañada Grande C23)

14.3 Archaeological Investigations in the Mound-Building Cultures of South-Eastern Uruguay: The 'Constructores de Cerritos'

Aerial photography has played a major role in the advancement of our knowledge of the pre-Columbian societies that inhabited south-eastern Uruguay, since these cultures are characterized by the construction of earthen mounds, which are highly visible on aerial photographs. The mound-building pre-Columbian cultures dating back to at least c. 4000 ^{14}C years BP are generally referred to as 'Constructores de Cerritos'('mound builders') (CDC hereafter) in Uruguay and are divided into the Umbu (Archaic Pre-ceramic) and Vieira (Ceramic) traditions in southern Brazil (Bracco 2006; Durán and Bracco 2000; Iriarte 2006a, 2007a; López 2001; Schmitz et al. 1991). They extend along the coastal and inland wetlands and grasslands that occur in the Atlantic coast between around 28° and 36°S. The region comprises a coastal plain along the Atlantic coast characterized by slight elevations (maximum 200 m above sea level), generated by the Late Quaternary marine oscillations (Montaña and Bossi 1995; Tomazelli and Villwock 1996; Bracco et al. 2000b). The region is characterized by coastal and inland wetlands encompassing large lagoons that are or were connected during the Late Quaternary to the Atlantic Ocean. Bordering these lagoons, half a million hectares of wetlands have been recognized as an environmentally diverse habitat (PROBIDES 2000). The study region, the southern sector of the Laguna Merin basin (Fig. 14.1), is characterized by a patchwork of closely packed environments including wetlands, wet prairies, grasslands, riparian forests, large stands of *Butia* palms and the Atlantic Ocean coast. It has a subtropical humid climate with high average temperatures of 21.6°C during the summer and low average temperatures of 10.9°C during the winter. Total annual rainfall averages 1,123 mm (PROBIDES 2000). In south-eastern Uruguay, the CDC are divided into two main periods: a Pre-ceramic Mound Period (hereafter PMP), which begins around 4200 BP and ends with the appearance of ceramics in the region around 3000 BP; and a Ceramic Mound Period (hereafter CMP), which extends from around 3000 BP to the Contact Period (Iriarte 2006a: 648, Fig. 2).

Recent research in south-eastern Uruguay had advanced previous interpretations of cultural development of these Early Formative groups in significant ways (Bracco 2006; Gianotti 2000, 2005; López 2001; Criado et al. 2006; Iriarte 2006a, 2007a; Iriarte et al. 2004). Our investigations at the Los Ajos site (Figs. 14.1b and 14.4) indicate that this area of south-eastern Uruguay was a locus of population concentration and emergent complexity not registered before in this region of southern South America. It has shown that the mid-Holocene was characterized by sea-level fluctuations and significant climatic changes marked by increased dryness and that these changes were associated with important cultural transitions (Bracco 1992) involving the adoption of a mixed economy and the development of more permanent mounded settlements situated within resource-rich, circumscribed wetlands during the Pre-ceramic Mound Period (PMP) (c. 4200–3000 ^{14}C cal years BP) (Iriarte 2006b). Our renewed community-focused archaeological programme at Los Ajos showed that large Pre-ceramic mound complexes in the region were not the result of random, successive short-term occupations of mobile hunter-gatherers (Schmitz et al. 1991), nor the burial mounds or monuments of complex hunter-gatherers as previously proposed (Bracco 1992; Bracco et al. 2000a; Gianotti 2000; López 2001), but well-planned plaza villages built by people who practised a mixed economy.

Available data also indicates that during the succeeding Ceramic Mound Period (CMP) (3000–500 ^{14}C cal years BP), there was a marked increase in the number of sites. It witnessed the appearance of collective cemeteries and a formalization and spatial differentiation of the earthen mound architecture that appears to represent an early and distinct civic-ceremonial architectural tradition in lowland South America (Iriarte et al. 2004; Iriarte 2006a, 2007a; Stalh 2004). The Early Formative cultures of south-eastern Uruguay are beginning to manifest the existence of a unique, independent and a more complex cultural trajectory than previously thought for the La Plata Basin. The unexpected cultural sequence at Los Ajos reveals an early expression of cultural complexity never before registered in this region of lowland South America, which clashes with the long-held view that it was inhabited by marginal, small groups of simple, highly mobile hunter-gatherers that had not experienced significant changes since the beginning of the Holocene (Meggers and Evans 1978; Steward 1946), and endorses previous views (Andrade and López 2001; Bracco et al. 2000a; López 2001).

14.4 The Use of Aerial Photography in the Investigation of the 'Constructores de Cerritos'

Aerial photography has been extensively used in the archaeology of south-eastern Uruguay, which is characterized by the presence of conspicuous earthen structures generally called mounds, for reconnaissance survey of new regions since the mid-1980s (Bracco and López 1987a, b; Durán and Bracco 2000; Gianotti 2000, 2005). Most of CDC mound sites have been discovered from the air using the 1966–1967 1:20,000 stereoscopic pairs. Archaeologists have used the latter combined with 1:50,000 topographic maps to discover and locate mounds and subsequently infer settlement patterns and community organization of the CDC cultures (López and Bracco 1992; Gianotti 2005; Bracco 2006; López 2001; López and Pintos 2000). The technique has also proved useful when integrating other natural scientists in the investigation of questions relevant to archaeology. For example, aerial photographs have also allowed geologists and geomorphologists to trace the location of Late Quaternary marine terraces (Montaña and Bossi 1995). More importantly, since the landscape of south-eastern Uruguay has been dramatically transformed by the drainage of wetlands for rice cultivation (Fig. 14.2), the use of historic aerial photographs is crucial in order to locate and obtain a complete inventory of mound sites. This is particularly so in some regions, like the Treinta y Tres province (Prieto et al. 1970) and Arroyo Yaguarí (Gianotti 2005), where numerous mounds have been literally flattened to prepare land for rice cultivation (for a detailed map of these regions see Iriarte 2006b: 646, Fig. 1).

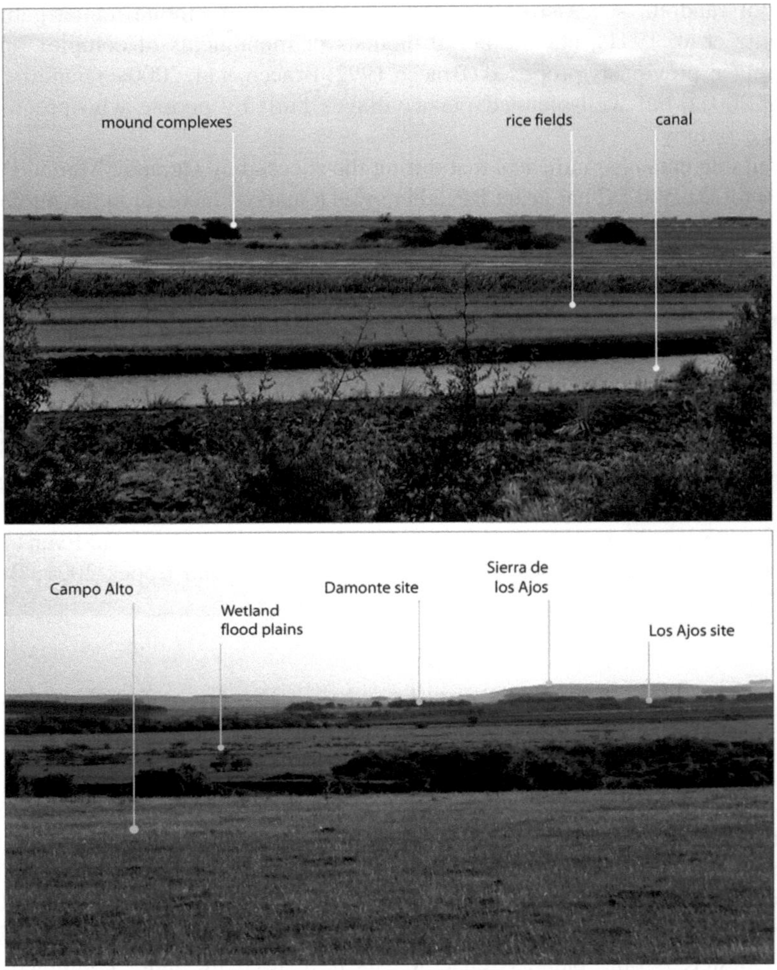

Fig. 14.2 *Above*. View of the India Muerta wetlands showing rice fields, irrigation canals and archaeological mound complexes. *Below*. Panoramic view of the Los Ajos mound complex sites from Campo Alto

Mounds are artificially constructed earthen structures with a circular, elliptical, oblong or quadrangular base, which ranges in diameter between 15 and 35 m, though they frequently fall between 20 and 30 m and between 0.3 and 7 m in height. Mounds can be found in various topographical positions from wetland floodplains to hilltops 160 m above sea level, and they are positioned at different locations within the landscapes, such as fossil sand dunes, marine terraces, lake margins, levees, flat knolls and prominent hilltops. They can be isolated, form small clusters of two to five mounds or aggregate in large mound complexes, such as the Estancia

Fig. 14.3 (**a**) *Left*. Notice radiating cattle trails leading to mound group. *Right*. Circular mound showing darker outer ring that stands out from clearer mound top surface. (**b**) 3 Islas mound group showing darker vegetation on ring surrounding mound where person is walking. (**c**) Nitrogen-loving vegetation growing on top of mound with trees growing on its upper slope

Mal Abrigo site which occupies an area of 60 ha and contains 77 mounds (Figs. 14.1b and 14.6; Iriarte et al. 2001: 64, Fig. 2). Finally, mounds can be uni- or multi-component, bearing only Pre-ceramic or Ceramic components, or containing both.

In the wetland floodplains CDC earthen mounds are highly visible using stereoscopic pairs of 1:20,000 aerial photographs. They can be identified based on a number of criteria: their dimensions, their shape, the distinct vegetation cover that grows on them, circular borrow pits that surround them; associated cattle trails; and the shadows they cast. In terms of shape and size, mounds can be distinguished by their spherical/sub-spherical shape that is about 1–4 mm in diameter on the photographs. In low areas where mounds are above 2.5 m in height, it is generally feasible to visualize a slightly darker semicircle surrounding part of the mound, which represents its shadow on one of its sides (Marozzi O., personal communication 2008). Mounds also stand out in the low wetland floodplains by their height, which ranges from 1 to 7 m. Additionally, they can be identified by the distinct vegetation that grows on them since they provide a dry *terra firma* habitat for the development of certain grasses, nitrogen-loving plants, herbs and forest. Cattle use mounds as dry sleeping platforms and their faeces unintentionally make certain type of nitrogen-loving herbs thrive, such as some non-native Asteraceae like 'cardo' (*Cynara cardunculus*) and 'cardo negro' (*Cirsium vulgare*), and some Solanaceae like 'chamico' (*Datura ferox*), among others (Fig. 14.3c) (Iriarte and Alonso 2009). Some mounds are also colonized by the aggressive Bermuda grass (*Cynodon dactylon*). In most cases, this distinct herbaceous vegetation makes the centre of mounds appear lighter in colour compared to the darker hydrophyte vegetation (sedges and reeds) growing in the seasonally flooded, surrounding waterlogged soils (Fig. 14.3a, b). Cattle trails that usually radiate from mounds are another important feature that helps to detect their presence (Fig. 14.3a). Mounds are also generally associated with darker, peripheral ring-shaped depressions surrounding them, which usually represent borrow pits (Fig. 14.3a).

In higher areas of the landscape, such as the flattened spurs of hills, mounds can also be distinguished by their size, height and shape, and also as distinct 'dots' of forest (Figs. 14.3a, 14.5b, and 14.6). These dots are readily distinguished from both the natural gallery forest that grows following the main streams, which are dense, continuous and linear, and from the hill forest that grows on the piedmont of hills, which usually consists of rather large, discrete patches along their base. In higher parts of the landscape, where large mound complexes occur, aerial photography allows us to estimate the overall size of the site, give some indications of their formal arrangements (linear, circular, horseshoe) and in many cases the size and shape of central plaza areas (Figs.14.3a, 14.5b, and 14.6). However, ground inspection is needed to determine the number and characteristics of the mounds hidden below the forest.

14.5 Settlement Patterns in the Wetlands of India Muerta

Our study region, the India Muerta wetlands (Fig. 14.1b), shows one of the major aggregations of multi-component mound complexes (Bracco 1993, 2006; Bracco et al. 2000a; Iriarte 2003, 2006a; López 2001). We have focused our archaeological programme in the wetlands of India Muerta and in particular at the Los Ajos site (33° 41′ 54″ S/ 53° 57′ 25″ W) (Figs. 14.1b and 14.4). Los Ajos is located on a flattened spur of the Sierra de Los Ajos, which overlooks the wetlands of India Muerta. The Los Ajos site, which covers about 12 ha, is one of the largest and most formally laid out sites in the study area (Fig. 14.4). Its Inner Precinct includes six flat-topped, quadrangular platform mounds (called 6, Alpha, Delta, Gamma, 4 and 7) closely arranged in a horseshoe formation and with a height above ground level of 1.75–2.5 m (Fig. 14.4). Two dome-shaped mounds (called Beta and 8) frame the central, oval plaza with a size of 75–50 m. The formal and compact Inner Precinct contrasts with more dispersed and informally arranged peripheral sectors, which include two crescent-shaped raised areas (named TBN and TBS), five circular and

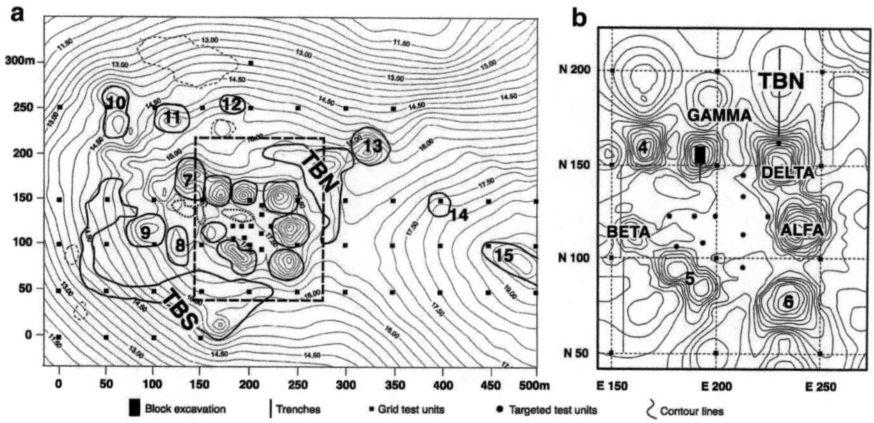

Fig. 14.4 (**a**) Los Ajos site planimetric and topographical map. (**b**) The Inner Precinct (Modified from Iriarte et al. 2004)

three elongated lower dome-shaped mounds, borrow pits and a vast off-mound area bearing subsurface occupational refuse. The TBN crescent-shaped raised area (14–25 m wide and 0.40–0.80 m high) extends over 150 m surrounding Mounds Alpha and Delta. At its base, it becomes wider, extending to the north-east and forming a rounded elongation facing Mound 13.

As part of our project, with the assistance of geomorphologist Juan Montaña, we carried out a detailed study of the locality constituting the wetlands of India Muerta, covering an area of 160 km^2. Building on the experience and knowledge accumulated by previous projects (Bracco and López 1987a, b; Bracco et al. 2000a; Gianotti 2000, 2005), we used 1966–1967 1:20,000 stereoscopic pairs of aerial photographs to discover, characterize and locate mound sites in our study region.

The following methodology was applied. We used stereoscopy to discover mounds on the aerial photographs. All sites identified in the field were then inserted onto the 1:50,000 topographic maps produced by the SGM (Fig. 14.1b), and their geographic coordinates were obtained. An earlier sketch map of the north-eastern sector of the Sierra de los Ajos done by Bracco (1993) was also incorporated. After that we carried out a pedestrian-targeted survey using a portable GPS to check all the information collected in the aerial photographs on the ground. Assisted by geologist Juan Montaña, we were also able to reconstruct many geomorphological features, such as palaeochannels (Fig. 14.5a).

The study paid particular attention to the relationships between mound sites and geomorphological units. Our survey showed that within the southern sector of the Laguna Merín basin, the wetlands of India Muerta contain one of the major aggregations of mound sites, confirming previous studies (Bracco et al. 1999). Our analyses also revealed a number of patterns. The first observable pattern is the association of all mound sites with wetland areas. More specifically, we noticed that mounds are strongly concentrated in two major geomorphological units, the wetland floodplains

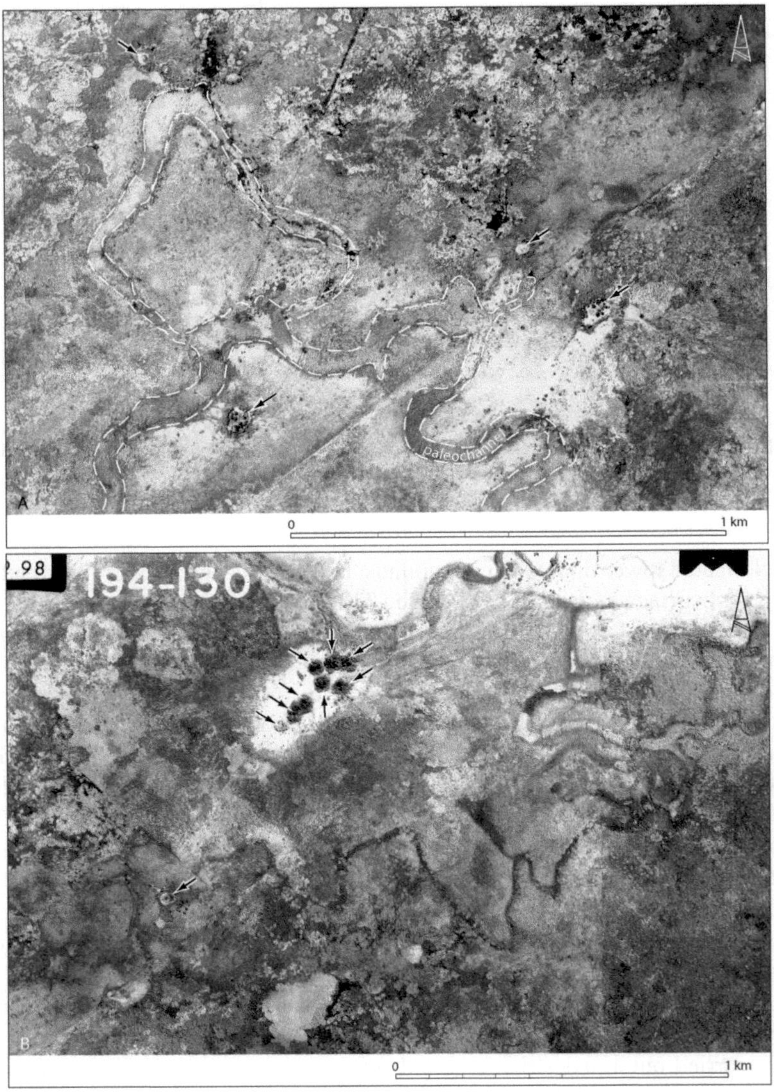

Fig. 14.5 (**a**) Aerial photograph showing palaeochannels (*dashed lines*) and isolated mounds located in the most prominent levees (*arrows*) (Aerial photos 1:20,000, No. 183–205, taken at 3,000 m above sea level). (**b**) Aerial photograph of mound group 5 Islas (Aerial photos 1:20,000, No. 194–130, taken at 3,000 m above sea level)

and the hills and knolls adjacent to these wetlands (see also Bracco et al. 1999; Gianotti 2000, 2005; López and Bracco 1994).

In the wetland floodplains, mounds are generally isolated or in small groups of two to three. In this sector of the landscape, mounds are positioned on top of the most prominent levees, following the courses of streams, displaying a linear or curvilinear pattern (Figs. 14.1b, 14.5a, and 14.6). Aerial photography has also allowed us to trace how mound sites are located along palaeochannels (Fig. 14.5a). Secondly, and in agreement with previous findings by Bracco et al. (1999) and Copé (1991), we found that mounds located in the low wetland floodplains are generally taller (some of them reaching 5–7 m) and smaller in diameter (less than 30 m). In contrast, the earthen structures aggregated in large complexes on the knolls and hills are relatively lower in height and larger in diameter. Thirdly, the largest, more numerous and spatially complex multi-mound sites occur in the more stable sectors of the landscape, such as the flattened spurs of the Sierra de los Ajos (e.g. Los Ajos, Damonte, Campo Alto) (Fig. 14.1b) and the topographical prominences in the wetland floodplains, such as Estancia Mal Abrigo (Fig. 14.6) and Cinco Islas (Fig. 14.5b).

The general location of sites in the region not only allows for immediate access to the rich and abundant wetland resources, but they are also in ecotonal areas. These transitional zones comprise a mosaic of wetlands, wet prairies, upland prairies and riparian and palm forest, which present a greater diversity and abundance of resources than the surrounding areas by themselves. These sites located in higher topographical positions are secure from the seasonal flooding and at the same time allow for immediate access to the resource-rich and fertile wetland areas. Unlike the more unstable locations in the wetland floodplains, which are at times subjected to unpredictable seasonal inundations, these locations were preferred sites for the establishment of larger and more permanent communities. Similar patterns have been documented in the Illinois River (Brown and Vierra 1983), the lower Mississippi valley (Gibson 1994), Amazonia (Denevan 1996) and the lowlands of the Gulf coast (Cyphers 1997), among several other case studies, where in general larger and more permanent communities are located on higher, more stable topographical areas adjacent to rich river floodplain resource zones.

Within these flattened spurs and knolls overlooking the wetlands, there are other regularities that could be observed. For example, large sites are generally located where the active channel of streams impinge on the slopes of the hills and where numerous lagoons or oxbow lakes occur (e.g. Los Ajos). These would have provided pre-Columbian populations with easily accessible water, the excellent fishing grounds that oxbow lakes supply and ready access to the riparian forest resources, which grow along the major lagoons in the area. In addition, this particular area comprises the distal sector of the wetlands of India Muerta alluvial fan. Thus, it was the least susceptible to water stress, most probably retained the highest water table during dry periods compared to other areas of the region, and its soils were periodically replenished with nutrients from the floodwater and overbank flows of the Cebollatí River that inundates the area (Juan Montaña 1999).

Like Los Ajos, these large multi-mound sites not only contain a central formal mounded area surrounding a communal space, but they also display a large peripheral outer sector bearing more dispersed and less formally integrated mounded architecture that may constitute large domestic outer precincts. For example, the Estancia Mal Abrigo site (Fig. 14.6), which was surveyed by the author, is one

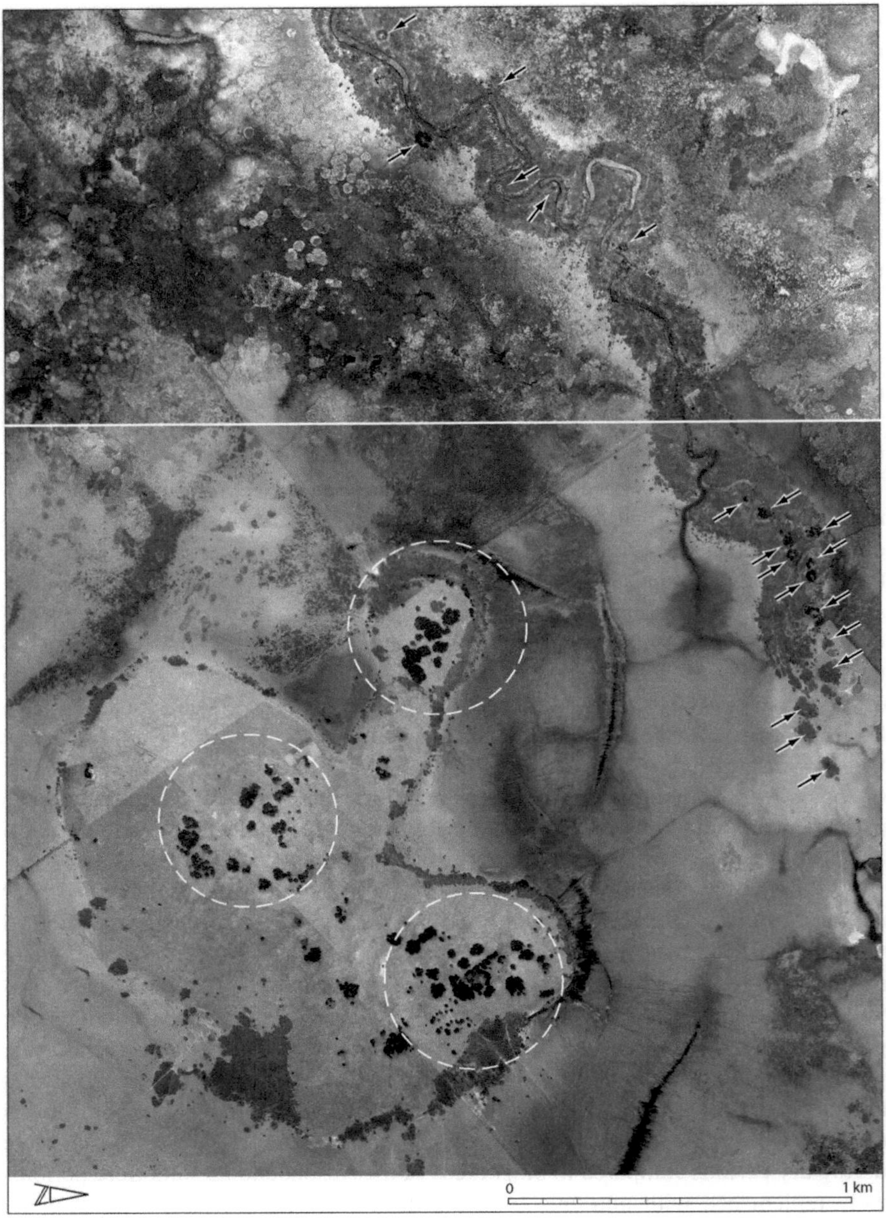

Fig. 14.6 Aerial photograph of Estancia Mal Abrigo mound complexes (*dashed circles*), small mound groups and isolated mounds along the stream. Composite of 1:20,000 aerial photographic stereoscopic pairs (Aerial photos 1:20,000, No. 183–203 and 19–131, taken at 3,000 m above sea level)

example of these large complex sites covering 60 ha and containing 77 mounds (Iriarte et al. 2001: 64, Fig. 6.2). In addition, through a more detailed inspection of the sites in the locality, it became clear that the circular, semicircular and horseshoe geometrical arrangement of mounds, like Los Ajos, represents one of the recurrent spatial groupings within multi-mound sites in the region. Although most of the large mound complexes in the region display the recursive geometrical layout (circular, elliptical and horseshoe), there is also considerable variability not only in the formal structure of the sites but also in the combination, dimensions and shapes of mounds (Bracco et al. 2000c; Gianotti 2000, 2005; López and Pintos 2000).

During our survey, we also discovered other site types. One other major site type represented in the region is the large platform mound that can reach 150 by 50 and 7 m in height, consisting of two well-defined platforms connected by a ramp, as is the case of site Isla de Alberto (Iriarte 2003). These sites are associated with two or three low dome-shaped mounds. Notably, most of these sites are oriented SE-NW with the highest platform located to the NW and the low mound positioned in the SE sector. Another site type is represented by the high conical mounds, as for example at Cerrito de la Viuda (Bracco and Ures 1999), which can attain 7 m in height. Other sites present combinations of circular and linear arrangements of mounds, such as at Cinco Islas (Fig. 14.5b). This type of variation in site plans and mound types has also been observed by Gianotti (2000: 98, Fig. 16) in the wetland of Arroyo Yaguarí. The presence of these distinct mound types, such as the large double-platform mounds and tall conical mounds (30 m in diameter and 7 m high), suggest distinctive site types serving specific functions within an integrated regional settlement system. Future work at a regional level will be able to clarify what is now a rather complicated picture of settlement variability, allowing a more precise understanding of the role that Los Ajos played in the emergence of Early Formative societies in the region.

Our study also corroborates previous patterns found by Bracco et al. (1999) showing a strong correlation between mound complexes and highly fertile floodplain soils, which is reflected in the remarkable correspondence between the larger archaeological sites and the most fertile agricultural lands in the region. This pattern is in agreement with our combined subsistence data showing that, starting shortly after c. 4190 ^{14}C years BP, the Los Ajos inhabitants engaged in a variety of subsistence activities, including hunting, fishing and collecting wild resources, in addition to the adoption of maize, squash and, possibly, domesticated beans (*Phaseolus* spp.) and tubers (*Canna* spp. and *Calathea* spp.) (Iriarte 2003, 2006a, 2007b, 2009; Iriarte et al. 2004). Our study indicates that the ancient use and manipulation of wetlands along lakes and river margins was a much more important and frequent activity than previously thought (see also Blake 1999; Siemens 1999). Wetlands represent an ideal setting for the adoption and intensification of agriculture (Niederberger 1979; Pohl et al. 1996; Sherratt 1980; Siemens 1999). Although the practice of flood-recessional agriculture remains hypothetical, it should be noted that the wetlands of India Muerta represent an ideal scenario for such practice. Unlike other continental areas in Uruguay that experience 25–30 days of frost, the influence of coastal lakes and lagoons and the extensive wetland areas in the southern sector of the Laguna Merin produce a buffer zone which dramatically reduces the occurrence of days of frost. In addition, the upper freshwater wetlands provide greater stability in water supply and, therefore, reduce risks during periods of drought. During the spring and summer months, organic soils are exposed on the wetland margins which constitute a privileged location for the practice of flood-recessional horticulture. The superficial peat horizons are highly fertile, hold moisture and are easy to till. Moreover, the floodwater and the overbank flow of the Cebollatí River that inundates the area periodically replenish these soils with nutrients (Juan Montaña personal communication 1999). Alternatively, floodplains, the tops of levees and, in particular, the back slopes of levees represent an ideal context for the development of a seasonally pulsing agriculture in the wetlands (Siemens 1999). Lastly, the numerous oxbow lakes and abandoned channels occurring in the wetlands floodplain constitute excellent fishing grounds, representing a prime protein resource. In sum, wetlands not only concentrate diverse and abundant wild resources, but also are a good setting for the practice of flood-recessional small-scale horticulture.

14.6 Concluding Remarks

The use of historic aerial photographs in Uruguay has provided a point of departure for the discovery of new sites and has helped recognize the sheer number and complexity of mound sites in the region. Most of the archaeological sites in south-eastern Uruguay were first discovered from the air. More importantly, in a region like our study area, which has been radically transformed by modern development, historical aerial photographs represent an archive containing unique information that is not otherwise accessible. The use of historic aerial photography to look at CDC regional settlement patterns along the mid-south Atlantic coast of South America has still to meet its full potential. Furthermore, reconstructing the past environment will allow us to map archaeological sites in the reconstructed landscape, and carrying out catchment composition analysis will facilitate comparisons of access to resource diversity among the different categories of mound sites.

Detailed topographical mapping and intensive excavation programmes in the region are sorely needed to gain a better understanding of settlement patterns. They are necessary to understand how the CDC people were organized at a regional level and what kind of sociopolitical organization they had (e.g. hierarchical, decentralized or heterarchical). The precocity and apparent distinctiveness of CDC people's settlement types and ceremonial architecture in relation to other Andean and Amazonian Formative processes will make an important contribution to the comparative study of the rise and dynamics of complex societies in lowland South America and elsewhere. In this task, the use of historic aerial photographs will continue to provide a crucial resource in reconnaissance of new regions.

Acknowledgements Research at Los Ajos and the India Muerta wetlands was funded by grants from the National Science Foundation, Wenner-Gren Foundation for Anthropological Research, Smithsonian Tropical Research Institute and the University of Kentucky Graduate School. I also received support from the Comisión Nacional de Arqueología, Ministerio de Educación y Cultura, Uruguay and the Rotary Club of Lascano, Rocha, Uruguay. Sean Goddard from the University of Exeter drafted all the figures. I am also particularly grateful to Oscar Marozzi and Juan Montaña for their insightful comments. I should also like to acknowledge Roberto Bracco, Leonel Cabrera and José López for their pioneering work in aerial photography in south-eastern Uruguay. Their work laid the basis of this chapter. All statements made herein, however, are my own responsibility.

Bibliography

Andrade, T., & López, J. M. (2001). La emergencia de la complejidad cultural entre los cazadores recolectores de la costa Atlántica meridional Sudamericana. *Revista de Arqueología Americana, 17*, 129–167.

Blake, M. (1999). *Pacific Latin America in Prehistory. The Evolution of Archaic and Formative Cultures*. Washington State University Press, Pullman, Washington.

Bracco, R. (1992). Desarrollo cultural y evolución ambiental en la región este del Uruguay. In C. Zubillaga, R. Pi, M. Sans, R. Bracco, J. M. López Mazz, L. Cabrera, C. Curbelo, M. Martínez, E. Mazzolini, S. Romero, S. Rostagnol, & A. J. Lezama (Eds.), *Ediciones del Quinto Centenario, Vol. 1: Estudios Antropológicos* (pp. 43–73). Montevideo: Facultad de Humanidades y Ciencias de la Educación, Universidad de la República.

Bracco, R. (1993). *Proyecto arqueología de la cuenca de la Laguna Merín*. Unpublished report presented to PROBIDES. Rocha: Uruguay.

Bracco, R. (2006). Montículos de la cuenca de la Laguna Merín: tiempo, espacio, y sociedad. *Latin American Antiquity, 17*, 511–540.

Bracco, R., & López, M. (1987a). *Prospección arqueológica y análisis de foto aérea (Bañado de la India Muerta y Bañado de san Miguel, Depto. de Rocha)*. Primeras Jornadas de Ciencias Antropológicas en Uruguay (pp. 51–56). Montevideo: Ministerio de Educación y Cultura.

Bracco, R., & López, M. (1987b). *Rescate Arqueológico en la Cuenca de la Laguna Merin: Informe de la Etapa de Prospección*. Unpublished manuscript, Montevideo.

Bracco, R., & Ures, C. (1999). Ritmos y dinámica constructiva de las estructuras monticulares: sector sur de la cuenca de la Laguna Merín-Uruguay. In J. M. López & M. Sans (Eds.), *Arqueología y Bioantropología de las Tierras Bajas* (pp. 13–33). Montevideo: Facultad de Humanidades y Ciencias, Universidad de la República.

Bracco, R., Marozzi, O., Orsi, L., & Castillo, A. (1999). *Suelos y "Cerritos". Nuevas Perspectivas en el Análisis de los Montículos del Este del Uruguay*. Manuscript on file. Montevideo: Departamento de Antropología, Facultad de Humanidades y Ciencias de la Educación.

Bracco, R., Cabrera, L., & López, J. M. (2000a). La prehistoria de las tierras bajas de la cuenca de la Laguna Merín. In A. Durán & R. Bracco (Eds.), *Arqueología de las Tierras Bajas* (pp. 13–38). Montevideo: Ministerio de Educación y Cultura, Comisión Nacional de Arqueología.

Bracco, R., Montaña, J., Bossi, J., Panarello, H., & Ures, C. (2000b). Evolución del humedal y ocupaciones humanas en el sector sur de la cuenca de la Laguna Merín. In A. Durán & R. Bracco (Eds.), *Arqueología de las Tierras Bajas* (pp. 99–116). Montevideo: Ministerio de Educación y Cultura, Comisión Nacional de Arqueología.

Bracco, R., Montaña, J., Nadal, O., & Gancio, F. (2000c). Técnicas de construcción y estructuras monticulares. Termiteros y cerritos: desde lo analógico a lo estructural. In A. Durán & R. Bracco (Eds.), *Arqueología de las Tierras Bajas* (pp. 285–300). Montevideo: Ministerio de Educación y Cultura, Comisión Nacional de Arqueología.

Brown, J. A., & Vierra, K. (1983). What happened in the Middle Archaic? Introduction to an ecologic approach to Koster site archaeology. In J. L. Philips & J. A. Brown (Eds.), *Archaic hunters and gatherers in the American Midwest* (pp. 201–230). Orlando: Academic.

Copé, S. M. (1991). A ocupação pré-Colonial do sul e sudeste do Rio Grande do Sul. In A. Kern (Ed.), *Arqueologia Prehistorica do Rio Grande do Sul* (pp. 191–211). Porto Alegre: Mercado Aberto.

Criado, F., Gianotti, C., & Manana, P. (2006). Before the barrows: Forms of monumentality and forms of complexity in Iberia and Uruguay. In L. Šmejda (Ed.), *Archaeology of burial mounds* (pp. 38–51). Plzen: Vlasta Králová.

Cyphers, A. (1997). Olmec architecture at San Lorenzo. In B. L. Stark & P. A. Arnold (Eds.), *Olmec to Aztec: Settlement patterns in the ancient gulf lowlands* (pp. 96–114). Tucson: University of Arizona Press.

Denevan, W. M. (1996). A bluff model of riverine settlement in prehistoric Amazonia. *Annals of the Association of American Geographers, 86*(4), 654–681.

Durán, A., & Bracco, R. (2000). *Arqueología de las Tierras Bajas*. Montevideo: Ministerio de Educación y Cultura, Comisión Nacional de Arqueología.

Gianotti, C. (2000). Monumentalidad, ceremonialismo, y continuidad ritual. In C. Gianotti (Ed.), *Paisajes Culturales Sudamericanos: De las Practicas Sociales a las Representaciones* (pp. 87–102). Galicia: Laboratorio de Arqueoloxia de Paisaxe.

Gianotti, C. (2005). *Desarrollo Metodológico y Aplicación de Nuevas Tecnologías para la Gestión Integral del Patrimonio Arqueológico en Uruguay*. Galicia: Laboratorio de Arqueoloxia de Paisaxe.

Gibson, J. L. (1994). Before their time? Early mounds in the lower Mississippi valley. *Southeastern Archaeology, 13*(2), 162–187.

Iriarte, J. (2003). *Mid-Holocene emergent complexity and landscape transformation: The social construction of early formative communities in Uruguay, La Plata Basin*. Unpublished PhD dissertation, University of Kentucky, Lexington.

Iriarte, J. (2006a). Landscape transformation, mounded villages, and adopted cultigens: The rise of early Formative communities in south-eastern Uruguay. *World Archaeology*. doi:10.1080/00438240600963262.

Iriarte, J. (2006b). Vegetation and climate change since 14,810 ^{14}C yr BP in southeastern Uruguay and implications for the rise of early Formative societies. *Quaternary Research*. doi:10.1016/j.yqres.2005.05.005.

Iriarte, J. (2007a). La construcción social y transformación de las comunidades del Formativo Temprano del sureste de Uruguay. *Boletín de Arqueología PUCP, 11*: 143–166. Lima: Pontificia Universidad Católica del Perú.

Iriarte, J. (2007b). Emerging food-production systems in the La Plata Basin. In T. Deham, J. Iriarte, & L. Vrydaghs (Eds.), *Rethinking agriculture: Archaeological and ethnoarchaeological perspectives* (pp. 256–272). Walnut: Left Coast Press.

Iriarte, J. (2009). Narrowing the gap: Exploring the diversity of early food producing economies in the Americas. *Current Anthropology*. doi:10.1086/605493.

Iriarte, J., & Alonso, E. (2009). Phytolith analysis of selective native plants and modern soils from southeastern Uruguay and its implications for paleoenvironmental and archaeological reconstruction. *Quaternary International*. doi:10.1016/j.quaint.2007.10.008.

Iriarte, J., Holst, I., López, J. M., & Cabrera, L. (2001). Subtropical wetland adaptations in Uruguay during the Mid-Holocene: An archaeobotanical perspective. In B. Purdy (Ed.), *Enduring records: The environmental and cultural heritage of wetlands* (pp. 61–70). Oxford: Oxbow Books.

Iriarte, J., Holst, I., Marozzi, O., Listopad, C., Alonso, E., Rinderknecht, A., & Montaña, J. (2004). Evidence for cultivar adoption and emerging complexity during the mid-Holocene in the La Plata Basin. *Nature*. doi:10.1038/nature02983.

López, J. M. (2001). Las estructuras tumulares (Cerritos) del litoral Atlántico Uruguayo. *Latin American Antiquity, 12*, 231–255.

López, J. M., & Bracco, R. (1992). Relaciones hombre-medio ambiente en las poblaciones prehistóricas del este del Uruguay. In O. Ortiz-Troncoso & T. Van der Hammen (Eds.), *Archaeology and environment in Latin America* (pp. 259–282). Amsterdam: Instituut voor Pre- en Protohistorische Archeologie Albert Egges van Giffen, Universiteit van Amsterdam.

López, J. M., & Bracco, R. (1994). Cazadores-recolectores de la Cuenca de la Laguna Merín. *Arqueología Contemporánea, 5*, 51–64.

López, J. M., & Pintos, S. (2000). Distribución espacial de estructuras monticulares en la Cuenca de la Laguna Negra. In A. Durán & R. Bracco (Eds.), *Arqueología de las Tierras Bajas* (pp. 49–58). Montevideo: Ministerio de Educación y Cultura, Comisión Nacional de Arqueología.

Meggers, B. J., & Evans, C. (1978). Lowland South America and the Antilles. In J. Jennnings (Ed.), *Ancient native Americans* (pp. 543–591). San Fransisco: Freeman.

Montaña, J., & Bossi, J. (1995). *Geomorfología de los Humedales de Humedales de la Cuenca de la Laguna Merín en el Depto. de Rocha*. Montevideo: PROBIDES.

Niederberger, C. (1979). Early sedentary economy in the basin of Mexico. *Science, 203*, 131–142.

Pohl, M. D., Pope, K. O., Jones, J. G., Jacob, J. S., Piperno, D. R., de France, S. D., Lentz, D. L., Giffford, J. A., Danforth, M. E., & Josserand, K. (1996). Early agriculture in the Maya lowlands. *Latin American Antiquity, 7*(4), 355–372.

Prieto, O., Alvarez, J., Arbenoiz, J., de Los Santos, A., Vesidi, P., Schmitz, P. I., & Basile Becker, I. (1970). *Informe Preliminar sobre las Investigaciones Arqueológicas en el Depto. de Treinta y Tres. R.O.U.* Sao Leopoldo: Instituto Anchietano de Pesquisas/UNISINOS.

PROBIDES Plan Director. (2000). *Reserva de Biosfera Bañados del Este*. Rocha: PROBIDES.

Schmitz, P. I., Naue, G., & Becker, I. (1991). Os aterros dos campos do sul: a tradicão vieira. In A. Kern (Ed.), *Arqueologia Prehistorica do Rio Grande do Sul* (pp. 221–251). Porto Alegre: Mercado Aberto.

Sherratt, A. (1980). Water, soil, and seasonality in early cereal cultivation. *World Archaeology, 11*, 313–328.

Siemens, A. H. (1999). Wetlands as resource concentrations in southeastern Ecuador. In M. Blake (Ed.), *Pacific Latin America in Prehistory. The Evolution of Archaic and Formative Cultures* (pp. 13–147). Pullman: Washington State University Press.

Stalh, P. (2004). Greater expectations. *Nature, 432*, 561–562.

Steward, J. (1946). *Handbook of South American Indians, Vol 1, The marginal tribes*. Washington, DC: Smithsonian Institution.

Tomazelli, L. J., & Villwock, J. A. (1996). Quaternary geological evolution of Rio Grande do Sul coastal plain, southern Brazil. *Annais da Academia Brasileira de Ciencias, 68*(3), 373–381.

Chapter 15
The Archaeological Exploitation of Declassified Satellite Photography in Semi-arid Environments

Anthony R. Beck and Graham Philip

Abstract Declassified satellite photographs are becoming an increasingly important archaeological tool. Not only are they useful for residue prospection and, when in stereo pairs, digital elevation model (DEM) generation, they can also provide large-scale temporal snapshots that provide essential information on landscape change. Importantly, in some instances, declassified photographs may be the only available record of archaeological residues that have subsequently been eradicated.

This chapter outlines a generic approach to accessing, digitising and processing declassified satellite photographs and utilising them in conjunction with modern fine-resolution satellite images. The methodological issues of acquisition and pre-processing are addressed. A number of potential archaeological applications are described and illustrated with examples from the Settlement and Landscape Development in the Homs Region, Syria (SHR) project. These examples demonstrate that there is no single approach to processing and image selection. Rather, processing is dependent upon the nature of the archaeological residues and their surrounding matrix, the type of analysis one wants to undertake and the range of ancillary datasets which can be used to 'add value' to the source data.

A.R. Beck (✉)
School of Computing, University of Leeds, LS2 9JT Leeds, UK
e-mail: A.R.Beck@leeds.ac.uk

G. Philip
Department of Archaeology, University of Durham, DH1 3LE Durham, UK
e-mail: graham.philip@durham.ac.uk

W.S. Hanson and I.A. Oltean (eds.), *Archaeology from Historical Aerial and Satellite Archives*, DOI 10.1007/978-1-4614-4505-0_15,

15.1 Introduction

Aerial photography is the oldest form of remote sensing in archaeology. Historically, bespoke archaeological remote sensing has been based on low-altitude aerial survey using handheld cameras with films sensitive to the optical and near infrared (Wilson 2000). However, traditional aerial photography is not without its problems. The reliance on a small component of the electromagnetic spectrum raises a number of issues. The small spectral window can induce a significant bias as only certain residues under specific conditions express contrasts in these wavelengths (e.g. aerial photography in the UK is broadly recognised not to work in clay environments). In areas that have been intensively studied, such as the UK, a point of saturation can be reached. This could mean that increasingly extreme environmental conditions may be required to detect new archaeological residues (with climate change, this may be an unfortunate reality). On a more pragmatic note, some countries operate a closed skies policy or impose severe bureaucratic challenges in obtaining the necessary permits. Finally, traditional archaeological aerial photography collects many localised photographs in a predominantly unsystematic way influenced by what is seen by the observer (Cowley 2002).

Photographs and digital images[1] collected from a satellite platform have the ability to address a number of these drawbacks. Satellites, by their very nature, represent open skies collection systems and can have fine resolving characteristics (Fowler 2004 and Chap. 4, this volume). Furthermore, in common with conventional aerial imaging, satellite images have a large synoptic footprint and are therefore more systematic and exhaustive approaches to survey.

Archaeologists were quick to spot the potential of the first public Earth observation satellites in the 1970s (e.g. Lyons and Avery 1977). Initially, however, imagery from publically available sensors, such as Landsat, was not only expensive but had a ground resolution which was too coarse for thorough archaeological prospection. Though archaeological features could be detected, the vast majority was already well documented. This situation changed in the mid-1990s for two important reasons: the declassification of high-resolution photographs by the American and Russian governments and the deregulation of commercial remote sensing systems allowing the collection of sub-metre resolution images. The availability of images with a ground resolution approaching that of traditional aerial photographs had the potential to revolutionise archaeological prospection. This is particularly true for areas where the archaeological resource is poorly understood or documented.

In the nearly 15 years since the first 'spy' satellite photographs were declassified, there have been several changes, including the declassification of other programmes, a better understanding of the potential of the resource and improvements in how the

[1] Photograph is used explicitly throughout this chapter to refer to a film (analogue) product. The term image refers explicitly to a digital product.

resource is accessed and disseminated. This chapter aims to provide an overview of the processing and applications of declassified satellite imagery for archaeological purposes, with descriptive exemplars from a semiarid environment.

15.2 Declassified Satellite Photography

At the start of the 'Cold war', both the American and Russian governments conducted photographic reconnaissance from manned 'spy' planes and unmanned adapted V-2 rockets. Between the 1960s and the early 1990s, researchers in each country developed dedicated military reconnaissance systems. Initially, these systems were camera based, requiring short-term missions (days to weeks) and an ability to redeploy captured photography back to Earth. From the mid-1970s, the US military adopted electro-optical imaging for new programmes. These sensors could be placed in stable orbits, and the digital images were electronically relayed to Earth. The level of secrecy was extremely high and all the resultant photographs and images were classified (Kramer 1996).

In 1992, Russia started selling selected photographs from their fine-resolution (2 m) KVR-1000 camera. In February 1995, President Clinton declassified the first generation of American reconnaissance satellites (programmes active between 1960 and 1972), with a second round of declassification in 2002 (Fowler 2004 and Chap. 4, this volume). Though the following sections focus on the use of American declassified satellite photography from the CORONA programme available from the United States Geological Survey (USGS), whose characteristics are summarised in Table 15.1, the majority of the processing techniques described are generic and would apply to other declassified image sets (such as the Russian KVR-1000 imagery) or to any photographic datasets that may be declassified in the future.

15.2.1 Digitising Declassified Photography

Although most users will purchase pre-digitised satellite photography through the USGS Earth Explorer web interface (http://edcsns17.cr.usgs.gov/EarthExplorer), some may want to digitise copies of film sourced from the National Archives and Records Administration (NARA). Alternatively, other 'spy' satellite programmes may be declassified in the future and only made available as film that requires digitising. As it is likely that the resultant digital facsimile will be employed in a variety of quantitative processes, the spatial and radiometric fidelity of the scanning process is of paramount importance. For example, if one wanted to generate stereo elevation models, then a high-quality initial scan is essential. Therefore, one should determine an appropriate scanning methodology appropriate for any subsequent analytical tasks (Philip et al. 2002a; Ur 2002).

Table 15.1 CORONA Keyhole camera mission characteristics

	KH-1	KH-2	KH-3	KH-4	KH-4A	KH-4B
Period of operation	27/6/59-13/9/60	26/10/60-23/10/61	30/8/61-13/1/62	27/2/62-24/3/64	24/8/63-22/9/69	15/9/67-25/5/72
Amount of frames	1,432	7,246	9,918	101,743	517,688	188,526
Mission life (days)	1	2–3	1–4	6–7	4–15	19
Lower altitude (estimated in km)	192	252	217	211	180	150
Higher altitude (estimated in km)	817	704	232	415	n/a	n/a
Successful missions	1	3	5	20	49	16
Targets	USSR	Emphasis on USSR		World-wide/emphasis on denied areas		
Aperture width	5.265°	5.265°	5.265°	5.265°	5.265°	5.265°
Pan angle	71.16°	71.16°	71.16°	71.16°	71.16°	71.16°
Stereo angle				30°	30°	30°
Lens	F/5.0 Tessar	F/5.0 Tessar	F/3.5 Petzval	F/3.5 Petzval	F/3.5 Petzval	F/3.5 Petzval
Focal length (cm)	61	61	61	61	61	61
Ground resolution (ft)	40	25	12–25	10–25	9–25	6–25
Film (lp/mm)	50–100	50–100	50–100	50–100	120	160
Nominal ground coverage image frame (km)	15.3×209 to 42×579	15.3×209 to 42×579	15.3×209 to 42×579	15.3×209 to 42×579	17×232	13.8×188
Nominal photoscale in film	1:275,000 to 1:760,000	1:275,000 to 1:760,000	1:275,000 to 1:760,000	1:300,000	1:305,000	1:247,500

After Kramer 1996 and Galiatsatos 2004

For example, on the project 'Settlement and Landscape Development in the Homs Region, Syria (SHR)', described below, the following approach was taken. The project team wanted to generate elevation models and conduct other quantitative analyses on the CORONA imagery and hence required a high-quality digitising process. As the CORONA photography has a nominal resolution of 160 lp/mm (see Table 15.1), a full-resolution scan would require a 3-μm (c. 8,000 dpi) scanner. The most appropriate available option was a high-resolution photogrammetric scanner (a Vexcel VX4000). This scanner has a high-resolution (7.5 μm) scan head without interpolation, a geometric accuracy of 1/3 pixel RMSE and a radiometric accuracy (in 8 bit) of 2 digital number RMSE. The resultant imagery had a nominal ground resolution of approximately 2 m.

15.2.2 Geo-referencing Declassified Imagery

Prior to any rectification or data collection procedure, a projection system must be determined. In most areas that have institutionalised cultural resource management (CRM) bodies, the regional or national projection system is easily accessible. It is advisable (and in some instances mandatory) that this projection system is used. This will ensure that any results will integrate seamlessly with the national CRM data and other datasets enabling subsequent data reuse and integration (Bewley et al. 1999).

Where such a system does not exist, then it is advisable to use one of the standard worldwide referencing systems such as Universal Transverse Mercator (UTM) or lat/long projections (both standard worldwide reference projections) and an appropriate datum (if in doubt, use WGS84).

UTM is more intuitive for in-field work than lat/long (units are metres as opposed to seconds of arc) and is in use within many CRM databases (cf. Palumbo 1992).

Rectification is the process of correcting systematic and random errors in imagery. Rectification procedures can either be spatial, to geolocate an image, or non-spatial, to remove scanning or camera aberrations. Spatial rectification relies on the ability to recognise objects within the imagery that have known coordinates. These objects are referred to as ground control points (GCPs) or tie-points. These points can be derived from topographic mapping, global navigation satellite systems (GNSS: such as GPS) or other remote sensing imagery. It is advisable to employ only a single data source as otherwise complicated error propagation issues could arise.

As declassified imagery is, by necessity, historic, there are likely to have been a range of different modifications that can make GCP identification ambiguous. In this respect, the use of remote sensing images as a rectification source has significant advantages: the interpreter is aware of the localised context of any GCP and can make a value judgement on quality and fitness for purpose.

Prior to rectification, the spatial accuracy requirements of the resultant image must be established. The accuracy is dependent upon the end use of the imagery. If the aim of the geo-referenced image is to facilitate detection, characterisation and approximate location, then high positional accuracy is not required. However, if mapping is derived from the imagery, then high positional accuracy is required in order to ensure that the mapping and field survey results correlate.

Finally, co-registration issues should be considered. Co-registration is the geo-referencing of two different images so that each overlaying pixel corresponds to the same location. Due to the errors associated with rectification, it is rare for this to occur by accident. Accurate co-registration is important for some time change analyses and photogrammetric extraction of elevation models from stereo pairs.

For the SHR project, the declassified photography was geo-referenced using 1-m ground resolution Ikonos imagery as a reference source. The comparable spatial resolution of the imagery allowed the confident determination of appropriate tie-points. The reported nominal accuracy for the Ikonos Geo-product™ was 23.3 m (Gerlach 2000): empirical GPS field measurements confirmed this value. Rectification trials using GPS tie-points derived from points of hard detail (such as bridge and road junctions) produced better accuracy rectifications. For stereo pairs, the second image was co-registered directly to the contemporaneous primary image.

15.2.3 The Archaeological Application of Declassified Photography

Fine spatial resolution declassified satellite photography has a number of archaeological applications. The primary one is that of archaeological prospection, for which, due to destructive modification, historic photography may be the only available resource. However, there are also other reasons for using such imagery. Many declassified satellite programmes had stereoscopic viewing capabilities. This allows the generation of elevation models. These may allow the identification of archaeological phenomena not directly visible in the photograph (e.g. hollow ways and other subtle topographic features). The fact that the imagery is historic can provide insights into any modifications that have occurred within the landscape and potentially quantify their impact on the archaeological resources. Time change analyses can also reveal other important environmental and anthropogenic factors that may impact on the management of the cultural resource. Finally, one should not underestimate the use of digital imagery for site navigation and field interpretation. When imagery is incorporated within a GIS-based recording and curation system, a number of synergies can be exploited. For example, when in the field, access to an overhead perspective of the residues can significantly clarify contextual ambiguities and improve recording and interpretation.

15.3 Archaeological Prospection

Unlike the majority of mainstream remote sensing specialists, archaeologists cannot rely on explicit spectral signatures to identify archaeological residues. Rather, it is hypothesised that archaeological residues produce localised contrasts in the landscape matrix which can be detected using an appropriate sensor under appropriate conditions. Although this statement sounds self-evident, it requires an understanding of the dynamics of both the nature of the residues and the landscape matrix within which they reside.

Once this data has been acquired, then physical, chemical and biological models can be developed to help understand how archaeological contrast may be expressed in different areas of the electromagnetic spectrum and under what conditions this contrast is most identifiable. With these models, one can then determine what type of sensors have the resolving capacity to detect identified contrasts. An appropriate sensor is one which has the following appropriate characteristics:

- Spatial resolution that allows the interpreter to identify the spatial structure of the object
- Spectral resolution that records reflectance in the area of the electromagnetic spectrum where the contrast is expressed
- Radiometric resolution that has the sensitivity to discriminate the contrast difference between the object and its surrounding matrix
- Temporal resolution that collects the imagery when the contrast is expressed

Spatial and spectral resolution is generally well understood by an archaeological audience. However, radiometric and temporal resolutions require further explanation. Radiometric resolution is particularly important for prospection as it describes the subtlety of the sensor measurements which, in part, determines whether an object can be detected. For example, if two panchromatic sensors with exactly the same spatial and spectral resolution, but different radiometric resolutions, take a digital image of the same object from the same location (within the shadow of a building) at the same time, only the sensor with the finer radiometric resolution can be used to differentiate the object from the shadow. However, because under normal viewing conditions the human eye can discriminate only between 20 and 30 shades of grey,[2] it is unlikely that the brain would be able to detect the object even though it exists numerically within the structure of the data. It is only with appropriate contrast manipulation of the finer radiometric resolution image that the object becomes apparent. This is analogous in trying to detect centimetre variations with one ruler that rounds measurements to the nearest millimetre and a different ruler that rounds measurements to the nearest decimetre. As archaeological residues commonly represent subtle shifts in reflectivity, much important archaeological information contained within an image can often go undetected.

Temporal resolution refers to how often a sensor system records a particular area. For all platforms except satellites in a fixed orbit, this value is likely to be infrequent. However, satellite images tend to cover the same area at the same time of day, whereas all other sensor platforms can cover an area at different times of day. This is particularly significant for some forms of contrast which occur at different times

[2] This figure is under debate. However, it is true to say that the brain can distinguish far fewer shades of grey than colours.

of day (such as shadow marks or diurnal temperature variations) or under specific, temporally constrained, conditions.

For declassified imagery purchased directly from USGS, all the axes of resolution are constrained. The temporal resolution is fixed: the majority of CORONA and GAMBIT mission were in sun-synchronous orbits with local collection times of between 10:00 and 14:00. However, some missions were scheduled at different times to exploit specific phenomena. The spatial resolution is dependent on the platform, camera system, orbital characteristics and scanning technique, but for CORONA KH-4b, it is nominally 2 m. The spectral resolution varies with the film which is generally agreed to be sensitive to the visual and near-infrared wavelengths. The radiometric resolution is a function of the film and scanning process. For USGS imagery, this is 8 bit or 256 distinct values. However, the film has a nominal 12 bit (4096 value) depth (USGS 2010, personal communication), and Leachtenauer et al. (1998) have scanned film at this detail. Hence, there is data loss in the digital USGS product due to the scanning process.

15.4 The 'Settlement and Landscape Development in the Homs Region, Syria' (SHR) Project

The previous sections have discussed some of the technical details surrounding the photographic sensors, data acquisition and data processing. The following section describes the use of declassified satellite photography on one project in Syria. Though the utility of declassified satellite photography in the region has been demonstrated by several authors (Kennedy 1998; Kouchoukos 2001; Stone 2003; Ur 2003; Challis et al. 2004; Challis 2007), the SHR project was one of the first projects to conduct in-depth research into the archaeological potential of fine (high) spatial resolution satellite imagery. Declassified photography, particularly CORONA imagery, has played a pivotal role in the detection, interpretation, management and long-term understanding of the different archaeological landscapes in the SHR environs.

In common with many other areas of the world, the SHR project is working in a data-poor environment. The extant archaeological inventory for the study area is biased towards large 'monumental' archaeological sites, such as tells (Rosen 1986), a settlement form characteristic of many parts of the Middle East. Other, generally smaller, settlement and land management components of the landscape are under-represented. Furthermore, although available, there was difficulty in acquiring appropriate datasets which are commonly used to contextualise interpretations, such as contour, topographic and soil maps. Following preliminary field visits, a remote sensing programme was introduced in 1999 to identify archaeological residues as part of a site prospection programme and to generate landscape themes that were previously unavailable to the project.

All reasonable quality CORONA KH-4b mission photographs intersecting the study area were purchased prior to analysis. As the photographs were going to be used for a range of quantitative applications including DEM extraction, the digitising strategy described earlier was adopted. The resultant CORONA images were used in conjunction with bespoke high-resolution panchromatic and multispectral Ikonos satellite images. The time frame for Ikonos data capture was determined for each environmental zone independently, based upon models of peak archaeological contrast. The pre-purchased CORONA imagery was particularly useful for testing model validity. Finally, it must be emphasised that, although CORONA is a useful tool in its own right, its full potential was only realised when used in conjunction with the more recent Ikonos images.

15.4.1 The Environmental Makeup of the Homs Region

The SHR project was designed to investigate long-term human-landscape interaction in adjacent but contrasting environmental zones, located in the upper Orontes Valley near the present-day city of Homs, Syria (Philip et al. 2002b, 2005; Beck et al. 2007a). Each zone is typical of a larger area, and initial study suggested that they differed substantially in both their settlement histories and in the nature of their archaeological records. There are two principal zones, basalt (140 km^2) and marl (370 km^2). The marl zone is a relatively flat landscape developed on lacustrine marls of Upper Miocene-Pliocene date (Wilkinson et al. 2006). It is an eroding terrace sequence sloping down to the river Orontes (Bridgland et al. 2003). Aggradation in this zone means that the majority of archaeological residues will be on or very near the surface (Wilkinson et al. 2006). The only buried deposits are likely to be under tells or other areas of long-term occupation. The basalt zone has a series of low boulder-strewn plateaus, interspersed with shallow colluvium-filled valleys and depressions. Since the mid-1970s, the region has experienced moderate expansion of settlement. Of greater impact has been the use of agricultural machinery: the deep plough in the marl zone and the bulldozer to clear fields in the basalt zone.

In the marl zone, the majority of the archaeological residues take the form of tells and low relief soilmark sites (Fig. 15.1: 197, 256, 308, 454). Tells are prominent landscape features and, unless heavily eroded, are easy to detect, so that the majority have already been mapped and recorded. These reflect mainly the settlement record of the Bronze through to Hellenistic periods. On the other hand, soilmark sites, many of which date between the Roman and Islamic periods are very difficult to spot on the ground and, when identified at all, have traditionally been located using intensive surface survey programmes.

In the basalt zone, the archaeological residues take the form of cairns, field walls and concentrations of rubble which constitute the remains of abandoned structures (Fig. 15.1: A, B, D). For an initial morphological classification of such structures, see Philip et al. (2005), subsequently modified by Philip and Bradbury (2010: 141–142) and Philip et al. (2011: 40–42), following more extensive field observation. The smallest of these features are stone alignments with a width of less than 1 m, which in some cases, may project only a few tens of centimetres above the present ground surface.

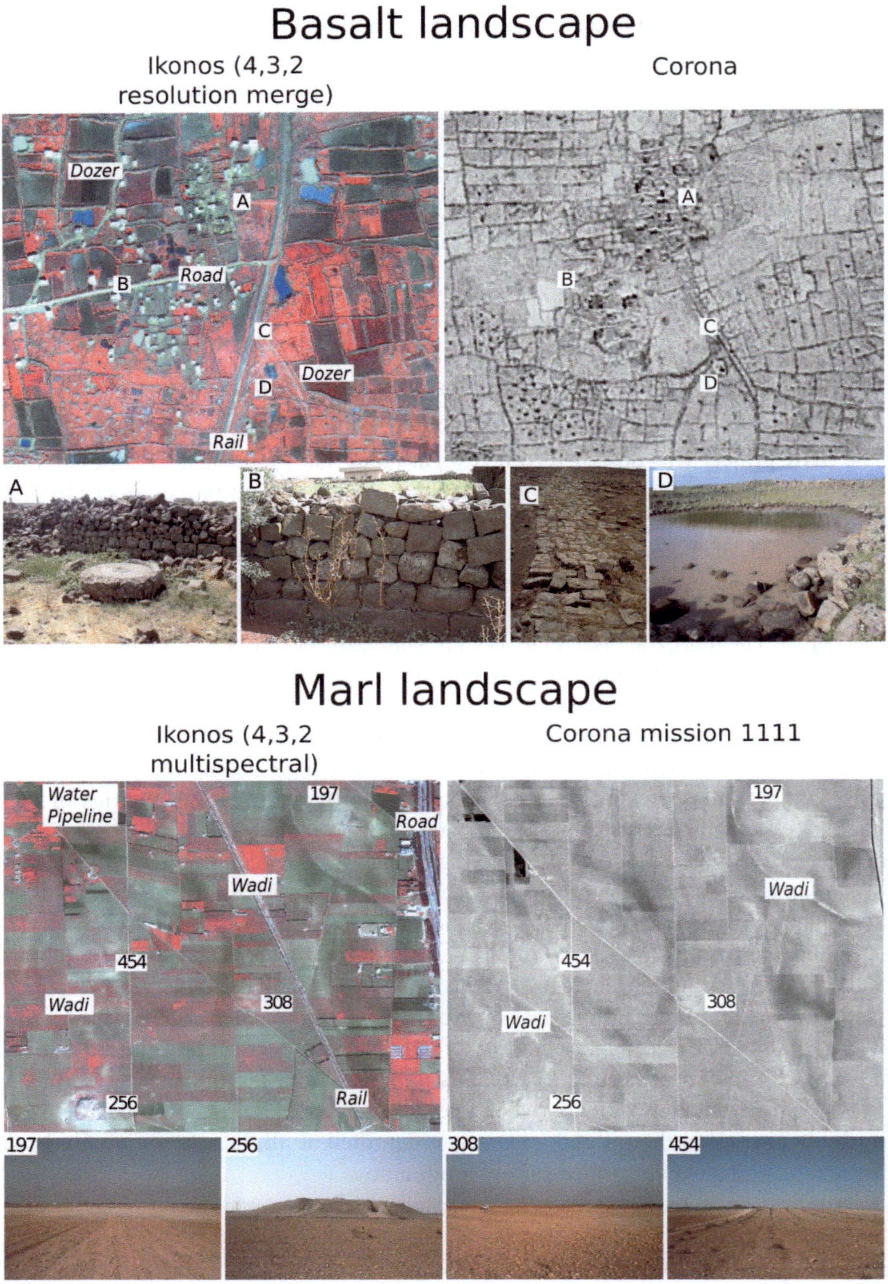

Fig. 15.1 Archaeological residues in the study area as observed in imagery from both Ikonos (*left-hand side*) and CORONA (*right-hand side*) satellites. The Basalt zone (*upper* half of the figure) containing fields, cairns and structures contrasts with the soil mark sites and tells in the Marl zone. Residues photographed in the field are illustrated below for both zones, the letters or numbers corresponding with locations on the satellite imagery

15.4.2 Prospection in the Basalt Zone

With comparatively stable soils with little indication for either soil erosion or sediment aggradation, the basalt zone contains a complex multi-period palimpsest of archaeological structures. Features are primarily constructed from locally sourced basalt (i.e. they have a similar spectral signature to the background basalt soils). In order to detect archaeological features in this environment, the following are relied upon: topographic effects, which might produce contrast through shadows; and spectral response, in the form of tone or texture differences between structures and soil or vegetation.

The width of the smallest archaeological features (c. 0.5 m) necessitated the use of fine spatial resolution imagery. For the mapping of small features, such as field walls, image fidelity needs to be high. Hence, the summer months (May to September) were to be avoided, when airborne particulates increase specular reflection and, therefore, decrease spatial resolution.

Residue detection in the basalt zone was relatively straightforward (Fig. 15.2). Even though the smallest feature size was much less than the resolving power of the sensors, the structural continuity of features and shadows meant that the features were readily detectable. Both the CORONA and Ikonos imagery were used for visual interpretation and mapping. Due to its finer spatial resolution, the pan-sharpened Ikonos imagery provided the best resource for mapping. Metric measurements and identification were also more accurate. However, there were fewer landscape modifications of the kind that were likely to hinder interpretation in the CORONA imagery. The synergies obtained by using both data sources together confirm that in combination they provided a better resource for archaeological interpretation than did either dataset alone.

15.4.3 The Benefits of Temporal Resolution in the Basalt Zone

The basalt zone is a landscape which is under significant threat. In the past 30 years, enhancements to the road and rail networks, and the concomitant increase in associated settlement activity (cf. Sever 1998), have removed archaeological features. Even more significant is the clearance of fields, walls and cairns by bulldozing as part of agricultural improvement schemes. Hence, historical imagery provides a view of the basalt zone prior to modern destruction (Fig. 15.1). The CORONA imagery predates the major phase of bulldozing and has recorded a landscape with minimal destruction, disturbance or masking of archaeological residues by present-day agricultural or settlement expansion. This is a potentially fortuitous set of circumstances as the application area lies in a region which was considered militarily sensitive and over which a relatively large number of CORONA missions had been flown.

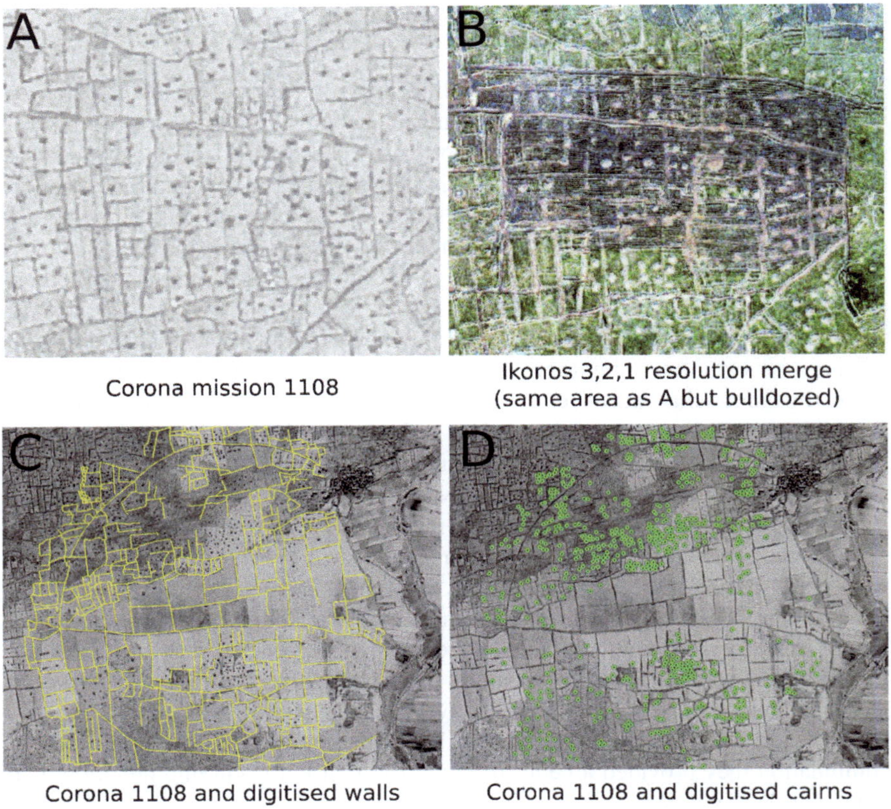

Fig. 15.2 Prospection evidence in the basalt zone. (**A**) and (**B**) show the same area in 1969 and 2002, respectively, indicating the impact of bulldozing on the cairns and field systems. (**C**) and (**D**) show the same area with overlays of walls and cairns, respectively

15.4.4 Improving Geo-referencing Accuracy

The basalt zone represents a complex palimpsest containing many archaeological residues in close proximity. The result was that even using handheld GPS and a printout of CORONA or Ikonos imagery, surveyors found it nearly impossible to establish a one-to-one correspondence between the majority of features appearing on the imagery and those visible on the ground. Thus, while CORONA offered a useful means of mapping the landscape as a whole, its value for more localised survey was constrained because of the difficulty of feature identification. A way needed to be found to improve the spatial accuracy of the Ikonos and CORONA rectification.

Fraser et al. (2002) demonstrated that the positional accuracy of the Ikonos Geo-product™ could be increased to sub-metre levels by using tie-points located by differential GPS. Using handheld GPS, the Ikonos imagery was re-geocorrected with a nominal accuracy of 5–8 m. The CORONA imagery was geo-referenced to this re-rectified Ikonos, retaining approximately the same error. The resulting greater degree of accuracy in the imagery allowed desk-based mapping and subsequent field navigation to be undertaken with improved confidence. It is clear from Fig. 15.2 that accurate geo-referencing is required in order to relocate digitised features due to their sheer number in the basalt zone.

15.4.5 Prospection in the Marl Zone

In order to detect the archaeological residues, it was necessary to understand not only how any contrast would be expressed but also the physical causation of that contrast. The rationale was to ensure that the imagery collected provides the maximum observable information for the phenomena of interest. It was postulated that these sites represent the decayed and thoroughly ploughed remains of abandoned settlements originally composed of mud-brick structures (Wilkinson et al. 2006). If this were the case, then the soil associated with each archaeological site could in theory be differentiated from the localised soil by some difference in grain size, structure, moisture content or chemical/biological composition due to the degraded building material. It seems reasonable to suggest, therefore, that this might give rise to differences in soil and/or crop properties that ought to be detectable using satellite imagery.

It was noticed that each site exhibited a subtle soil colour difference. When compared against a Munsell chart, it was established that, when dry, archaeological residues were significantly lighter in colour (reflecting an increase in chroma) than the surrounding off-site soils, but that on- and off-site soils were indistinguishable by eye when wet (demonstrating that both soils share the same parent regolith). The inspection of CORONA imagery from different seasons revealed that the colour differences between archaeological and non-archaeological soils were most evident during peak aridity (September and January), although sites were also readily detectable during periods of drying out following rainfall. This presumably reflects differences in the moisture retention capacity of archaeological soils which will be a function of grain size and organic content. This simple observation provided enough information to determine that for optical satellite imagery the archaeological residues in the marl zone would exhibit the most contrast during periods of peak aridity. Further in-situ and laboratory analysis was undertaken which confirmed this hypothesis (Wilkinson et al. 2006; Beck et al. 2007b; Beck 2007).

Hence, the ideal time for image collection would be between September and January when the soil is either arid or hyper-arid. The problem with this time frame is that there are a range of airborne particulates which could decrease image fidelity. These particulates are, however, reduced during winter rains which tend to start in November/December.

Archaeological residues in the marl zone take the form of discrete settlement sites that are easy to identify as colour or textural variations in soil (Fig. 15.1: flat sites 197, 308 and 454). Both the Ikonos MS and CORONA are particularly useful resources for displaying changes in soil colour. These residues are an order of magnitude larger than those found in the basalt zone. Hence, there is not such a reliance on high spatial resolution data. There is more scope for interpretation by proxy in this zone. Some sites are associated with kinks in the road network, where the road respects the archaeological site. These are useful indicators when interpreting the satellite imagery. Since the CORONA photographs were obtained, this zone has been subject to a range of landscape modifications, but due to the nature of the residues, few sites have been eradicated. Rather, deeper ploughing has removed some of the surface textural components in the Ikonos imagery and brought subsurface marl deposits to the surface, creating a number of potential but negative features. Hence, using CORONA and Ikonos in combination generates a number of synergies which result in a more confident interpretation of the marl landscape.

15.4.6 Generating DEMs from CORONA for the Homs Region

Galiatsatos and co-workers (Galiatsatos 2004; Galiatsatos et al. 2008) extensively discuss the creation of DEMs by photogrammetry using a CORONA – CORONA stereo pair. Galiatsatos extracted the DEM using traditional photogrammetric techniques. Like Altmaier and Kany (2002), an empirical non-metric camera model was employed.

Even considering the distortions introduced by the panoramic KH-4b camera system, Galiatsatos achieved DEM accuracies approximating to 5 m in all three dimensions and a ground resolution of c. 17 m. The increased accuracy and resolution means that DEMs derived from CORONA imagery can be applied to more sophisticated archaeological problems, such as the identification of wadi channels by modelling surface deformation. Interestingly, Galiatsatos (2004) proposes that stereo models using historic and modern images as stereo pairs can be used for time change analysis. He postulates that if one were to analyse the error surface associated with the DEM, then locations with large errors will be due to changes (such as house construction). Finally, it should also be noted that the declassified KH-7 GAMBIT and KH-9 HEXAGON have stereo capabilities and camera systems that introduce fewer distortions than CORONA. However, the archaeological applicability of DEMs from these sensors has yet to be evaluated.

15.5 Discussion and Conclusions

CORONA photographs in conjunction with Ikonos images have provided a wealth of new archaeological information for the area around Homs, Syria. The differences between the two environmental zones and the nature of the archaeological residues mean that different approaches are required for image capture and processing in

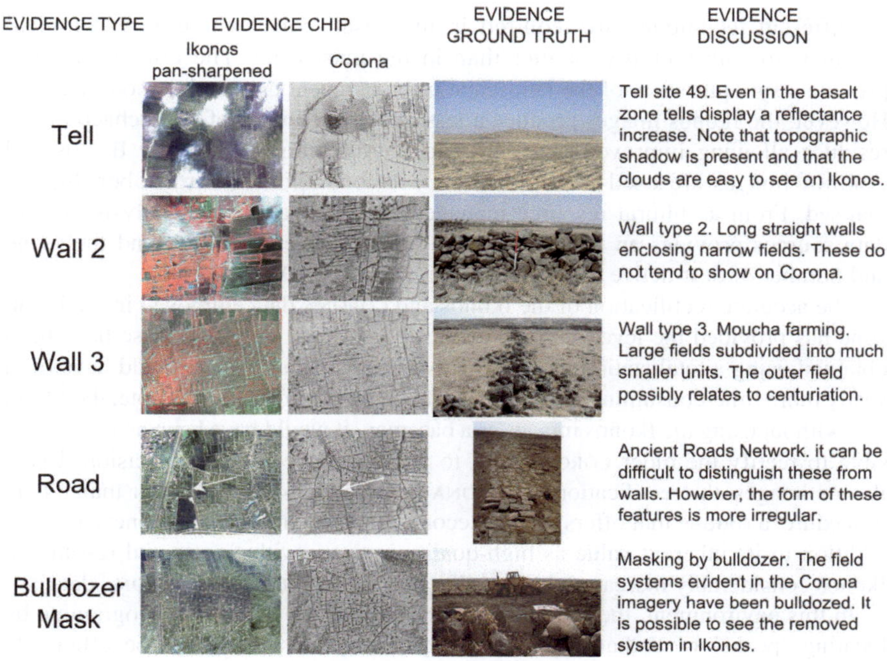

Fig. 15.3 Positive image interpretation key from the Basalt zone

order to extract the maximum archaeological information. To aid image interpretation, keys for the different zones have been produced (e.g. Fig. 15.3).

In the marl zone, the reincorporation of degraded mud-brick building material gave rise to changes in the moisture content, grain size and structure of the soil at the site. Having an understanding of the physical nature of this archaeological deformation process allows one to determine what types of sensing device, or other detection technique, and what conditions are appropriate for identifying this contrast.

In this instance, the localised reflectance difference expressed in the optical wavelengths was used. In order to achieve the maximum contrast for the archaeological residues, dry soils with limited crop cover were required. This choice of conditions and contrast type was determined by the detecting sensor – an optical sensor was employed so contrast differences expressed in the optical region are required. However, the modelling suggested that sensors in the short wave infrared (SWIR) may be more sensitive to variations in mineralogy and structure.

This knowledge can be used to enhance the visualisation. Once it was understood that in the optical wavelengths sites did not produce a specific spectral signature, but rather a relative shift to the spectral curve, bespoke enhancement algorithms were developed (Beck et al. 2007b).

Different techniques are required in the basalt zone, where archaeological residues are substantially smaller than in the marl zone. The CORONA images provide a synoptic view of the landscape prior to recent destructive modifications. However, the Ikonos images produce a less generalised view of the archaeological residues allowing improved detection and interpretation. When the Ikonos and CORONA images are used in conjunction with one another, further benefits are realised. From a cultural resource management perspective, the analysis of both data sources provides an overview of the archaeological residues and the range and number of destructive modifications over the past 30 years.

The accurate rectification of the Ikonos and CORONA data employed in the basalt zone has provided the level of spatial control that could only otherwise have been obtained using a total station or DGPS survey, a technique which would have been vastly more time-consuming for an area of this size. It is important to note, therefore, that without using the Ikonos images as a basemap, it would have been nearly impossible to rectify the older CORONA data to an acceptable level of precision. Using Ikonos images, the rectification of CORONA became a desk-based rather than a field procedure, a routine that offers obvious economies of both time and money. Thus, in addition to its inherent value as high-quality imagery with fine spatial resolution, Ikonos considerably increased the usability and value of the older CORONA data.

In this environment, the satellite imagery has framed the survey programme by locating 'peaks' of archaeological activity. Hence, resources can be efficiently deployed during field seasons resulting in improved modes of data collection and analysis. Importantly, the project team determined that satellite imagery detected the majority of surface residues. That said, a degree of 'off-site' sampling is required to provide some control over classes of feature which may not be readily detectable using imagery or to identify any landscape types in which the presence of archaeological material does not generate the kinds of indicators discussed above.

This chapter has outlined a generic approach to accessing and digitising declassified satellite photographs, highlighted some of the potential archaeological issues to which these photographs can be applied and illustrated this with examples from the SHR project. These examples demonstrate that there is no single approach to processing and image selection. Rather, processing is dependent upon the type of analysis one wants to undertake and the range of ancillary datasets, such as present-day imagery, or devices, such as GPS, which can be used to 'add value' to the source photographs. Image selection is a critical part of the process and requires an understanding of the nature of the archaeological residues, the localised contrasts they may exhibit and how these contrasts may vary over time.

Declassified photographs are a good resource in their own right, but their value is enhanced when they are utilised with other datasets. Particular synergies are observed when declassified photography is used with modern fine-resolution satellite images (such as Ikonos or Quickbird). At a practical level, modern images provide a more robust reference source for geo-referencing. From a prospection perspective, the combination of modern and historic images offers many benefits. The value of historic data in areas where recent change has removed or obscured archaeological evidence is obvious.

Declassified satellite photographs are becoming an increasingly important archaeological tool. Not only are they useful for residue prospection and, when in stereo pairs, digital elevation model (DEM) generation, they can also provide large-scale temporal snapshots that provide essential information on landscape change. Importantly, in some instances, declassified photographs may be the only available record of archaeological residues that have subsequently been eradicated. Future declassification of other 'spy' satellite programmes, some with even higher spatial resolution, will provide greater granularity to this temporal sequence.

Acknowledgements The authors gratefully acknowledge the support provided by the Natural Environment Research Council to Beck through Award Ref. GT0499TS53 and for the purchase of the Ikonos imagery by their Earth Observation Data Centre. Thanks are due to Nikolaos Galiatsatos for help provided during the writing of this paper. The Ikonos imagery includes material © 2003, European Space Imaging GmbH, all rights reserved. CORONA and GAMBIT data compiled by the US Geological Survey. We also wish to thank the British Academy and the Council for British Research in the Levant for their financial and logistical support of our fieldwork. All illustrations have been produced by the first named author. Thanks are also due to the directors and staff of the Damascus and Homs offices of the Directorate General of Antiquities and Museums, Syria, for all their help and assistance during the field seasons, with particular thanks due to our collaborators: Dr. Michel al-Maqdassi, Director of Excavations DGAM Damascus, and engineers Farid Jabbour and Maryam Bshesh of the DGAM office in Homs.

Bibliography

Altmaier, A., & Kany, C. (2002). Digital surface model generation from CORONA satellite imagery. *ISPRS Journal of Photogrammetry and Remote Sensing, 56*, 221–235.

Beck, A. R. (2007). Archaeological site detection: The importance of contrast. In *Proceedings of the Remote Sensing and Photogrammetry Society Annual conference 2007*, TS6, Newcastle.

Beck, A. R., Philip, G., Abdulkarim, M., & Donoghue, D. (2007a). Evaluation of Corona and Ikonos high resolution satellite imagery for archaeological prospection in western Syria. *Antiquity, 81*, 161–175.

Beck, A., Wilkinson, K., & Philip, G. (2007b). Some techniques for improving the detection of archaeological features from satellite imagery. In E. Manfred & M. Ulrich (Eds.), *Remote sensing for environmental monitoring, GIS applications, and geology VII* (Proceedings of SPIE, Vol. 6749, no. 674903). Bellingham: Society of Photo-Optical Instrumentation Engineers.

Bewley, R., Donoghue, D., Gaffney, V., Van Leusen, M., & Wise, A. (1999). *Archiving aerial photography and remote sensing data: A guide to good practice*. Oxford: Oxbow.

Bridgland, D. R., Philip, G., Westaway, R., & White, M. (2003). A long Quaternary terrace sequence in the Orontes River valley, Syria: A record of uplift and occupation. *Current Science, 84*, 1080–1089.

Challis, K. (2007). Archaeology's Cold War Windfall – The CORONA programme and lost landscapes of the Near East. *Journal of the British Interplanetary Society, 60*, 21–27.

Challis, K., Priestnall, G., Gardner, A., Henderson, J., & O'Hara, S. (2004). Corona remotely-sensed imagery in dryland archaeology: The Islamic city of al-Raqqa, Syria. *Journal of Field Archaeology, 29*, 139–153.

Cowley, D. C. (2002). A case study in the analysis of patterns of aerial reconnaissance in a lowland area of Southwest Scotland. *Archaeological Prospection, 9*, 255–265.

Fowler, M. J. F. (2004). Archaeology through the keyhole: The serendipity effect of aerial reconnaissance revisited. *Interdisciplinary Science Reviews, 29*, 118–134.

Fraser, C. S., Baltsavias, E., & Gruen, A. (2002). Processing of Ikonos imagery for submetre 3D positioning and building extraction. *ISPRS Journal of Photogrammetry and Remote Sensing, 56*, 177–194.

Galiatsatos, N. (2004). *Assessment of the* corona *series of satellite imagery in landscape archaeology: A case study from the Orontes valley, Syria.* Unpublished PhD thesis, Department of Geography, University of Durham.

Galiatsatos, N., Donoghue, D. N. M., & Philip, G. (2008). High resolution elevation data derived from stereoscopic CORONA imagery with minimal ground control: An approach using Ikonos and SRTM data. *Photogrammetric Engineering and Remote Sensing, 74*, 1093–1106.

Gerlach, F. (2000). Characteristics of space imaging's one-meter resolution satellite imagery products. *International Archives of Photogrammetry and Remote Sensing, 33*(B1), 128–135.

Kennedy, D. L. (1998). Declassified satellite photographs and archaeology in the Middle East: Case studies from Turkey. *Antiquity, 72*, 553–561.

Kouchoukos, N. (2001). Satellite images and Near Eastern landscapes. *Near Eastern Archaeology, 64*, 80–91.

Kramer, H. J. (1996). *Observation of the earth and its environment.* Berlin/Heidelberg/New York: Springer.

Leachtenauer, J., Danniel, K., & Vogl, T. (1998). Digitizing satellite imagery: Quality and cost considerations. *Photogrammetric Engineering and Remote Sensing, 64*, 29–34.

Lyons, T. R., & Avery, T. E. (1977). *Remote sensing: A handbook for archaeologists and cultural resource managers.* Washington, DC: Cultural Resources Management Division National Park Service U.S. Dept. of the Interior.

Palumbo, G. (1992). JADIS (Jordan Antiquities Database and Information System): An example of national archaeological inventory and GIS applications. In J. Andresen, T. Madsen, & I. Scollar (Eds.), *Computing the past: Computer applications and quantitative methods in archaeology* (pp. 183–188). Aarhus: Aarhus University Press.

Philip, G., Donoghue, D. N. M., Beck, A. R., & Galiatsatos, N. (2002a). Corona satellite photography: An archaeological application from the Middle East. *Antiquity, 76*, 109–118.

Philip, G., Jabour, F., Beck, A. R., Bshesh, M., Grove, J., Kirk, A., & Millard, A. R. (2002b). Settlement and landscape development in the Homs Region, Syria: Research questions, preliminary results 1999–2000 and future potential. *Levant, 34*, 1–23.

Philip, G., Abdulkarim, M., Beck, A. R., & Newson, P. G. (2005). Settlement and landscape development in the Homs region, Syria: Report on work undertaken 2001–2003. *Levant, 37*, 21–42.

Philip, G., & Bradbury, J. (2010). Pre-classical activity in the basalt landscape of the Homs region, Syria: the development of "sub-optimal" zones in the Levant during the Chalcolithic and Early Bronze Age. *Levant, 42/2*, 136–169.

Philip, G., Bradbury, J., & Jabbur, F. (2011). The Archaeology of the Homs Basalt, Syria: the main site types. *Studia Orontica, 9*, 38–55.

Rosen, A. M. (1986). *Cities of clay: The geoarcheology of tells.* Chicago: University of Chicago Press.

Sever, T. L. (1998). Validating prehistoric and current social phenomena upon the landscape of the Peten, Guatemala. In D. Liverman, E. F. Moran, R. R. Rinfus, & P. C. Stern (Eds.), *People and pixels: Linking remote sensing and social science* (pp. 145–163). Washington, DC: National Academy Press.

Stone, E. (2003). Remote sensing and the location of the ancient Tigris. In M. Forte & P. R. Williams (Eds.), *The reconstruction of archaeological landscapes through digital technologies: Proceedings of the 1st Italy-United States workshop, Boston, Massachusetts, USA, November 1–3, 2001* (British Archaeological Reports International Series 1151, pp. 157–162). Oxford: Archaeopress.

Ur, J. (2002). Settlement and landscape in Northern Mesopotamia: The Tell Hamoukar Survey 2000–2001. *Akkadica, 123*, 57–88.

Ur, J. (2003). CORONA satellite photography and ancient road networks: A Northern Mesopotamian case study. *Antiquity, 77*, 102–115.

Wilkinson, K. N., Beck, A. R., & Philip, G. (2006). Satellite imagery as a resource in the prospection for archaeological sites in central Syria. *Geoarchaeology, 21*, 735–750.

Wilson, D. R. (2000). *Air photo interpretation for archaeologists.* Stroud: Tempus.

Chapter 16
Uses of Declassified CORONA Photographs for Archaeological Survey in Armenia

Rog Palmer

Abstract Photographs taken of Armenia on 20 September 1971 during CORONA KH-4B mission 1115 provided a 'first look' at the landscape of a small research area. The fortress of Amberd provided keys through which to guide on-screen interpretation of the CORONA photographs, and this led to the identification of more than 200 'sites' (objects that were not of natural origin) in a 400 km^2 area. Selected sites were examined on the ground, and the area was further recorded on oblique aerial photographs taken at low altitude from a paramotor.

16.1 Introduction

Until the end of the twentieth century, archaeology in Armenia was earthbound: most archaeological work had been excavation and most known sites were monumental or funereal. In October 2000, following a request to the Aerial Archaeology Research Group from Professor Hayk Hakobyan of the Institute of Archaeology in Yerevan, I began to help get aerial survey off the ground in Armenia (Hakobyan and Palmer 2002). During my first visit there, I was told that no aerial photographs or maps were available and that civil aviation had been banned by the president. This seemed likely to make things a little more difficult than I had anticipated. However, one way of making a beginning was to acquire CORONA photographs that covered the 400 km^2 area we had selected for preliminary examination. This area lay south of Mt Aragats and was bounded on the west and east by two gorges – Amberd and Kasach, respectively (Fig. 16.1).

R. Palmer (✉)
Air Photo Services, 21 Gunhild Way, Cambridge CB1 8QZ, UK
e-mail: rog.palmer@ntlworld.com

W.S. Hanson and I.A. Oltean (eds.), *Archaeology from Historical Aerial and Satellite Archives*, DOI 10.1007/978-1-4614-4505-0_16,

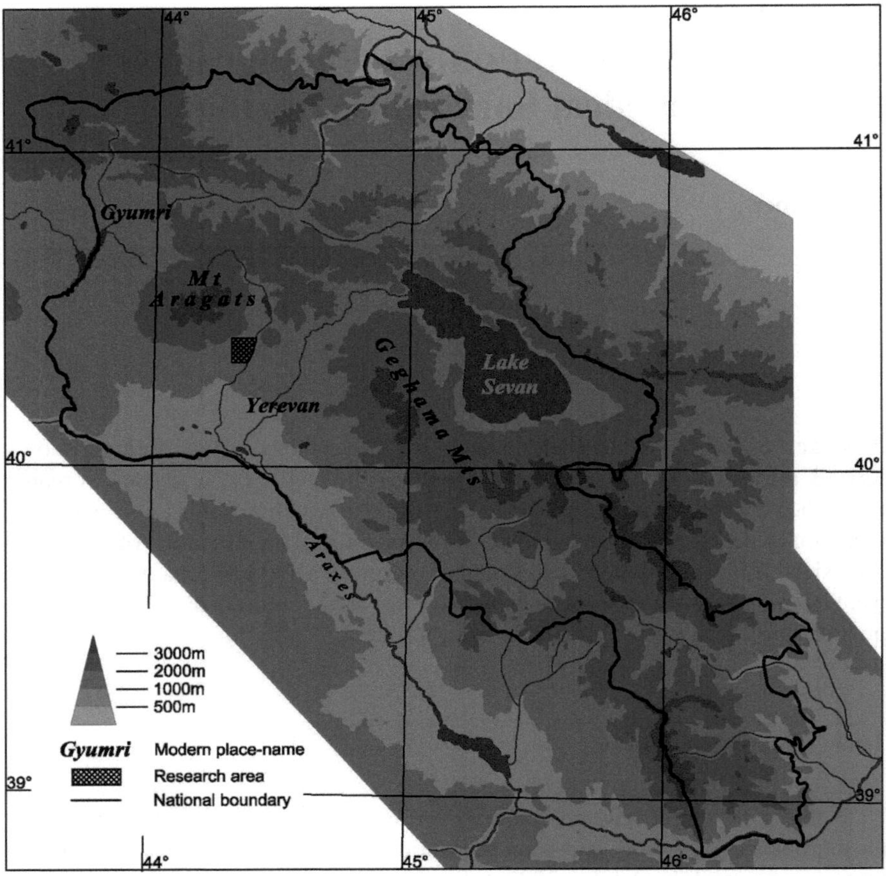

Fig. 16.1 Armenia showing (approximately) the final research area

My working hypothesis was that there should be, or may be, 'rural' sites scattered between the known monumental ones, and that examination of the CORONA photographs may allow identification of some of these as well as providing aerial views of sites that were already known. Some confirmation of the existence of such sites has come from recent field investigations, led by Adam Smith, University of Chicago, to survey land north of Mt Aragats (Smith 2009). The suitability of CORONA images for this type of survey had recently been demonstrated in Syria by the team from Durham University, UK (Philip et al. 2002; Chap. 15 by Beck and Philip, this volume).

What follows includes a brief description of a method of photo examination that was developed in 2001 for the study of extracts from CORONA photographs. Advances in technology in the past few years have made some aspects of this slightly antiquated, although the basis of the method and the definition of keys for known features in the area remain sound. Examples are given of features identified on CORONA photographs with follow-up field visits that have allowed photo interpretations to be confirmed or refined. Many sites have now been given added detail from oblique photographs that we have been taking since 2003.

16.2 Method

Photographs taken of Armenia on 20 September 1971 during CORONA KH-4B mission 1115 were cloud-free – a rarity among the Armenian cover – and five consecutive strips of negatives were purchased. KH-4B missions carried two cameras – fore and aft – to provide overlap for stereoscopic examination of the photographs (see Chap. 4 by Fowler, this volume). Unfortunately, the aft camera appears to have ceased functioning soon after it reached my area, so stereoscopic cover is restricted to two overlaps. However, the negatives have the greatest resolution and clarity of any CORONA material I have seen.

I had no sophisticated technology for scanning 70-mm negatives so I decided to examine the images in much the same way as I would with conventional vertical photographs. The image area of each negative was 60 mm deep by about 1 m in length and covered a ground area of roughly 15 km by 150 km. Our research area was a small part of this, and photographic prints of it were made on 25×20-cm paper where the 25 cm dimension was the 6 cm width of the negative. This gave about four times magnification and our area was covered by 13 high-quality prints, including stereoscopic overlap where it existed. An advantage of making enlargements is that each frame can be dodged appropriately to produce an optimum print. These were scanned at 1,200 dpi on a Microtek X12 flatbed scanner. This resolution was decided after visual experimentation. Higher resolution gave bigger files but no appreciable difference on screen; at lower resolution, some information was lost or indistinct. The resulting files were slightly larger than 100 MB and were easy to manage, without recourse to compression, on a Pentium III computer with 512 MB RAM – the peak of my technology in 2001.

In practical terms for viewing, this meant that a negative with a nominal contact scale of 1:247,500 was enlarged photographically to about 1:60,000. Scanned copies of those prints were examined using ER Viewer at an initial enlargement of about four times – giving a scale in the region of 1:15,000. Work on aerial photographs in Britain and Europe has shown this scale is adequate to identify settlement sites, especially those that survive as upstanding monuments, but not to depict much detail within them.

My aim was to examine each photograph as systematically as possible using a constant basic scale, but zooming in as necessary to check my identifications. Experience gained from work with aerial photographs in England and parts of Europe during the previous 30 years gave me the confidence to expect to identify features that were not natural. Among these should be archaeological sites. All digital copies were rotated so that the shadows fell towards the viewer as is recommended practice when examining and interpreting vertical photographs. Thus, all

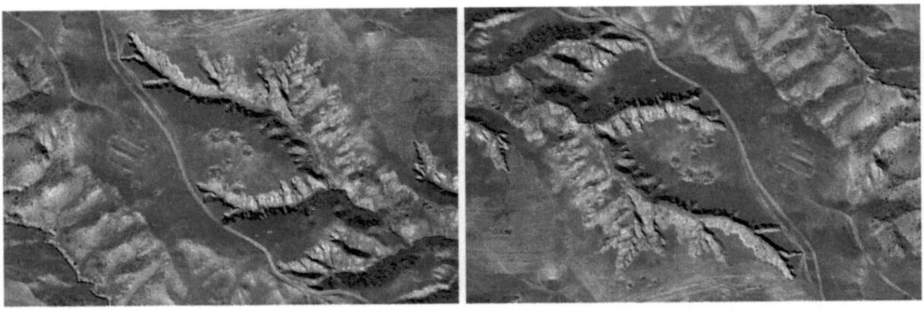

Fig. 16.2 A single view cropped from a vertical aerial photograph taken in the northern hemisphere and used to illustrate the effect of the direction of sunlight on a viewer's perception of height. *Left*: With north to the top and the shadows falling away from the viewer. *Right*: With south to the top so that shadows fall towards the viewer. The effect of the rotation on the shape and understanding of the topography and the types of archaeological (or other) features visible should be obvious (Photo: OFEK, June 2001)

photographs were viewed on- and off-screen (and are presented here) with south to the top to avoid the reversal of topography that can occur when shadows do not fall towards the viewer (Fig. 16.2).

Since the date of this work, CORONA photographs have been available only in digital form, making the printing and scanning part of my work no longer necessary. Visually, there is little apparent difference between the two sources to judge by the duplicate copies I purchased of Armenia. In some places, the digital image appears to be a little sharper; in others, the photographic prints seem to show more detail – and there is definitely more 'noise' (scratches and dust) on the digital copies. But when using either source, it is important to be systematic in examining these data and the above may provide a method for doing this. There are advantages to having good-quality photographic prints as they can be examined stereoscopically where overlap allows this, and they are useful in the field where they can act as a guide to location and to be annotated. A photographic print is also considerably lighter to carry than a notebook computer and is cheaper to replace if it gets wet.

The first example given below is of one site among several that I had visited on the ground in 2001. Detailed examination of these on the CORONA photographs provided a key of sorts to guide my examination of the rest of the images. Ground photographs taken at Amberd show the castle and church and a selection of low walls remaining from former buildings (Fig. 16.3). Thus, the fortress at Amberd includes examples of a range of features of different heights that can be used to illustrate the way in which light and shadow define such objects on a CORONA photograph and so indicates the ways in which other unknown features may be identified elsewhere during systematic examination of the photographs. Amberd is a long promontory with steep, deep, gorges on its north and south sides. This is apparent in Fig. 16.3 (inset) and, at the time of writing (January 2009), can be viewed at high resolution in Google Earth. The wall flanking the north edge of the promontory is the clearest feature and is made so through the combination of highlight, on the side the sunlight

Fig. 16.3 *Above*: Amberd castle and church (Note the low house walls behind the church in the southern part of the fortress (seen in detail below). (Photos: Rog Palmer, October 2001). *Inset*: Extract from a CORONA photograph showing the fortress and its environs. South is to the top and the scale is approximate (Photo: USGS, DS115-2154DF094, 20 September 1971)

strikes the wall, and shadow, on the opposite side. This feature – a wall – is also the type of structure that may be identified elsewhere on the CORONA photographs and noted as 'possibly archaeological'. The castle walls form three sides of a rectangular shape which has, so the shadows suggest, considerable height. This, however, is of similar shape and size to some obvious natural features and is unlikely to have been recognised as a castle without prior knowledge. The church is less distinct and is little more than a dark blob although, by comparison with other dark blobs, it is more likely to be a shadow than a stain on the ground surface as can be seen immediately south of it in Fig. 16.3 (inset). Despite the confidence of that description, this would not have been recognised as anything of interest on the basis of this photograph alone. The house walls are even less distinct but can be said to show as a slightly 'rough' area. Again, pre-knowledge was essential in this identification but, as will be seen below, the Amberd example does help confidence of interpretation of other 'rough areas', in particular those that indicate recently abandoned villages.

During the systematic examination, features identified – whether archaeological or not – were tabulated using the on-screen coordinates as an initial location. This table, and the indication of locations on a 1:100,000 map, has provided a guide for field examination of a sample of the CORONA sites. That scale of map was the largest available to me at the time (2001). Since then, digital copies of 1:50,000 maps have been acquired. At that date, maps in Armenia were very much in the province of the military, and virtually no previous use has been made of them for archaeological purposes.

16.3 Results

Two days of on-screen examination of the images resulted in the identification of more than 200 'sites'. These included 'fields', cultivation terraces or those enclosed by walls and what appeared to be lynchets, sometimes with stone clearance cairns; 'enclosures', some of which included internal or external detail; and 'dark areas', some of which were associated or within walled enclosures that most likely remain from recent shepherds' camps. All were located by eye on the 1:100,000 map and given Pulkovo coordinates (the local grid system) that can be refined using GPS during field visits. Correlation between the map and satellite images was surprisingly good, and the site locations are reasonably accurate. Certainly, they have been good enough to allow them to be found on the ground.

The unavailability of large- or medium-scale maps of Armenia means that the CORONA photographs may serve this purpose by transforming them to match available maps or GPS coordinates. A test transformation was made using the specialist software, AirPhoto (Scollar 2002). Some 45 control points were identified on an image extract and the 1:100,000 map allowing use of a fifth-order polynomial transformation algorithm (Scollar et al. 1990: 235–238). The result was a good visual correlation and achieved mean mismatch values for control points of less than ±15.0 m – probably as good as the 1:100,000 map survey accuracy. With the use of field-collected GPS control, this is likely to improve. Recently, but beyond the scope of this contribution, extracts from these CORONA photographs have been transformed to match an Ikonos image using about 90 control points. This provides excellent accuracy of position for the 1971 data, and the advantages of this method of using earlier and non-georeferenced material have been discussed elsewhere (Beck et al. 2007: 163–167; Chap. 15 by Beck and Philip, this volume).

Field visits made since June 2002 provide ground comparison for the examples below (Figs. 16.4, 16.5, and 16.6). These visits, along with examination of oblique aerial photographs that we have taken from a paramotor, have been continuing, although research grants ran out in 2005.

As soon as we began using the paramotor in 2003, it became apparent that we had to reduce the extent of our study area for the simple reason that the western side was above the maximum altitude at which our machine could fly. The smaller area could be classed as 'lowland' even though it ranges in altitude from about 1,400 to 1,850 m above sea level. However, it includes a range of easily-identifiable sites plus others of previously-unknown form or type and many that require ground investigation before they become much more than 'possible' status.

Within the reduced area, the site that showed the most obvious archaeological elements on the CORONA photographs was at Ushi, some elements of which were previously known. There, just beyond the west side of the present village, is a small hill topped by an enclosure from which walls of a field system appeared to radiate (Fig. 16.4). These features are as described and named by my UK-trained eye and, in the UK, there would be little doubt of their antiquity. Without excavation, dating some of these structures can be proposed in the field from the style of their stone construction. From ground inspection, my Armenian colleagues suggested a date for the defensive site as late Bronze Age and/or early Iron Age – in fact, our oblique photographs show it clearly to have two main phases of construction. Some of the field walls on the south side include cyclopean construction and thus may be of similar date, while others on the north-west of the hill show Hellenistic characteristics. Oblique aerial photographs add details to the field walls and, in places, show small features, possibly houses, incorporated in them. Confident recognition of such small detail is beyond the resolution of the CORONA photographs.

Another of the sites identified on a CORONA photograph appeared to be a walled structure that included straight-line elements and enclosed the top of a small hill (Fig. 16.5 inset). Following my numerical system, this was initially site 141, but ground visits and further research proved it to be a known defended site named Shahward. On the ground, as in the CORONA photograph, the wall can be traced almost all the way round the hilltop, and it has a possible entrance on the west side by an internal tower. Rectangular walled structures, possibly houses, were identified inside the walled enclosure and possible walls were outside (Fig. 16.5). Surface pottery from inside the enclosing wall was of eleventh–ninth century BC date.

CORONA site 119 was described as 'possible walled enclosures' and appeared to comprise two conjoined rectangular walls. Ground inspection showed it to be an entirely natural formation of large boulders on the side of a small gorge. This was one of the sites that helped refine subsequent interpretations as my knowledge of the local topography increased and helped to inform subsequent examination of CORONA or other photographs.

A recurring type of feature on the CORONA photographs was a cluster of small dark blobs many of which appeared square or rectangular. These differed in appearance to modern villages, but clearly were something similar and were usually in

Fig. 16.4 *Above*: CORONA extract of Ushi. The dark feature is the previously-known stone-walled Bronze Age-Iron Age hilltop settlement with medieval features on its south side. Around these are former fields defined by walls and terraces. South is to the top and the scale is approximate (Photo: USGS, DS115-2154DF093, 20 September 1971). *Below*: Ground view of Ushi showing part of the field system on the south side of the hill. The walled hilltop settlement is central on the skyline behind the spoil heaps raised from excavation of the medieval site (Photo: Rog Palmer, October 2001)

Fig. 16.5 *Inset*: CORONA extract of Site 141 that was initially identified as a 'walled enclosure within modern fields' and subsequently confirmed to be a known defended site called Shahward. South is to the top and the scale is approximate (Photo: USGS, DS115-2154DF093, 20 September 1971). *Main image*: Part of the site photographed from the ground showing a hilltop crested with a stone wall within which are rectangular stone-built features and outside of which are other walls. Clearance cairns are downslope (Photo: Rog Palmer, June 2002)

remote locations. On the 1:100,000 map, they were sometimes identified as 'ruins'. One of these, Buravet, lay between two gorges that carried water from Mt Aragats (Fig. 16.6 inset) and comprised a number of buildings, some with attached 'paddocks', a 'village pond' and areas of cleared ground in the immediate environs. We overflew the village to take oblique photographs on at least two dates, and considerable detail was recorded that can be compared with the information in the CORONA photograph (Fig. 16.6). Ground inspection showed it to be one of the villages that were forcibly deserted in Soviet times and then partly destroyed to prevent people reoccupying the houses. House walls remained more than a metre high in places and rutted village roads showed evidence of long use. A cemetery area lay close to the village and included graves of pre-Christian type (i.e. before second–third century AD), suggesting that the location had been occupied, not necessarily continuously, for some 2,000 years.

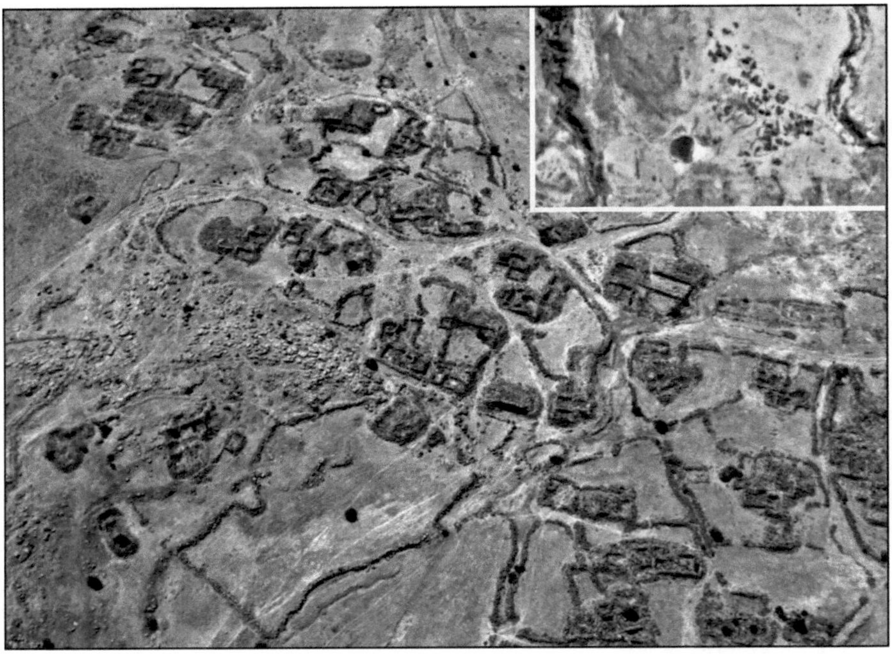

Fig. 16.6 *Inset*: Buravet from CORONA. South is to the top (Photo: USGS, DS115-2154DF093, 20 September 1971). Between the two gorges, the cluster of houses is clearly visible as are some of the walled enclosures and compounds. The pond is the large pear-shaped feature, and some stone heaps are apparent in cleared ground. Many of the features can be identified in one of our oblique photographs (*Main image*) which has similar orientation (Photo: Karen Martirosyan, August 2004)

Many other sites that were first identified on the CORONA photographs have been confirmed on the ground, and our field visits have also allowed us to delete 'modern' and natural features from the original list. Sites thus identified tend to be large – defended hilltops, fields and terraces and deserted villages – as is to some extent dictated by the 2-m spatial resolution of the original photographs. Field visits have suggested that many smaller sites, or parts of sites, remain between presently cultivated land on small hillocks of ground that were perhaps too stony to cultivate. These include a number of small buildings within which some Hellenistic pottery has been found. The CORONA photographs enable us to identify such areas of uncultivated ground and so can assist in planning field visits.

One final use that we have made of the CORONA photographs returns the theme to my first need for them – to act as a map in a country where maps were not available. Since that date, we have acquired maps of 1:50,000 scale and smaller, but the detail is considerably greater in the CORONA photographs. As described above, by transforming the CORONA photograph to match the 1:100,000 map, a rough geo-rectified image was produced. Comparison of the landscape recorded in 1971 by CORONA and that on our present-day obliques showed there to be no change in places and minimal change in others (it also showed considerable village expansion, but that is another story). These closely-similar landscapes meant that the geo-rectified CORONA photograph could be used as a 'map' on which to locate precisely all of the oblique photographs that we had taken from the paramotor – something that would not have been possible using only the conventional maps.

16.4 Potential

There are two immediately apparent ways in which uses of CORONA photographs may help Armenian archaeology. The first of these is through their ability to add new information to the known archaeological record. This would follow a process from the identification of possible archaeological sites on CORONA photographs to their confirmation (or not) from a ground visit or by examination of targeted oblique photographs. When, or if, observer-targeted aerial photography becomes more feasible and is undertaken by more sophisticated methods than at present, it may eliminate the need for all sites to be visited on the ground. Second, CORONA photographs can be used to try to locate sites that are known and are listed in the Armenian National Record. The location of these sites may already be marked on maps of 1:200,000 scale, but accurate location of sites does not seem to be a routine component of Armenian archaeology. The Record describes locations on a 'parish' basis in terms such as '3 km north-west of the village'. We were not always successful in correlating those locations with recognisable features on the CORONA photographs, nor did we always succeed in finding them on the ground. Furthermore, if we did find something on the ground close to the Record's location, we had no definite way of knowing if it was, in fact, the site that had been recorded. The people responsible for the Armenian Record hope to bring it up to European standards. One step towards this will be to provide coordinate locations for their sites, and specialist examination of geo-located CORONA images may prove a cost-effective way to achieve certain elements of this.

The success of our uses of CORONA photographs in Armenia shows their potential in other parts of the world where local archaeologists are ready to expand from excavation to survey and want to begin their examination of the landscape. Examination of CORONA photographs offers high archaeological potential to an experienced photo-interpreter who is prepared, in cases, to stick his/her neck out a bit. Such a person, with some familiarity of the local terrain, should be able to make well-judged identifications of likely archaeological features that can be followed by closer inspection. As with aerial photographs, the ability to find the relevant objects, to sense changes in the topography and to understand what is being seen is dependent on the experience of the photo-interpreter. This will be improved considerably by making field visits in the area being studied. In Armenia, I was very aware of the improvements to my perception of CORONA photographs and our oblique photographs after each field season.

CORONA photographs also have historical value in their own right as they record ground conditions in the 1960s and 1970s in many places that have since been changed by development of various kinds. This value has previously been noted in the Homs environs of Syria (Beck et al. 2007; Chap. 15 by Beck and Philip, this volume), and there are certainly places in Armenia where, for example, villages have expanded to cover archaeological features whose only record now is that recorded on CORONA photographs and other (still secret or inaccessible) earlier lower-altitude aerial photographs that lurk, unknown to archaeologists, in archives in Moscow or Yerevan.

Acknowledgements Thanks for help, advice and for sharing their knowledge of Armenia go to Professor Hayk Hakobyan, without whom the project would not have begun, and to Tigran and Vardan Hovhannisyan. Generous funding for the project that has been named *Wings over Armenia* has been given by the Association for Cultural Exchange (2001, 2002) and the British Academy (2002–2003).

Bibliography

Beck, A., Philip, G., Abdulkarim, M., & Donoghue, D. (2007). Evaluation of Corona and Ikonos high resolution satellite imagery for archaeological prospection in western Syria. *Antiquity, 81*, 161–175.

Hakobyan, H., & Palmer, R. (2002). Prospects for aerial survey in Armenia. In R. H. Bewley & W. Rączkowski (Eds.), *Aerial archaeology: Developing future practice* (NATO Science Series, Vol. 337, 140–146). Amsterdam: IOS Press.

Philip, G., Donoghue, D., Beck, A., & Galiatsatos, N. (2002). CORONA satellite photography: An archaeological application from the Middle East. *Antiquity, 76*, 109–118.

Scollar, I. (2002). Making things look vertical. In R. H. Bewley & W. Rączkowski (Eds.), *Aerial archaeology: Developing future practice* (NATO Science Series, Vol. 337, 166–172). Amsterdam: IOS Press.

Scollar, I., Tabbagh, A., Hesse, A., & Herzog, I. (1990). *Archaeological prospecting and remote sensing*. Cambridge: Cambridge University Press.

Smith, A. T. (2009). *University of Chicago: Project Aragats*. http://aragats.net/field-projects. Accessed Jan 2009.

Chapter 17
Pixels, Ponds and People: Mapping Archaeological Landscapes in Cambodia Using Historical Aerial and Satellite Imagery

Damian Evans and Elizabeth Moylan

Abstract Over the last 20 years, two particular factors have contributed to a renewed focus on archived images of Cambodia and an increasing recognition of their importance. On the one hand, it has become clear that the great urban complexes of the Angkor era have left subtle traces of their existence everywhere on the surface of the landscape, and that remote sensing affords us the opportunity to uncover, map and analyse the various elements of medieval urban form. On the other hand, since the cessation of three decades of civil conflict, rapid urbanisation in Cambodia and the expansion of modern cities into rural areas have endangered and in some cases obliterated many of those remnant archaeological features. In addition to Second World War-era aerial photo archives and modest collections produced by the colonial authorities, the Second Indochina War has left behind a particularly rich legacy of now-declassified spy satellite imagery from the 1960s and early 1970s. These images, in particular those from the KH-4 CORONA missions, provide extremely valuable coverage of an almost pristine archaeological landscape immediately prior to the radical restructuring of agricultural systems during the Khmer Rouge period. In this chapter, we describe how various collections of archived imagery have been used not only to reconstruct the medieval landscapes of the Khmer Empire but also as a tool for evaluating the complex relationships between contemporary populations at Angkor and the archaeological landscape they have inhabited for generations.

D. Evans (✉)
Department of Archaeology, University of Sydney, Sydney, NSW, Australia
e-mail: damian.evans@sydney.edu.au

E. Moylan
School of Geosciences, University of Sydney, Sydney, NSW, Australia
e-mail: e.moylan@usyd.edu.au

W.S. Hanson and I.A. Oltean (eds.), *Archaeology from Historical Aerial and Satellite Archives*, DOI 10.1007/978-1-4614-4505-0_17,

17.1 Introduction and Background

In French Indochina in the 1920s, as in Europe and the Middle East, a revolution in archaeological research methods was quietly taking place. Aviation had developed very quickly during the Great War, and aircraft had been deployed very extensively over the battlefields of Europe. During the course of that conflict, a number of archaeologists, both professional and amateur, had noted the potential for discovering sites from the air. As with the great pioneer of aerial archaeology O.G.S. Crawford, the archaeologist Victor Goloubew of the *École française d'Extrême-Orient* (EFEO) had developed an appreciation for the value of aerial prospection through his experiences in the trenches of the Western Front (Crawford 1955: 117–122; Goloubew 1936; Malleret 1967: 339). There, in addition to reconnaissance, aerial survey had been used to generate continuously updated maps of enemy trenches and to develop precise and detailed topographic maps to assist with the ranging of artillery fire. In 1921, the Director of the EFEO, Louis Finot, made the first request of the nascent *Aéronautique d'Indochine* for an archaeological survey. A military organisation, the *Aéronautique*, nonetheless had a specific mandate to cooperate with civilian organisations, including the *Cadastre* and the *Service géographique*, and aerial photographs were duly provided to Finot. They covered the Plain of Reeds area of the Mekong Delta, an area that was to figure large in the later history of remote sensing in Indochina. The results were very impressive, and thus began a long period of cooperation with the *Aéronautique d'Indochine* throughout the 1920s, in which both aerial photographs and daring feats of low-level survey – which on at least one occasion resulted in the death of an aviator at an archaeological site – played a prominent role in the archaeology of France's Vietnamese possessions (Claeys 1951: 92–96). In the late 1920s, Victor Goloubew, a frequent participant in these projects, began to imagine the potential for an application of these methods in Cambodia, and in particular at Angkor, the great capital of the medieval Khmer Empire from c. 802 AD to c. 1431 AD (Fig. 17.1).

Goloubew arrived in the Siem Reap region of Cambodia early in 1932, via Saigon, where he had convinced a ship's captain within the colonial navy to collaborate on archaeological prospection in the Angkor region. In August that year, two sea planes were dispatched from the French fleet and conducted operations from the waters of medieval reservoirs and the moat of Angkor Wat, freeing Goloubew from the limitations of elephant-back surveys with his colleagues (Anonymous 1933: 520–521; Malleret 1967: 351). From the air, Goloubew noticed many subtle topographic features, for example, small elevated areas that indicated occupation mounds and depressions that indicated medieval ponds, which had previously gone unnoticed in the course of ground surveys. Using analyses that are now a stock-in-trade of aerial archaeology, he observed patterns of soil moisture and differential vegetation growth that were indicative of subsurface remains such as urban enclosures, extinct field wall systems and even small temples. Subtle linear traces were identified as previously undiscovered canals and roadways. As Pottier (1999) has pointed out, it marks essentially the first time at Angkor that these micro-level landscape features

Fig. 17.1 Angkor and Banteay Chhmar in their modern geographical context

had been considered in detail; previous field investigations had concentrated almost entirely on the great monuments and their immediate enclosures, and the largest of the medieval reservoirs.

The original field season of 1932, covered in a publication by Goloubew the following year (Goloubew 1933), pioneered the tandem use of aerial survey and ground-based investigations in Cambodia and provided the foundation for a typology of landscape artefacts (occupation mounds, excavated ponds, traces of medieval rice field walls, the moated mounds of long-disappeared 'village shrines', etc.) still in use by landscape archaeologists today. In the 80 years since then, the methodology has become a cornerstone of archaeological work in Cambodia and has revolutionised our understanding of the nature of early urbanism in the region (Evans 2007). From a theoretical and historical point of view, one of the most enduring legacies of Goloubew's landscape perspective has been the notion that Angkor had a vast and elaborate water management system that was (at least in part) designed to guarantee and improve annual rice yields (Goloubew 1941). This idea was later popularised by another member of the EFEO, Bernard-Philippe Groslier, who also expanded it to include the idea that Angkor's eventual collapse as the capital of the

Khmer Empire could be attributed to a failure of that system. Like Goloubew before him, Groslier was a keen aerial archaeologist who recognised the importance of gaining the widest possible perspective of the landscape and of mapping the vast quantity of physical traces of medieval civilisation etched upon it in order to assess his hypothesis. Once again, like Goloubew in the 1920s and 1930s, Groslier and his colleagues at the EFEO frequently enlisted the assistance of colonial civil and military services and their pilots in pursuit of archaeological research programmes from the 1940s to the 1960s, even if the primary focus of most researchers at Angkor remained – as ever – on the temples and on the inscriptions (Evans 2007).

Groslier was forced to depart from Angkor near the beginning of a lengthy civil conflict in the 1970s, never to return; his ambitious programme to survey and map the archaeological landscape of Angkor from above remained only partially achieved. In the next four decades, the social, cultural and physical landscape of Cambodia would undergo a series of radical changes caused by war, genocide, mass displacement of human populations and – at least in the last decade or two – extremely rapid development and urbanisation in many areas. During the Khmer Rouge regime from 1975 to 1979, for example, a key component of the Maoist revolutionary agenda was the radical restructuring of the agrarian landscape in order to implement irrigation systems with a view to greatly increasing Cambodia's rice yields (Himel 2007; Nesbitt 1997: 5–6; Pijpers 1989). This programme involved extensive clearance of forest that had shrouded – and to a certain extent protected – important archaeological landscapes for centuries. Even worse, from an archaeological point of view, is that it involved widespread destruction of pre-existing field wall systems and topographic features in order to implement a standard, national grid of 1 ha fields. Thus, vast and intricate networks of field walls, ponds and dams that had evolved over centuries and that were, in many cases, a direct legacy of the medieval period were either erased from the surface or modified beyond all recognition. Cambodia remained all but inaccessible to researchers until the last 20 years, in which time the remnant traces of the Angkorian landscape have faced an entirely new threat to their preservation: the explosion of the tourism industry in Siem Reap and the resulting deforestation and urban intensification and extensification (Gaughan et al. 2008; Winter 2007).

Essentially, therefore, by the 1990s historical circumstances had produced a situation where for decades modern archaeology had almost completely passed Angkor by, even though it is perhaps one of the world's richest archaeological landscapes, and in spite of the fact that it has been – and indeed remains – under tremendous pressure from development. Furthermore, much of this development took place prior to the availability of affordable, commercial high-resolution satellite imagery in the last 5–10 years. In this context, therefore, the corpus of archival aerial imagery from the twentieth century has taken on a particular importance. Unfortunately, however, in spite of the numerous aerial archaeology campaigns undertaken over Cambodia between the 1930s and 1990s, very few of the resulting images have survived due to the various periods of political upheaval, combined with the humid tropical environment and the lack of adequate facilities for the proper storage and care of negatives and prints. In spite of these challenges, three major collections of high-resolution stereo imagery have been preserved, scanned and made available in digital form: the Williams-Hunt collection from 1945 to 1946 at 1:25,000 scale (Moore 1985, 2009); a series of images acquired by the French *Institut Géographique National* at around the time of Cambodia's independence from France in 1953–1954 at 1:40,000 scale (Pottier 1999: 243); and a country-wide coverage of Cambodia acquired in 1992–1993 by the FINNMAP company at 1:25,000 scale (Evans 2007: 159–160; Pottier 1999: 243), following the Paris Peace Accords of 1991 that nominally ended nearly two decades of civil conflict. All three collections have good coverage of the major Angkor-period archaeological sites in Cambodia and have proven exceptionally useful for uncovering and mapping archaeological landscapes.

Another extremely valuable source of archival medium-resolution imagery is the United States Geological Survey (USGS) and in particular their collection of declassified spy satellite imagery from the 1960s and 1970s (see also Chap. 4 by Fowler, this volume). Perhaps because of the strategic importance of Indochina during this period, Cambodia is reasonably well covered by collections of declassified imagery from KH-4 CORONA and KH-7 GAMBIT missions. Both of these satellites provide imagery of adequate scale and resolution to be of use for archaeological analysis. Although cloud cover is a major challenge and tends to render the majority of the images useless for research purposes, a number of datasets were used successfully for archaeological applications at Angkor and beyond, in particular for determining the structure of hydraulic networks immediately prior to the radical, Khmer Rouge-era reorganisation in the early 1970s.

17.2 The Archaeology of Angkor

Following the 'return' of Angkor from the Kingdom of Siam to French Indochina in 1907, colonial authorities in Cambodia immediately commissioned a detailed topographic map of an extended region around the central temples of Angkor, with a view to defining the complete distribution of archaeological remains in the area (Evans 2007: 38–39). This very ambitious programme, however, was interrupted by the early onset of the rainy season in 1908 and was never completed; only the central temple area was mapped. Nonetheless, the totally arbitrary cartographic 'boundaries' of Angkor that were thus created (Fig. 17.2) became a de facto definition of the extent of archaeological remains in the area and would constrain virtually every archaeological survey and mapping project undertaken at Angkor until the twenty-first century. For example, no detailed archaeological map of the extended northern reaches of the Angkor temple complex was completed until 2002 (Evans 2007: 39). The mapping projects of the last decade have coincided with (and to some extent precipitated) a general re-thinking of the nature of early urbanism in Cambodia. Increasingly, scholars are questioning the traditional model of 'bounded settlements' contained within city walls and temple enclosures and moving towards a more holistic view of settlements that includes the low-density urban/agrarian landscapes that often stretch between, and also far beyond, the massive enclosures (Fletcher 2001, 2009; Pottier 2000a). Methodologically, the key component of this redefinition has been to uncover, map and analyse the distribution of

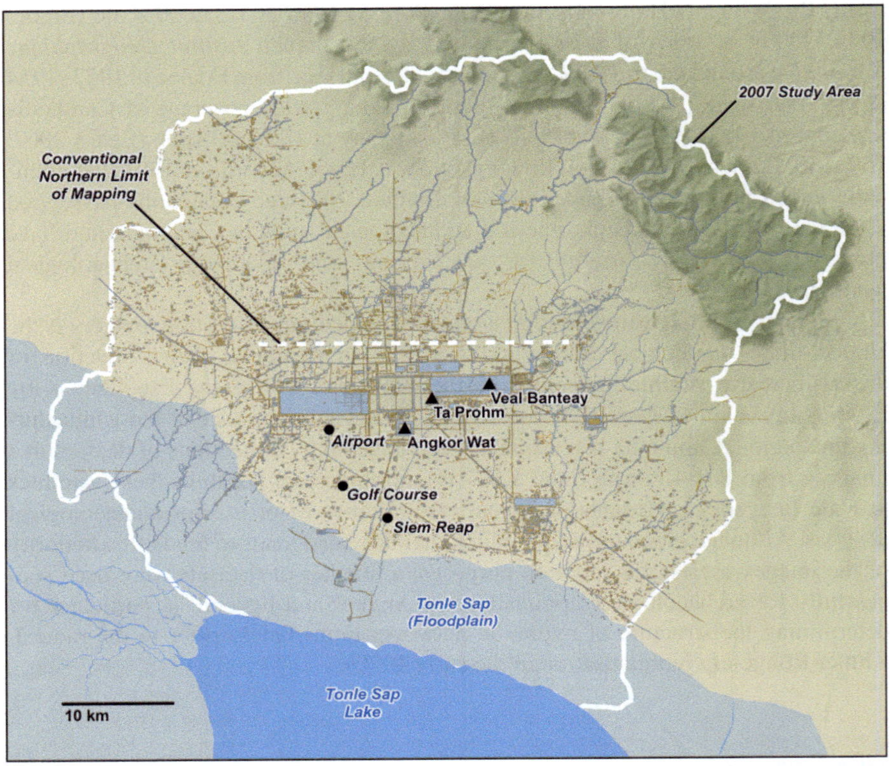

Fig. 17.2 The 1907–1908 cartographic project, compared to the most recent archaeological map of Angkor (After Pottier 1999; Evans 2007)

the small-scale elements of urban form that were first identified by French archaeologists like Goloubew in the 1930s and largely ignored thereafter due to the relentless focus on temples and inscriptions: the subtle topographic traces of occupation mounds, moated 'village shrines', canals and ponds.

As mentioned above, it is precisely these less tangible traces of the archaeological landscape that are the most under threat from modern development, particularly in the Siem Reap-Angkor area. In this context, therefore, the extensive aerial photography collections of the twentieth century provide an important window into the landscape history of the area. The images permit the identification of archaeological features that have been erased both from view and from memory, and allow the creation of a relatively comprehensive and accurate picture of medieval urban form that would otherwise be distorted by blank spaces under sprawling hotel zones and massive infrastructural projects. Perhaps the most important of these elements of early urban form are the small temples known as 'village shrines'. In contrast to the well-known monuments of stone that characterise the major state temples of the Angkor era, these shrines were modest structures consisting of a single, small brick tower, or possibly even of less durable material such as wood; in most cases, there

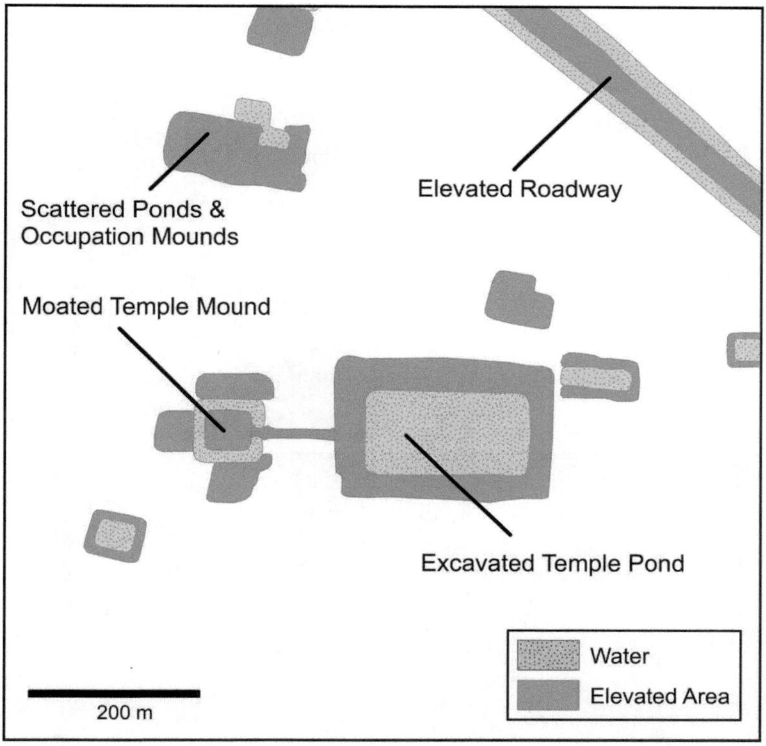

Fig. 17.3 Typical 'village shrine' configuration

is little or no trace of the shrine itself remaining on the surface. Nonetheless, determining their location and distribution is critically important for settlement pattern studies, because not only do they define a place of Hindu-Buddhist worship, they also define the nucleus of an extended residential community in much the same way as a Buddhist pagoda or *wat* commonly does in contemporary Cambodia.

Fortunately, in spite of the lack of architectural remains, these small temples can generally be identified from the air by the subtle traces they have left upon the landscape. The typical horizontal configuration of a village shrine consists of a small, square elevated mound of approximately 1–2 m in height, surrounded by a moat excavated to a depth of approximately 1 m. An earthen causeway leading to the east from the temple mound intersects the moat and gives it a characteristic east-facing 'horseshoe shape'. In general, the causeway leads to an excavated rectangular pond positioned to the east of the moated temple mound; the pond has x:y dimensions of 2:1 and is aligned east–west. Often, elevated occupation mounds and secondary ponds (also aligned east–west) are scattered around the immediate area. Taken together, this whole 'village shrine' assemblage forms the classic building block of low-density Khmer urbanism in medieval Cambodia (Fig. 17.3). In general, the temple mound and pond banks are substantially eroded, and the excavated moats and ponds largely infilled; nonetheless, they can usually be discerned in stereo-

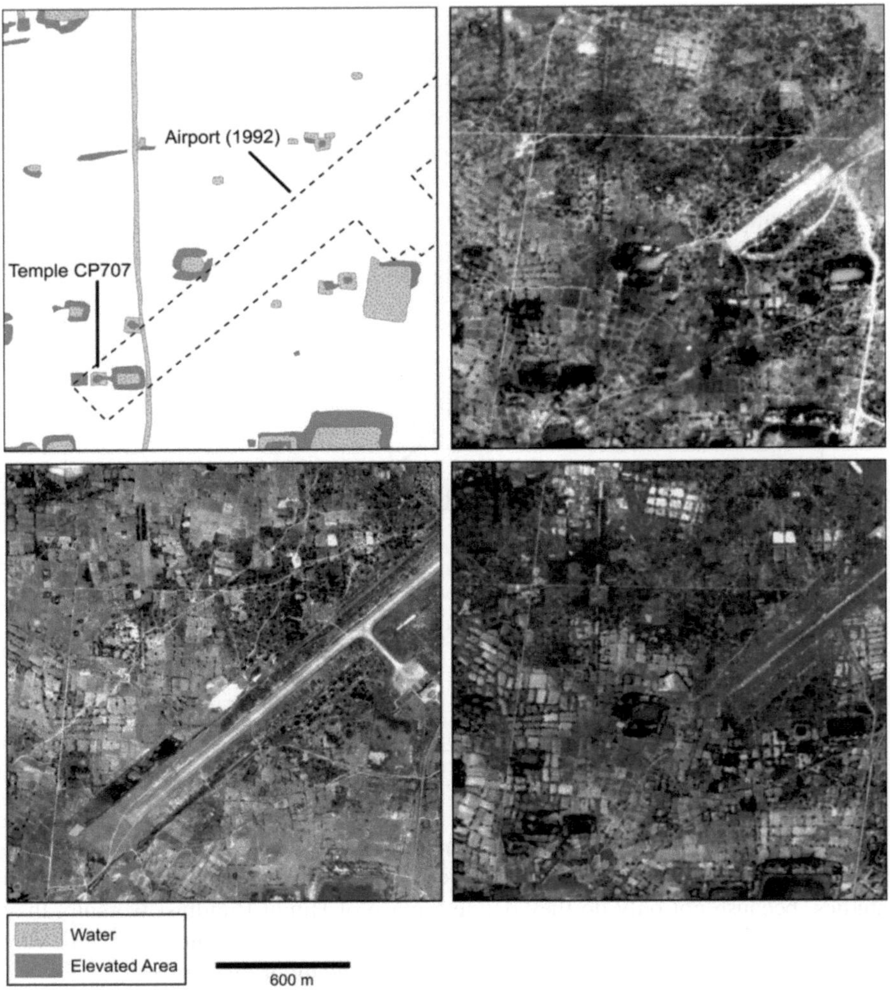

Fig. 17.4 Clockwise from *top left*: Temple CP707 as identified by Pottier (1999); in the Williams-Hunt imagery from 1945; in the IGN imagery from 1954; and finally covered by the airport in the FINNMAP imagery from 1992

scopic imagery by subtle, patterned variations in height, or in soil moisture, or in differential growth of crops or other vegetation. Identifying these village shrine complexes has been the foundation of most landscape-scale archaeological mapping projects in Cambodia over the last decade and a half (Evans 2007, 2010; Evans and Traviglia 2012; Pottier 1999).

One example of the destruction of these temples on the landscape is the historical expansion of Siem Reap airport. Perhaps four temples in total now lie beneath the current airport development; were it not for archives imagery from the 1940s and 1950s, it is likely that at least one of them would remain completely unknown, having been flattened and buried beneath the pavement of the main runway when it

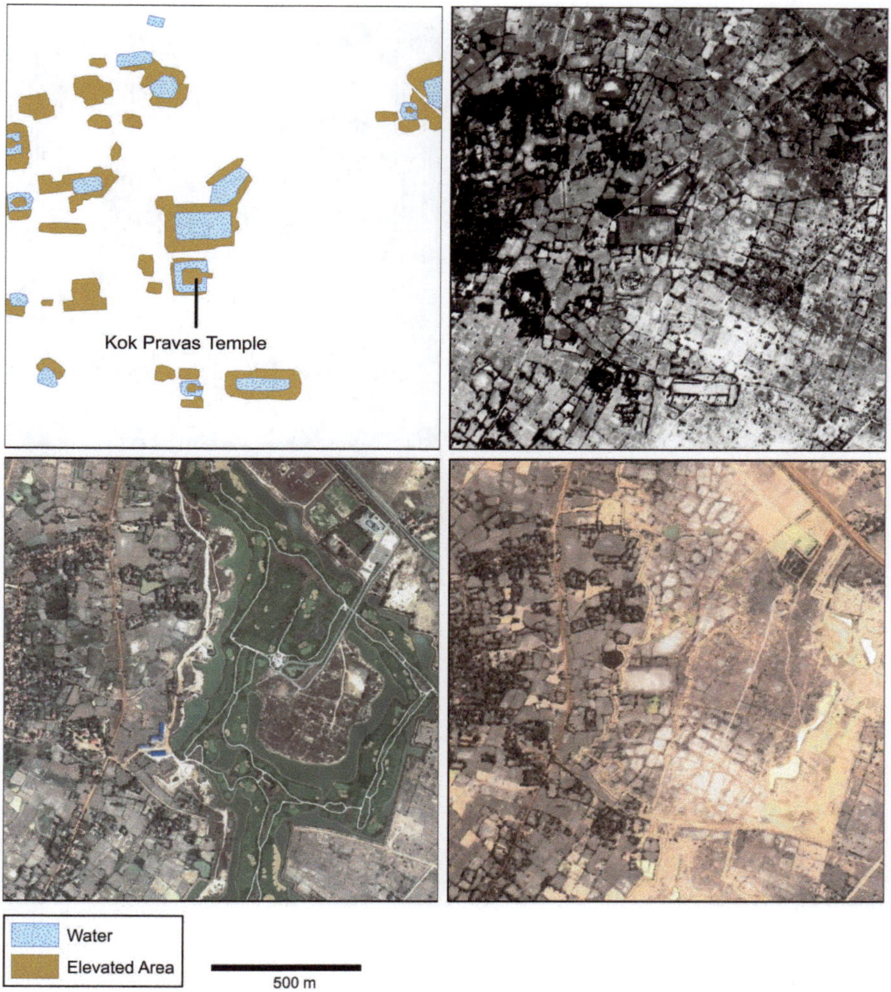

Fig. 17.5 *Top left*: the temple of Kok Pravas, as mapped by Pottier (1999); *top right*: in the FINNMAP imagery from 1992; *bottom right*: under threat from the construction of a golf course, as seen in QuickBird imagery from 2004; and finally *bottom left*: destroyed by the golf course in GeoEye-1 imagery from 2010

was extended in the mid-1960s (Fig. 17.4) (Baty 2005: 10–12; Pottier 1999: 85). It is likely that the FINNMAP series of stereo pairs from 1992 to 1993 will become an incredibly important archaeological dataset for researchers in the future, considering that the process of destruction continues apace: a recent survey of high-resolution GeoEye-1 data from 2010 revealed that more temples had been flatted and destroyed in the first few years of the twenty-first century, including at least one by a major golf course development (Fig. 17.5) (Christophe Pottier, personal communication 2011).

The fact that there are several archives of imagery for Angkor, illuminating different stages of landscape development over the last 70 years, is also of particular benefit

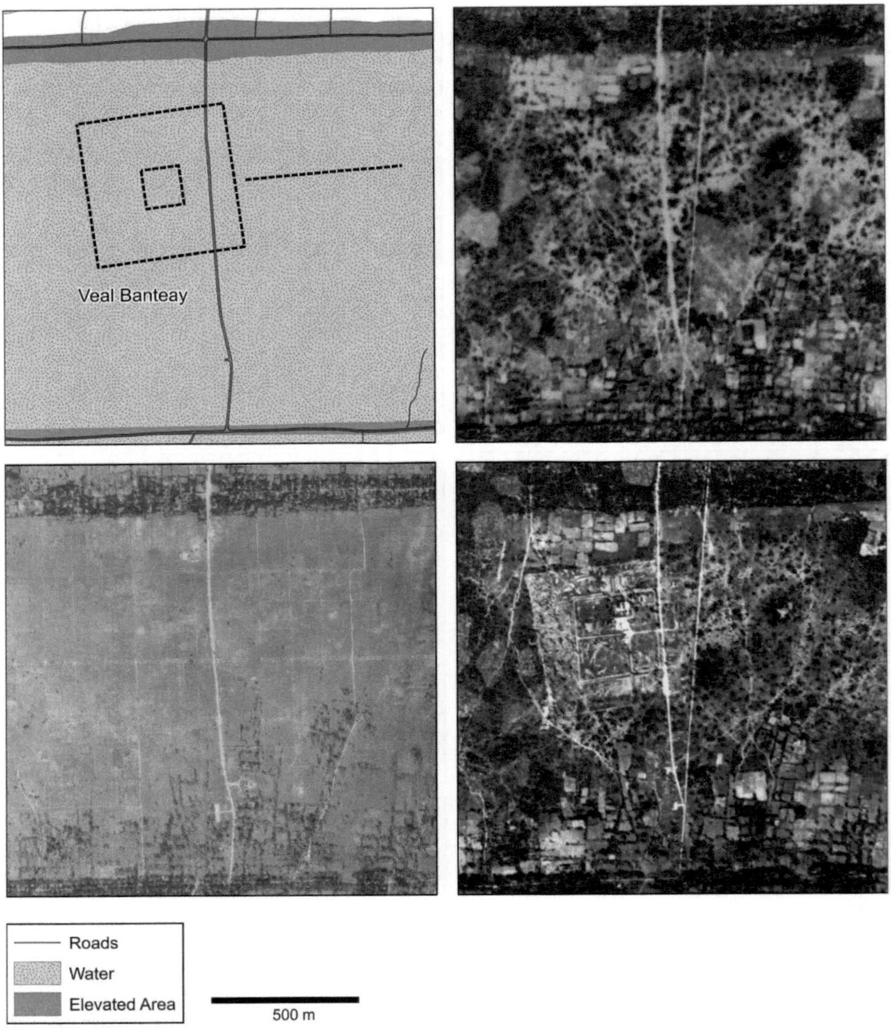

Fig. 17.6 *Top left*: the central enclosure and causeway of Veal Banteay, as tentatively identified in 2010; *top right*: visible traces of those features in the Williams-Hunt imagery from 1945; *bottom right*: IGN imagery from 1954 showing the military encampment built in the late 1940s or early 1950s; and *bottom left*: FINNMAP imagery from 1992 with traces of Veal Banteay erased from the landscape

to archaeological research. The resulting temporal resolution allows us to do more than simply define what is 'modern' and 'Angkorian' and arrive at a more nuanced view of landscape change and site morphology. A case in point is the site known as Veal Banteay ('the field of the citadel'), located in the eastern area of the Angkor Archaeological Park (Fig. 17.6) in the bed of a vast tenth-century reservoir known as the East Baray. Traces of the site were first identified in 2010 on the basis of the 1954

IGN imagery, in which at least three sides of a large rectangular enclosure, and also a causeway leading to the east, can clearly be seen. These traces had been completely erased from the surface by the time of the 1992 FINNMAP coverage. The identification of the site sparked particular interest because the ‘enclosure’ shares the same orientation as another highly unusual, and quite mysterious, pre-modern feature newly identified in 2007: a grid-like arrangement of 100 small mounds, contained within an earthen enclosure (Evans et al. 2007). Taken together, Veal Banteay and the ‘100 mounds’ ensemble appear to be a variation on the standard moated temple-mound and reservoir configuration (Fig. 17.6). Moreover, this ensemble aligns almost exactly with the nearby temple complex of Banteay Samré, of early twelfth-century construction, suggesting perhaps that Veal Banteay dates from the same period. This in turn would provide a convenient *terminus ante quem* for the abandonment and permanent drying out of the East Baray, a critically important milestone in the evolution of Angkor’s hydraulic infrastructure whose date remains obscure.

Ground verification of the site of Veal Banteay in 2011, however, revealed only ambiguous evidence for Angkor-period occupation: ceramic sherds were abundant, but the only diagnostic sherds were of clearly modern origin (Carter et al. 2011). A local informant advised the field team that the area was actually the site of a military encampment established in the 1940s by the forces of Dap Chhuon, a Siem Reap-based leader of the post-war resistance to French colonial rule (Thompson and Adloff 1953), although this could not be verified from written historical sources. Moreover, after a closer inspection of the 1954 aerial photos, the most prominent rectilinear features within Veal Banteay did indeed appear to be relatively recent constructions and were consistent with the spatial patterning of a modern military installation. Taken together, these findings called into question the antiquity of the site and offered a plausible alternative explanation for the Khmer-language description of the site as a ‘citadel’ as being a product of recent history, rather than a legacy of the Angkor period.

A closer inspection of the Williams-Hunt archival coverage of the same area in 1945, however, revealed that while the most obvious elements of Veal Banteay were certainly constructed at some stage between 1945 and 1953, traces of an outer rectilinear enclosure are clearly visible prior to these Dap Chhuon-era developments (Fig. 17.6). These traces are located in an area that was undeveloped and covered by forest throughout the historical period, suggesting that Dap Chhuon’s forces simply took advantage of an existing, medieval citadel wall rather than building an entirely new defensive fortification. This kind of re-use of Angkor-period topographic features is not unknown within the archaeological record: there is some stratigraphic evidence from excavations, for example, that the pattern of rice field walls visible on the surface today is the result of centuries of continuous renovation of medieval field systems (Baty 2005; Pottier 2000b). In this case, however, the archival imagery has provided a rare opportunity to both evaluate and add a certain depth of archaeological understanding to oral histories of landscape change.

17.3 Beyond Angkor: The Archaeology of Banteay Chhmar

The same lack of documentary evidence for landscape change is strikingly evident in the case of Banteay Chhmar, a twelfth- to thirteenth-century provincial temple centre constructed during the reign of Cambodia's greatest king, Jayavarman VII. Located in a remote, relatively dry and sparsely populated area some 100 km from Angkor (Fig. 17.1), the site has received relatively very little attention from scholars over the last century in spite of the great architectural and art historical significance of the temple ruins. The aridity of the area has been a common theme in the archaeological literature since the nineteenth century (Aymonier 1901: 335–6; Delaporte 1880: 143), and much of the region around the temple today consists of little more than scrubland and large expanses of almost desert-like plains of sand and rock. Foreshadowing aspects of the 'hydraulic city' hypothesis that his son Bernard-Philippe would later canonise in relation to Angkor, the French archaeologist George Groslier proposed as early as the 1930s that the hydraulic system was necessary for settlement in the region to be viable, ensuring a water supply for the population of Banteay Chhmar and irrigating an area that would otherwise be entirely barren (Groslier 1935: 161–3); moreover, he argued that it was impossible to maintain a population large enough to build and maintain the temple complex without such a hydraulic network and suggested that after the baray and canals were no longer maintained, the region reverted once again to a state of barrenness and desert.

The issue was not pursued until nearly 70 years later, when Pottier (2004) suggested that George Groslier's 'hydraulic city' hypothesis for Banteay Chhmar may in fact have some merit: in the 1992 FINNMAP coverage of the area, large linear features could be seen extending in at least two directions from the large, stone-lined reservoir associated with the central temples at the site. Unfortunately, however, it was also clear from the same images that the area around Banteay Chhmar had seen particularly intensive redevelopment during the Khmer Rouge regime. The 1945 Williams-Hunt aerial coverage revealed faint traces of several features such as canals and dams that were likely Angkorian; however, in the 1940s the area was so remote that it was almost entirely uncleared and obscured by vegetation, and in any case these features had been modified, extended and renovated beyond all recognition in the 1970s. It was, therefore, virtually impossible to identify which features – or which parts of them – dated to antiquity, much less to reconstruct the functioning of the original hydraulic network in detail. No coverage of the area survives in the IGN archive from 1953 to 1954.

In the 1950s and 1960s, however, the area around Banteay Chhmar was subject to an extensive programme of resettlement and land clearance; this process is not well documented historically, but the resulting deforestation is dramatically revealed in declassified, high-resolution KH-4 CORONA images from the late 1960s. Of particular importance from an archaeological point of view is a high-quality, almost cloud-free acquisition taken over the temple complex on 21 May 1967 during mission 1041-2 (Fig. 17.7) that captures the archaeological landscape with great clarity, just prior to the drastic Khmer Rouge-era remodelling. We can note, for example, that a reservoir just north of Banteay Chhmar called Trapeang Rohal Trach, commonly held to be

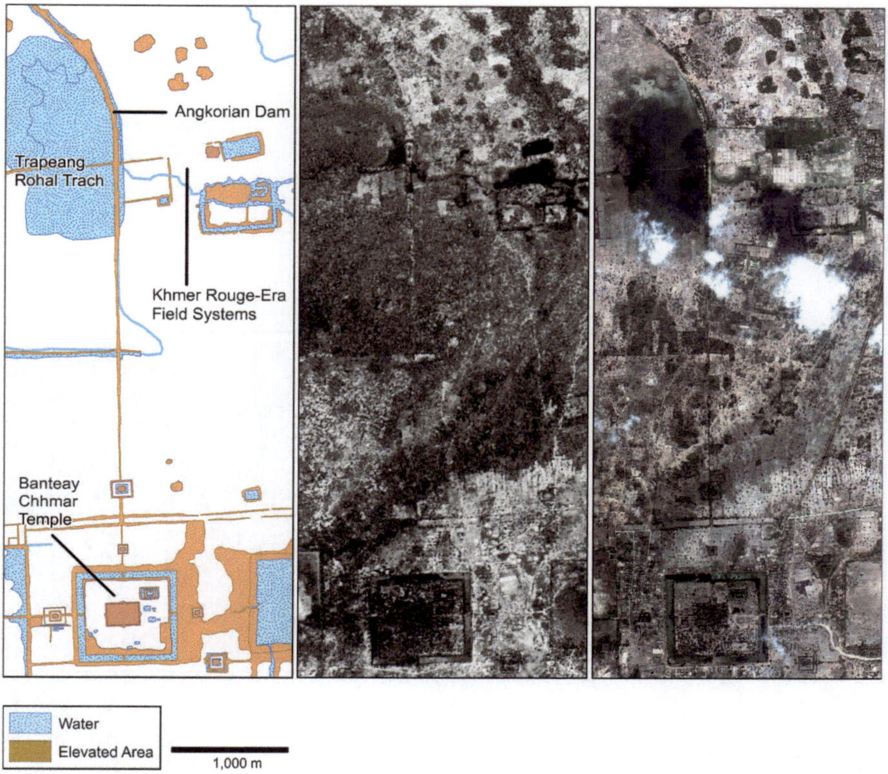

Fig. 17.7 From *left to right*: archaeological map of Banteay Chhmar; CORONA imagery from 1967, prior to Khmer Rouge hydraulic engineering, showing remnant Angkorian features; and GeoEye-1 imagery from 2010 showing Khmer Rouge-era adaptations to the Angkorian system

Khmer Rouge in origin, is in fact simply an ancient dyke that has been renovated in the 1970s and was at one time clearly part of an extended and quite elaborate Angkor-era water management system centred on Banteay Chhmar. The Khmer Rouge-era modifications were for the purposes of irrigation and were accompanied by the installation of a typical square-block field system immediately downslope from the reservoir, but in the declassified CORONA imagery from the 1960s, a pre-existing (and likely Angkorian) field system is faintly visible beneath the vegetation.

This Angkorian field system is dramatically clear in many other parts of the CORONA image, for example, around the satellite temple of Prasat Kbal Krabei, just to the south-east of Banteay Chhmar. Here, a comparison of the 1992 FINNMAP coverage and recent satellite imagery seems to suggest that the rice field system has been developed only in the last 20 years; closer inspection of the area in the CORONA imagery from 1967, however, reveals that a very distinctive pre-modern field pattern has been exposed due to partial clearing of the vegetation in the 1950s and 1960s (Fig. 17.8). Dozens of previously undocumented temples have been discovered in the Banteay

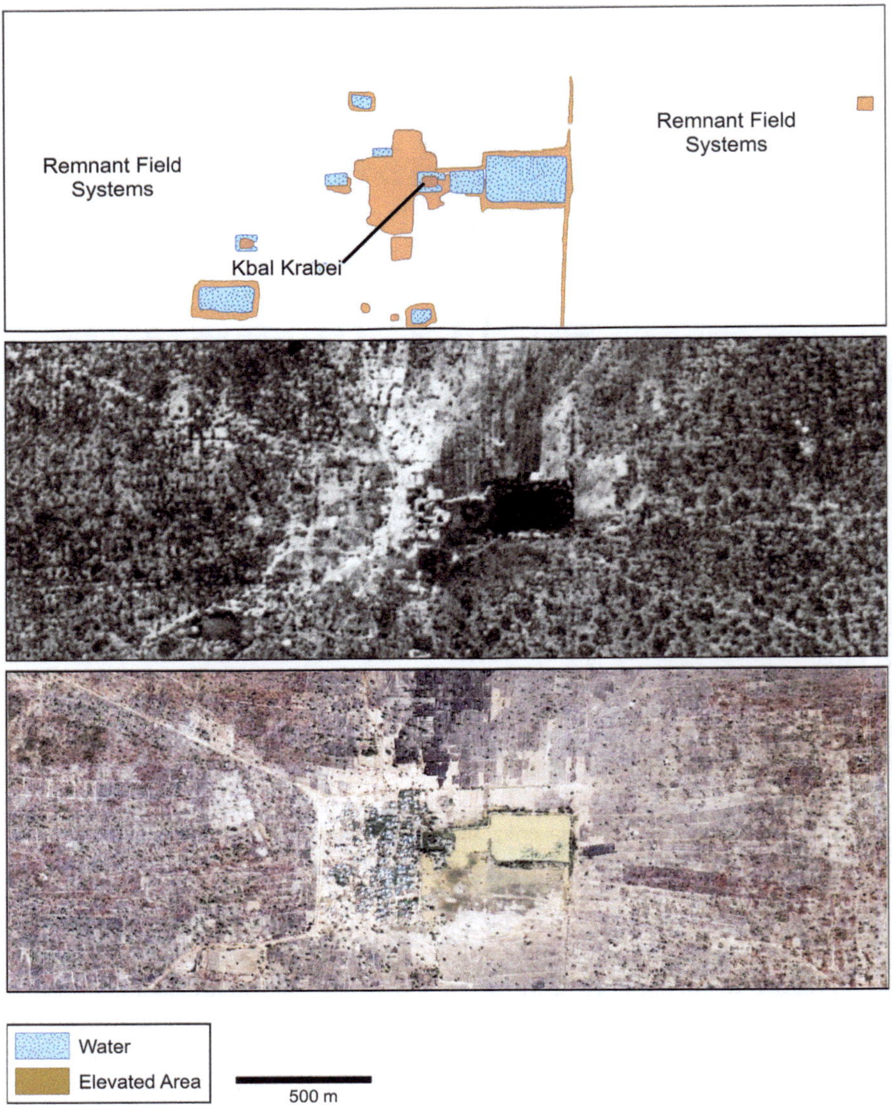

Fig. 17.8 From *top to bottom*: archaeological map of Kbal Krabei Temple at Banteay Chhmar; CORONA imagery from 1967, prior to habitation of the site, showing remnant field systems beneath the vegetation; and QuickBird imagery from 2004, after the development of a large town on the temple site and the clearance of all vegetation in the surrounding area

Chhmar area in recent years, with most showing evidence of an extended agricultural system around them. Overall, the evidence from CORONA imagery indicates that a highly elaborate state-sponsored hydraulic system around the main temple complex was complemented by a vast and intricate network of walled fields at satellite temples, and this new data provides crucial insight into Angkor-era subsistence and urban sustainability in the harsh environment of far north-western Cambodia. It now appears that the 'hydraulic city' of Angkor was not at all a unique phenomenon in the history of Khmer urbanism (Evans 2010).

17.4 Historical Geography and Heritage Management at Angkor

17.4.1 Background

The discovery and mapping of the low-density urban landscape of Angkor has drawn attention, in turn, to a series of important issues related to the conservation and management of the archaeological landscape. Even within the 400-km^2 zone defined by UNESCO as a World Heritage Site, several rural communities live in close proximity to the temples, and in this area there has been a historically uneasy relationship between the competing demands of mass tourism, heritage management and the traditional beliefs and practices of local people (Gillespie 2009; Kasiannan 2011). The revised archaeological map based on the work of Pottier (1999) and Evans (2007) further complicates the issue insofar as it redefines Angkor as a dispersed, low-density settlement complex stretching far beyond the borders of the World Heritage Site and including the burgeoning tourism hub of Siem Reap city. In 2005, a joint Cambodian-Australian programme called the 'Living with Heritage Project' was initiated with view to addressing some of these issues. In particular, the aim of the project was to explore ways in which spatial information systems might be used as a tool for site management, change monitoring and participatory planning, and also to provide a common mapping platform through which local communities can engage in management processes and assist in developing an understanding of diverse heritage values at local, national and international scales (Fletcher et al. 2007).

Recent case studies from Australia have shown that developing spatial representations of cultural landscapes can lead to a range of useful and important outcomes both for heritage managers and for communities within the area in question (Byrne and Nugent 2004; Moylan et al. 2009). Through the process of engaging with local communities and mapping their cultural landscapes, 'spatial representation of heritage is set within a cultural landscape framework, acknowledging that all parts of the landscape have inter-connected cultural histories, associations and meanings resulting from long-term and ongoing human-environmental interactions' (Moylan et al. 2009). As with the archaeological research that precipitated it, archival aerial imagery – in particular the Williams-Hunt collection – played an important role in achieving the goals of the 'Living with Heritage Project' in this particular domain.

In the first stage of the project, the archival images were simply used to familiarise researchers with broad, long-term change in the Angkor area. The photographs revealed crucial information about landscape history, allowed researchers to make a rapid appraisal of the major processes driving long-term change and provided important historical context for the finer-grained spatial and historical data that were eventually gathered as part of the project. Complex analytical techniques were not applied to the photographs and were by and large not necessary at this stage: even through simple visual analysis of the imagery, it was possible to note changes to the spatial extents of villages, the impact of new roads on patterns of development, areas of vegetation clearance and regrowth and – remarkably, given the tremendous growth in population in the area over a tumultuous half a century – large swathes of the landscape where no discernible change had occurred.

As with the archaeological mapping undertaken by the Greater Angkor Project, the Williams-Hunt collection was particularly valuable due to the scale and coverage of the imagery and the fact that there were overlapping prints so that individual trees and dwellings were visible using a simple stereoscope. Moreover, once again, the archival imagery proved to be tremendously important in light of the scarcity of alternative historical sources. Although much has been written about the political and economic history of modern Cambodia (Chandler 2000), very little of the literature deals specifically and directly with the environment. Some information on the modern history of Angkor's landscape can be gleaned from site-specific studies (e.g. Ebihara 1968; Kalab 1968), but it is difficult to extrapolate a broader landscape view from these limited sources. Cambodia's lengthy civil war from the 1970s to the 1990s resulted in the loss of historical documents and militated against effective record-keeping; meanwhile, French colonial archives are widely distributed, are in most cases not thoroughly catalogued or particularly easily accessible and (as with the secondary historical sources) tend not to be particularly informative on issues such as landscape change.

Thus, as in many other cultural contexts (Fox et al. 2005; Tobias 2000), the most valuable body of information about landscape history at Angkor resides in the collective memory of the local community and in particular in the various oral traditions passed down through generations. Acknowledging this, researchers in the 'Living with Heritage Project' conducted extensive interviews within local communities that revealed a rich web of connections, past and present, to the landscape within the World Heritage zone. During these interviews, hardcopy prints of remotely sensed data were presented to the interviewees to allow them to identify and define specific features and areas of interest and importance to them, in both the past and present. These included ancestral worship sites (*neak ta*), old temple sites, ponds, canals and channels, bridges, roads and paths, rice field areas and boundaries, community halls, forested areas, and the spatial extents of land ownership and villages. These data were marked by hand over the top of the prints while in the field and later digitised and recorded in a GIS. Recent, high-resolution, full-colour satellite imagery, such as QuickBird, proved to be the most useful dataset for the field interviews, as landscape features were immediately recognisable to interviewees. The primary use of the archival historical imagery, therefore, was to cross-check and validate the historical information provided by interviewees and to provide a basis for mapping these data in the GIS. Specific case studies illustrate how the archival imagery was used as a tool within this process.

17.4.2 Case Studies

Archival imagery was used to generate detailed maps of land use and land cover and of change over time, from the Second World War-era coverage of Williams-Hunt to the year 2005. The study area was classified into three broad landscape types: agricultural land, vegetated areas and inhabited areas. Individual rice field boundaries were delineated, and differences between vegetation patches observed. In addition to the broad patterns of landscape change, the analysis was able to reveal significant changes at a more detailed level.

In several instances, for example, informants reported that the spatial structure of villages had changed substantially over the course of previous generations due to the development of infrastructure, changing regulatory regimes and the degree to which villages depended on subsistence agriculture for survival. Reportedly, the construction of roads by French colonial administrators in the early twentieth century led to increased development of linear communities along the sides of the new routes and at the same time fundamentally altered long-standing patterns of subsistence and settlement in villages even at some distance from the new roads. Informants reported that this shift was characterised by an increasing focus on the commercial opportunities afforded by the road traffic and, at least to a certain extent, led to the abandonment of traditional modes of subsistence such as walled rice field agriculture. However, during the period of civil strife from the early 1970s to early 1990s (in which the tourism industry evaporated and roads were poorly maintained, if at all), there was a renewed emphasis on the development of water management infrastructure rather than roads and an increased reliance on traditional modes of subsistence such as rain-fed rice cultivation. The relative peace between the mid-1990s and the present day, however, in combination with more stringent regulation of the Angkor as a World Heritage Site, has led to yet another reversal of the process: livelihoods are once again focussed on taking advantage of the renovated thoroughfares that connect the temples and the million or more tourists who travel them every year, and regrowth covers much of the land previously given over to rice cultivation.

This oral history is supported, to a large extent, by the archival historical imagery. The cyclical process of growth and decline in agricultural areas over the twentieth century is illustrated in an area just to the north of the famous temple of Ta Prohm in central Angkor. In 1992, immediately after the Paris agreements that restored relative peace to the temple zone, much of the land in the area was cleared for cultivation, but 13 years later, it is clear from regrowth that many of these areas have been abandoned, and the spatial structure of land use and land

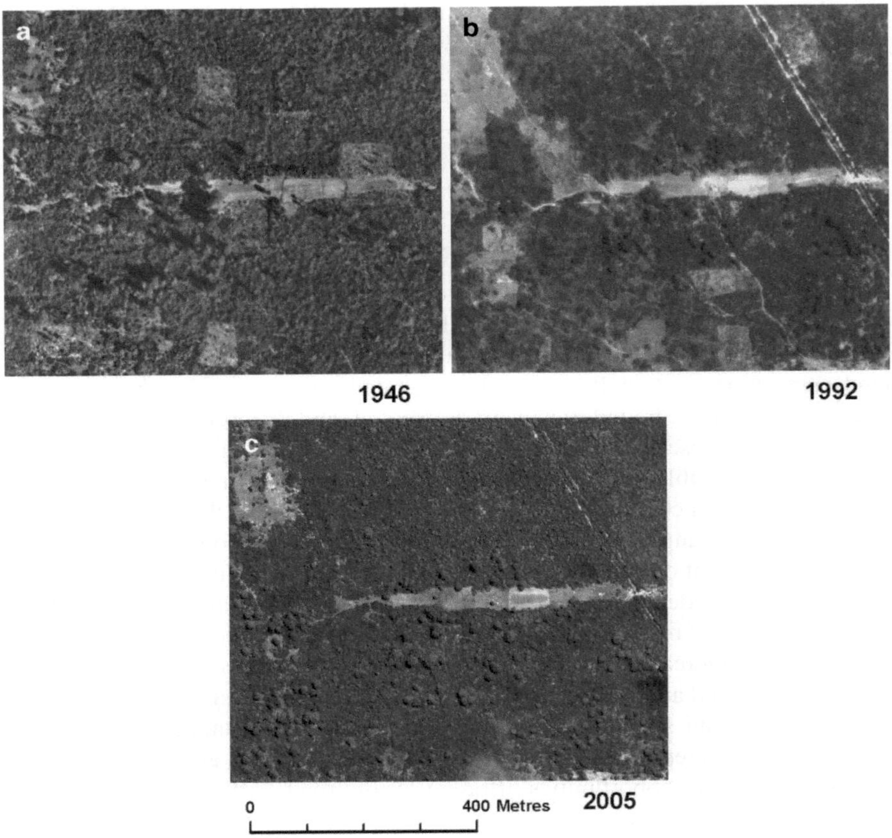

Fig. 17.9 Rice fields falling into disuse. (**a**) Williams-Hunt imagery from 1946. (**b**) FINNMAP imagery from 1992 showing large-scale clearance of the forest for roadways and cultivation. (**c**) QuickBird imagery from 2005 showing substantial regrowth of vegetation over previously cleared roadways and fields

cover more closely resembles that of the 1940s (Fig. 17.9). Significantly, in spite of these changes to the structure of the landscape, local knowledge of the specifics of the old land tenure system persists – even specific toponyms for individual mounds and fields – indicating that connections between modern communities and historical and archaeological landscapes persist and remain strong. The same was true at the scale of the village, where an oral tradition of toponymy preserves knowledge of the location and structure of settlements that have long been moved or abandoned; these villages, in turn, could also be identified in the archival imagery. The uses and applications of this verification process are not limited to establishing the longevity and continuity of land tenure but also have a potentially very important historical aspect: in some cases, ancient toponyms mentioned in the corpus of Khmer inscriptions have been preserved over the course of centuries and are reflected in modern place names, and can therefore assist in

Fig. 17.10 Persistence of traditional systems of land tenure: World War II-era land parcelling as visible in the Williams-Hunt imagery (*left*) compared to post-Khmer Rouge-era land parcelling as visible in QuickBird imagery from 2005 (*right*)

the identification of important archaeological sites (Ang 2007; Chandler 1996: 39; Vickery 1996).

In many areas, however, land parcelling and field boundaries did not change at all over the time period in question, and it became clear that the patterns visible on the landscape today are representative of at least 50 years of continuous use, including the period of civil upheaval. The strength of the rice field boundaries at this site was highlighted by fields that had been physically divided into two when a Khmer Rouge period canal had cut diagonally across them. The historical aerial photography shows clearly the field boundaries before and after the creation of the canal (Fig. 17.10). Interviews with the rice field owners confirmed that in many cases the divided fields on either side of the canal are still owned by a single person, and that informal systems of land ownership have persisted for decades, in spite of various attempts to abolish them and create entirely new systems of land tenure.

The identification of historical pathways through the archaeological park is another example of continuity in the landscape. As with the network of ancient roadways and highways stretching through and beyond Angkor (Evans 2007; Hendrickson 2007), the routes can be considered physical evidence of the historical ties that bind specific communities to areas of archaeological, cultural and economic significance. Once again, the archival imagery showed a remarkable degree of continuity within the network of small pathways through fields and forests over the last half a century; often, the paths mirrored the ancient road network and in some cases even made use of ancient embankments to create modern routes elevated above the floodwaters of the wet season. To take one example, again from the area just north of Ta Prohm temple, one pathway that had particular significance to local people was a route that delineated the boundary between two villages (Fig. 17.11). This pathway was considered 'ancient' by the community, and the historical photography helped to confirm

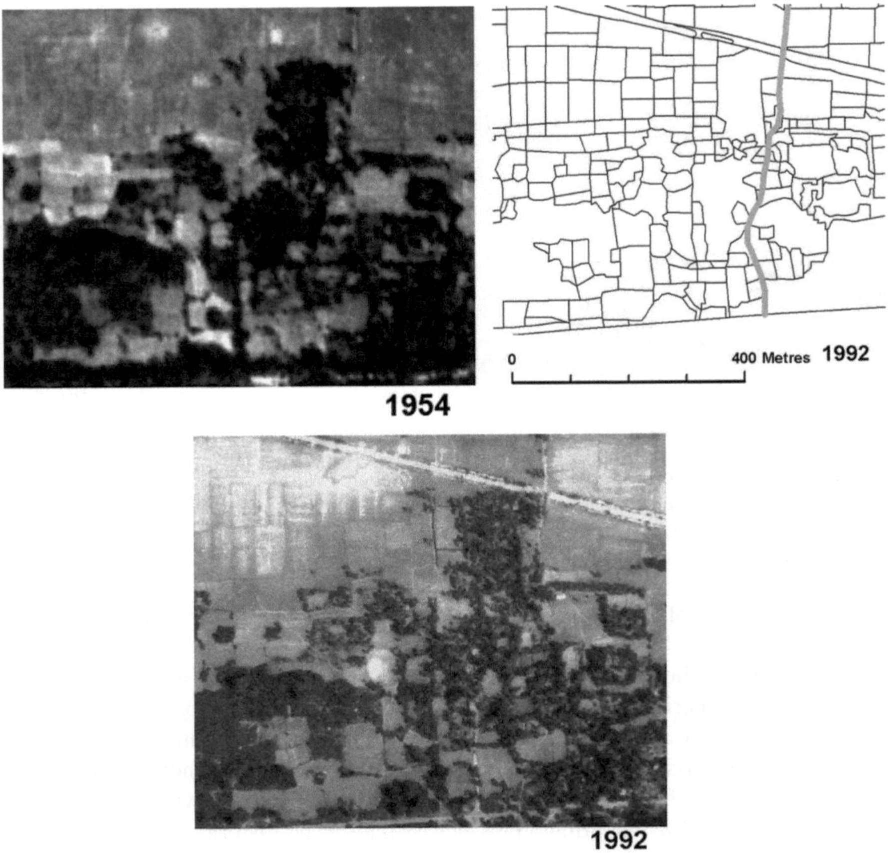

Fig. 17.11 Pathways as persistent elements of a village cadastre: digitising from historical sources such as the IGN coverage from 1954 (*left*) and the FINNMAP coverage of 1992 (*middle*) to create a historical cultural atlas of landscape features (*right*)

its continued use over at least 60 years not only as a pathway but as a key element in a kind of informal, 'folk cadastre' so typical of rural Cambodia.

17.5 Conclusions

Although the benefits of using historical aerial photographs for landscape change assessment have been mentioned here, there should also be a caution associated with their use. Interpretation must be done with care, and this means understanding the context in which the photography was taken. Given the sharp seasonality of weather in monsoon Southeast Asia the time of year at which imagery was acquired is of particular importance when trying to compare landscapes across time. It is important to remember, for example, that comparing aerial photography taken before and after harvest may result in radically different changes in the landscape than if they were taken at the same of the year. Beyond this seasonal variation, the high degree of annual variability in rainfall also needs to be considered. Years of early or late rain will strongly influence vegetation regrowth and rice production, and this will be reflected in the aerial photography and have consequences for the visibility and interpretation of historical and archaeological features.

Another difficulty in using archival imagery for the analysis of cultural and archaeological landscapes is that of linking dynamic processes of landscape change to the patterns captured in the landscape at quite specific moments in time. It is imperative that other sources of data are used to confirm suspected influences, and that programmes of ground verification and excavations are used to confirm initial archaeological interpretations. In addition to community knowledge, other sources can include economic data, census information and historical research. As mentioned above, this kind of independent verification remains particularly problematic in Cambodia, where the quality of record-keeping has been generally very poor, so the temptation to draw broad conclusions without additional sources needs to be understood, acknowledged and avoided wherever possible.

As mentioned above, in spite of the numerous aerial campaigns over Cambodia during the colonial period, very few negatives or prints remain. This can be ascribed in part to the tumultuous recent history of the nation, but also to the challenges presented by tropical conditions and the lack of adequate resources to care for these extremely valuable historical archives. The locally-stored negatives of the 1992 FINNMAP aerial coverage, for instance, have deteriorated in the last 20 years to such an extent that prints made from them are now virtually useless for archaeological investigation, and it has fallen to individual research teams who have partial holdings of high-quality prints to scan and archive them digitally. Considering the sheer number of archaeological sites in Cambodia, then, and the pace of modern development in the region, it is clear that the medium- and high-resolution imagery from the 1960s to the 1990s held by foreign governments, such as the CORONA imagery so successfully used for research in Cambodia, will become an even more important and valuable resource in the years to come. This is particularly true when we consider that perhaps only a small fraction of that spy satellite imagery has been declassified, and that the resolution of the best commercially available imagery today is roughly equivalent to the resolution of the best spy satellite imagery of a half century ago.

Bibliography

Ang, C. (2007). In the beginning was the Bayon. In J. Clark (Ed.), *Bayon: New perspectives* (pp. 362–77). Bangkok: River Books.

Anonymous (1933). Chronique: Cambodge. *Bulletin de l'École française d'Extrême-Orient, 33*(1), 514–29.

Aymonier, E. (1901). *Le Cambodge: Les Provinces Siamoises*. Paris: Ernest Leroux.

Baty, P. (2005). *Extension de l'aéroport de Siem Reap 2004: Rapport de fouille archéologique*. Siem Reap/Paris: APSARA/INRAP.
Byrne, D., & Nugent, M. (2004). *Mapping attachment: A spatial approach to aboriginal postcontact heritage*. Hurstville: Department of Environment and Conservation (NSW, Australia).
Carter, A., Chhay, R., Heng, P., Fehrenbach, S. S. , & Stark, M. T. (2011). *Greater Angkor project: July 2011 field season report*. Honolulu: University of Hawai'i-Manoa.
Chandler, D. P. (1996). *Facing the Cambodian past: Selected essays 1971–1994*. Sydney: Allen and Unwin.
Chandler, D. P. (2000). *A history of Cambodia*. Chiang Mai: Silkworm Books.
Claeys, J.-Y. (1951). Considérations sur la recherche archéologique au Champa et au Indochine depuis 1925. *Bulletin de l' École française d' Extrême-Orient, 44*, 89–96.
Crawford, O. G. S. (1955). *Said and done: The autobiography of an archaeologist*. London: Weidenfeld and Nicholson.
Delaporte, L. (1880). *Voyage au Cambodge: l'architecture khmer*. Paris: Delagrave.
Ebihara, M. (1968). *Svay, a Khmer village in Cambodia*. New York: Columbia University.
Evans, D. (2007). *Putting Angkor on the map: A new survey of a Khmer 'Hydraulic City' in historical and theoretical context*. Sydney: University of Sydney.
Evans, D. (2010). Applications of archaeological remote sensing in Cambodia: An overview of Angkor and beyond. In M. Forte, S. Campana, & C. Liuzza (Eds.), *Space, time, place: Third international conference on remote sensing in archaeology* (pp. 353–66). Oxford: Archaeopress.
Evans, D., Pottier, C., Fletcher, R., Hensley, S., Tapley, I., Milne, A., & Barbetti, M. (2007). A comprehensive archaeological map of the world's largest preindustrial settlement complex at Angkor, Cambodia. *Proceedings of the National Academy of Sciences of the United States of America, 104*, 14277–82.
Evans, D., & Traviglia, A. (2012). Uncovering Angkor: Integrated remote sensing applications in the archaeology of early Cambodia. In R. Lasaponara & N. Masini (Eds.), *Satellite remote sensing: A new tool for archaeology* (pp. 197–230). New York: Springer.
Fletcher, R. (2001). A.R. Davis Memorial Lecture. Seeing Angkor: New views of an old city. *Journal of the Oriental Society of Australia, 32–33*, 1–25.
Fletcher, R. (2009). Low-density, Agrarian-based urbanism: A comparative view. *Insights, 2*, 1–19.
Fletcher, R., Johnson, I., Bruce, E., & Khuon, K.-N. (2007). Living with heritage: Site monitoring and heritage values in Greater Angkor and the Angkor World Heritage Site, Cambodia. *World Archaeology, 39*, 385–405.
Fox, J., Suryanata, K., & Hershock, P. (Eds.). (2005). *Mapping communities: Ethics, values, practice*. Honolulu: East–West Center.
Gaughan, A. E., Binford, M. W., & Southworth, J. (2008). Tourism, forest conversion, and land transformations in the Angkor basin, Cambodia. *Applied Geography, 29*, 212–23.
Gillespie, J. (2009). Protecting world heritage: Regulating ownership and land use at Angkor archaeological park, Cambodia. *International Journal of Heritage Studies, 15*, 338–54.
Goloubew, V. (1933). Le Phnom Bakhèn et la ville de Yaçovarman. Rapport sur une mission archéologique dans la région d'Angkor en août-novembre 1932. *Bulletin de l'École française d'Extrême-Orient, 33*(1), 319–44.
Goloubew, V. (1936). *Collaboration de l'Aéronautique et de la Marine indochinoises aux travaux de l'Ecole Française d'Extrême-Orient*. Hanoi: Imprimerie d'Extrême-Orient.
Goloubew, V. (1941). L'hydraulique urbaine et agricole à l'époque des rois d'Angkor. *Bulletin Économique de l'Indochine, 1*, 9–18.
Groslier, G. (1935). Troisièmes Recherches Sur Les Cambodgiens. *Bulletin de l'École française d' Extrême-Orient, 35*, 159–206.
Hendrickson, M. (2007). *Arteries of Empire: An operational study of transport and communication in Angkorian Southeast Asia*. Sydney: University of Sydney.

Himel, J. (2007). Khmer Rouge irrigation development in Cambodia. In *Searching for the truth*. http://www.genocidewatch.org/images/Cambodia_11_Apr_07_Khmer_Rouge_Irrigation_Development_in_Cambodia.pdf. Accessed 15 Mar 2012.

Kalab, M. (1968). Study of a Cambodian village. *The Geographical Journal, 134*, 521–37.

Kasiannan, S. (2011). *Cultural connections amidst heritage conundrums: A study of local Khmer values overshadowed by tangible archaeological remains in the Angkor World Heritage Site*. Unpublished PhD thesis, University of Sydney.

Malleret, L. (1967). Le vingt-cinquième anniversaire de la mort de Victor Goloubew (1878–1945). *Bulletin de l'École française d'Extrême-Orient, 53*, 331–73.

Moore, E. H. (1985). The Williams-Hunt collection of aerial photographs of the archaeological research trust for Thailand. *Journal of the Siam Society, 73*, 252.

Moore, E. H. (2009). The Williams-Hunt collection: Aerial photographs and cultural landscapes in Malaysia and Southeast Asia. *Sari – International Journal of the Malay World and Civilisation, 27*, 265–84.

Moylan, E., Brown, S., & Kelly, C. (2009). Toward a cultural landscape atlas: Representing all the landscape as cultural. *International Journal of Heritage Studies, 15*, 447–66.

Nesbitt, H. J. (Ed.). (1997). *Rice production in Cambodia*. Manila: International Rice Research Institute.

Pijpers, B. (1989). *Kampuchea: Undoing the legacy of Pol Pot's water control system*. Dublin: Trocaire.

Pottier, C. (1999). *Carte Archéologique de la Région d'Angkor. Zone Sud*. Unpublished Ph.D thesis, Université Paris III – Sorbonne Nouvelle.

Pottier, C. (2000a). À la recherche de Goloupura. *Bulletin de l'École française d' Extrême-Orient, 87*, 79–107.

Pottier, C. (2000b). Some evidence of an inter-relationship between hydraulic features and rice field patterns at Angkor during ancient times. *Journal of Sophia Asian Studies, 18*, 99–120.

Pottier, C. (2004). À propos du temple de Banteay Chmar. *Aséanie, 13*, 131–50.

Thompson, V., & Adloff, R. (1953). Cambodia moves toward independence. *Far Eastern Survey, 22*, 105–11.

Tobias, T. (2000). *Chief Kerry's Moose: A guidebook to land use and occupancy mapping, research design and data collection*. Vancouver: Union of British Columbia Indian Chiefs and Ecotrust Canada.

Vickery, M. (1996). Review article: What to do about The Khmers. *Journal of Southeast Asian Studies, 27*, 389–404.

Winter, T. (2007). *Post-conflict heritage, postcolonial tourism: Culture, politics and development at Angkor*. London: Routledge.

Chapter 18
Integrating Aerial and Satellite Imagery: Discovering Roman Imperial Landscapes in Southern Dobrogea (Romania)

Ioana A. Oltean and William S. Hanson

Abstract This chapter demonstrates the value of analysing a range of remotely sensed imagery in order to study the development of the historic landscape in southern Dobrogea (Romania). The methodology involves integrating within a GIS environment low-altitude oblique aerial photographs, obtained through traditional observer-directed archaeological aerial reconnaissance; medium-altitude historical vertical photographs produced by German, British and American military reconnaissance during the Second World War; and high-altitude declassified US military satellite imagery (CORONA) from the 1960s. The value of this approach lies not just in that it enables extensive detailed mapping of large archaeological landscapes in Romania for the first time, but also that it allows the recording of previously unrecognised archaeological features now permanently destroyed by modern urban expansion or by industrial and infrastructural development. Various results are presented and illustrated, and some of the problems raised by each method of data acquisition are addressed.

I.A. Oltean
Department of Archaeology, University of Exeter, Laver Building, North Park Road, Exeter, EX4 4QE, UK
e-mail: I.A.Oltean@exeter.ac.uk

W.S. Hanson (✉)
Department of Archaeology, Centre for Aerial Archaeology, University of Glasgow, Glasgow, G12 8QQ, UK
e-mail: william.hanson@glasgow.ac.uk

W.S. Hanson and I.A. Oltean (eds.), *Archaeology from Historical Aerial and Satellite Archives*, DOI 10.1007/978-1-4614-4505-0_18,

18.1 Introduction

Over the last decade or more, the authors have been largely responsible for the reintroduction of aerial survey and a renewed appreciation of the value of the aerial perspective in Romanian archaeology (Hanson and Oltean 2001, 2003; Oltean and Hanson 2001, 2007a, b), where in common with most countries of Eastern Europe, there is little or no tradition of aerial survey. They have been involved in a major assessment of the archaeological landscape of southern Dobrogea since 2004,[1] in a research project which seeks to define the nature of the trends in occupation of the landscape from later prehistory to the Roman period. This work represents the first major landscape-focused survey ever to be undertaken in the region. More widely, however, the methodology applied, which is the primary focus of this chapter, breaks new ground. It involves the integration of a range of remotely sensed aerial and satellite data, combining both newly acquired and historical archival material. This chapter puts particular emphasis on the latter, in keeping with the theme of the volume, and provides several examples of its successful application, demonstrating its value in both recording features which are no longer visible and enhancing the ability to map the wider archaeological landscape.

Dobrogea in southeastern Romania is geographically defined by the Danube River to the west and north and by the Black Sea to the east. The present study covers its southern part, between the Bulgarian border and the canal which links the Danube and the Black Sea (Fig. 18.1). Southern Dobrogea has been a focus of human settlement since the Neolithic, but was also a major zone of movement and interaction, an area of constant ebb and flow of populations and of cultural interference between the classical Mediterranean and prehistoric Eurasian cultures (Batty 2007: 97). The indigenous north Thracian population (the *Getae*) experienced early exposure to Mediterranean civilization through the Greek colonisation of the Black Sea shores. The subsequent occupation of the area by the Roman empire from the

[1] This work began as part of a British Academy-funded research programme *Contextualizing change on the Lower Danube: Roman impact on Daco-Getic landscapes* under the direction of the first author, mentored by the second. That overarching project sought to study the effect of shifting Roman imperial politics of power, provincial administration and colonisation on shaping the traditional Daco-Getic settlement pattern of the Lower Danube, employing a wide-scale comparative analysis of a representative sample of three landscapes in modern Romania with a similar Daco-Getic late Iron Age ethnic backgrounds, but contrasting experience of Roman contact, namely, southern Dobrogea, southwestern Transylvania (see Oltean 2007) and northern Crişana. The first of these areas has since developed into the primary focus of ongoing research and become a collaborative project with the second author, under the auspices of the ArchaeoLandscapes Europe Project, funded by the European Commission.

The project also involves a number of Romanian collaborators and beneficiaries. These include the National History Museums of Romania (Bucharest) and of Transylvania (Cluj-Napoca); the history and archaeology museums in Constanţa and Oradea; the Institute for Cultural Memory (cIMeC-Bucharest); and the Babeş-Bolyai University (Cluj-Napoca).

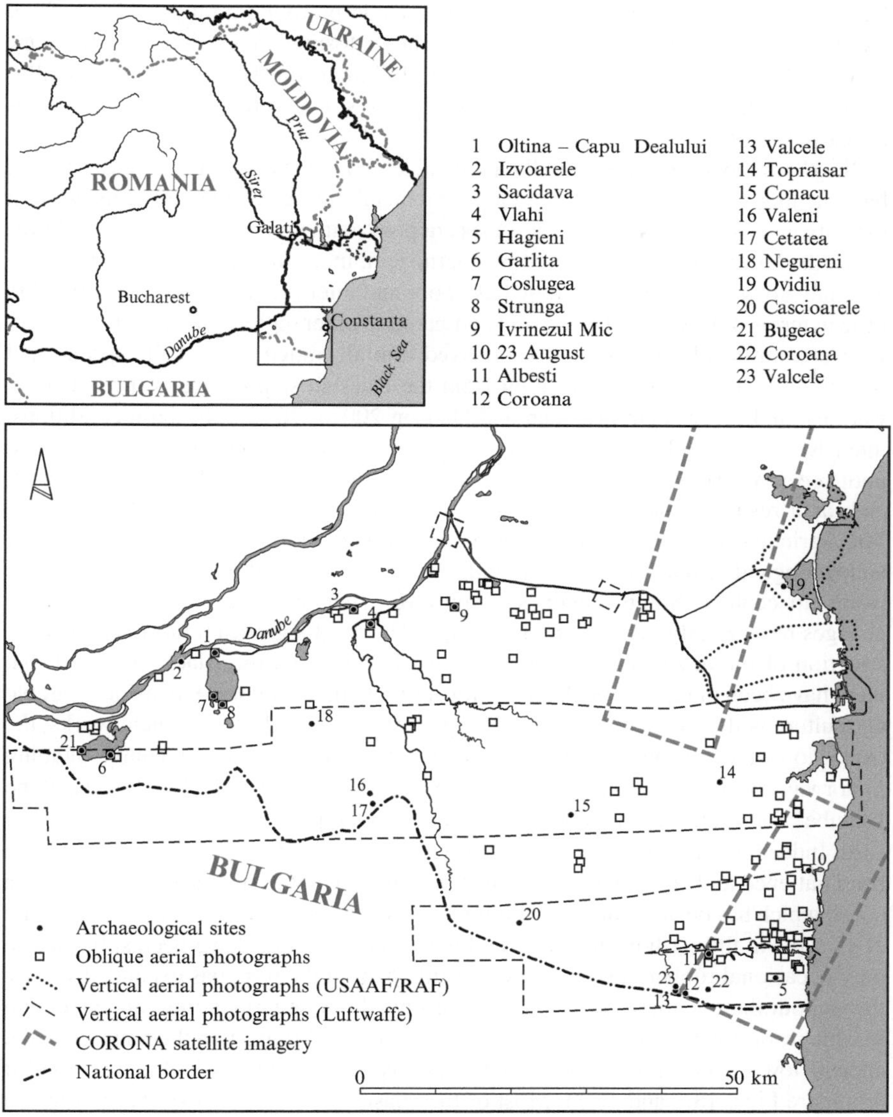

Fig. 18.1 Map of southern Dobrogea indicating the extent of archival imagery and the location of sites photographed during aerial reconnaissance

first century AD eventually dissolved into Byzantine, then Ottoman imperialism. Apart from making this an ideal case for investigating settlement and landscape change arising from colonisation and imperialism, the study area presents archaeological investigators with the challenge of recognising a wide array of man-made features. Within this archaeological landscape can be found a range of sites, from those of larger dimensions, such as settlement enclosures, roads and large funerary barrows dating from at least the sixth century BC to the third century AD, to smaller features, including sunken-floored structures or stone-walled buildings in unenclosed settlements.

Southern Dobrogea is essentially a low tableland, wide and flat in its eastern half, but reaching altitudes of 200 m and topographically more fragmented in its western half. Its geology consists of a thick loess deposit with some chalk/limestone intrusions, the latter sometimes covered by chernozem, more frequent towards the south-east and along the few rivers. These have poor and intermittent outflow, which offers little in terms of fresh water provision compared to other regions in Romania. Indeed, with predominantly intense and short-lived rainfall, much of it in the spring, and occasional hot dry winds emanating from the Russian steppe, this is the country's warmest and driest region (Oltean and Hanson 2007a: 76–77). In these conditions, intensive agricultural use of the landscape for cereal cultivation has been constrained until recent decades, when extensive irrigation made this possible on a wide scale, and, as a result, arable cultivation has successfully superseded animal husbandry. This agricultural history created reasonably good conditions for the preservation of archaeological remains in the area, with numerous sites still extant until late in the twentieth century. Some remain so even today, but over recent years, numerous changes have greatly affected the landscape. The most significant has been the construction of the navigable canal between the Danube and the Black Sea itself, but there have been many others: the intensification of quarrying; the expansion of arable agriculture with its concomitant ploughing erosion and land improvement works; the expansion of urban areas and infrastructure development, most recently ongoing motorway construction; and finally, erosion along the shores of the Black Sea and the Danube. These all present substantial dangers to the preservation of archaeological sites. Indeed, thousands of funerary barrows and numerous settlements have already been flattened and are now under cultivation. The town of Mangalia on the Black Sea coast, which has been occupied continuously since its foundation as the Greek colony of *Callatis* in the sixth century BC, and whose ancient ruins have been subjected to only occasional robbery and to accreted settlement in the past, has expanded greatly throughout the twentieth century from a sleepy, dusty settlement into an expanding seaside resort town. This expansion has had detrimental impact on the remains of the ancient town itself and on much of the extensive necropolis which surrounded it (compare Figs. 18.2 and 18.3). Most of the coastal area has been similarly affected through the development of the Romanian Black Sea Riviera for tourism.

Fortunately, in terms of aerial survey, moisture stress can be a regular occurrence in the area because of the local climate, geology and cultivation patterns, resulting in very favourable conditions for the formation of archaeological cropmarks (Hanson and Oltean 2007a, b). These may occur on much of the modern landscape, which is now predominantly given over to arable cultivation of cereals, with pasture land occurring where water shortages restrict cultivation. The peak period in crop development for revealing cropmarks is throughout June and very early in July,

Fig. 18.2 Ancient funerary barrows and road network appearing as lighter marks in the fields outside Mangalia (*Callatis*) visible on CORONA imagery in April 1966

though excesses of precipitation over the spring months, combined with the natural conditions, may produce occasional local variations. Accordingly, excessive rainfall may delay cropmark formation by a few weeks, while drought could prevent germination altogether. In the latter case, however, the geology of the study area with its often contrasting subsoil and topsoil makes possible further detection of some (though not all) archaeological remains as soilmarks.

Fig. 18.3 The area around Mangalia (*Callatis*) as shown on Google Earth in 2007, showing the impact of agricultural development and urban expansion on the survival of archaeological remains visible on Fig. 18.2 (© 2012 Google Earth; © 2012 TerraMetrics; © 2012 Digital Globe)

18.2 Imagery: Sources and Methodology

The specific conditions of this region and the appreciation that cropmarks, especially in cereal cultivation, best reveal the shape and layout of buried archaeological remains in ploughed landscapes (Hanson 2007), determined that particular attention was devoted to the identification and mapping of buried remains detectable primarily via cropmarks, with a secondary focus on the soilmark evidence. The mapping of archaeological remains in the study area involves interpreting and integrating data from a range of remotely sensed imagery of varying acquisition dates, technical parameters and reasons for acquisition (Fig. 18.1). All these characteristics had a radical influence on the quality of archaeological information contained in each data set and on the technical problems which needed to be addressed in order to integrate them within a coherent system of management and interpretation. In excess of 950, low-altitude oblique aerial photographs of archaeological sites were obtained by the authors through traditional aerial reconnaissance involving several flights at an altitude of approximately 500 m. These surveys were performed from a Cessna 4-seater light aircraft in separate programmes of work planned to coincide with the cropmark season in June 2005 and 2006 and again in June 2011. The aerial survey involved the long-established archaeological method of observer-directed reconnaissance, with both authors scanning the ground for any possible sites which were then recorded using handheld cameras, in this case a digital SLR camera (Canon EOS 300D) and two 35-mm single-lens reflex cameras (Nikon and Pentax) for panchromatic negative and colour slide film (Oltean and Hanson 2007b). Post-reconnaissance processing of these oblique photographs involves digital rectification using control point data, usually from base maps, in order to transform them into orthophotos which provide a sufficiently accurate basis for the transcription and mapping of archaeological features.[2]

In addition to these oblique photographs taken by the authors, the acquisition of historical archival aerial photographs and satellite imagery provides an opportunity to examine the landscape at different stages throughout its evolution over the past century and offers the potential to identify sites now lost to agricultural erosion or urban development. Over 200 aerial photographs of the area were identified and acquired from The Aerial Reconnaissance Archive (TARA), then located in the University of Keele but now housed in RCAHMS, Edinburgh. The research potential of this resource is extremely high but has been barely used by academic researchers, particularly for areas of Eastern Europe where the absence of readily available finding aids or keys makes accessing the data both difficult and time-consuming (see Chap. 1 by Hanson and Oltean and Chap. 2 by Cowley et al., this volume). Virtually all this material consists of vertical photographic block coverage acquired during the Second World War through medium-altitude reconnaissance flights for military intelligence gathering throughout April, May and June 1944 by the German and British-American Air Forces (Figs. 18.1, 18.5, 18.6, and 18.7b).[3] Additional single-frame photomosaics were taken by the German Luftwaffe over key strategic objectives, such as the marshalling yards at Medgidia on 18 April 1940 at an approximate scale of 1:10,000 (Fig. 18.7a) or railway bridges, most notably that of Cernavoda on 24 April 1943 at an approximate scale of 1:7,000. All were made available to us in digital format as high-resolution raster (.tiff) scans of the original prints, films and glass slides.

The subsequent Cold War coverage utilised in this study consists of declassified CORONA high-altitude military satellite photography acquired from the USGS (see Chap. 4 by Fowler, this volume). Although a basic search for imagery within the study area returned some 1,000 data base hits (951 declassified in 1996 and 45 declassified in 2002), only two provided sufficient data quality in terms of cloud

---

[2] Mapping of the data from the most recent flights in 2011 is still ongoing.

[3] The most extensive coverage was by the Luftwaffe in April and May 1944 (sortie reference numbers: GX 22248, 22249, 22251 and 22252); suitable Allied coverage was restricted to limited areas of the Black Sea coast (sortie reference number: MAPRW 60-PR-460).

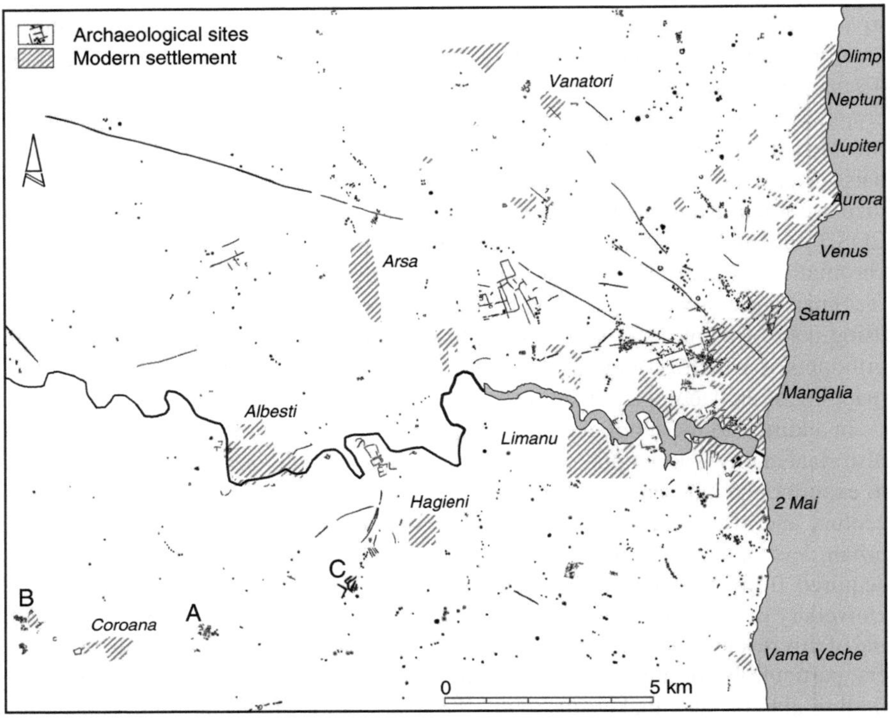

Fig. 18.4 Distribution of archaeological features recovered from aerial and satellite imagery in the area of Mangalia

coverage and ground resolution to be considered here (Figs. 18.1 and 18.2). Taken on 21 April and 21 March 1966, respectively,[4] both are photographic products of the CORONA satellite, taken by the KH-7 high-resolution surveillance camera with a high ground resolution (2–4 ft) on black-and-white 9-in. negative film (see Chap. 4 by Fowler, this volume). High-resolution satellite imagery (primarily QuickBird and GeoEye) of recent date, freely available via Internet-based resources, was also examined. Google Earth coverage for this area is excellent and frequently updated, providing extremely useful surveys obtained in different seasons over several years. This is extremely useful both for enhancing coverage generally and for comparison with older coverage to assess the impact of developmental change (compare Figs. 18.2 with 18.3 and 18.7 with 18.8). Finally, five QuickBird coverages acquired between 2002 and 2008 were obtained as separate multispectral and panchromatic data sets through collaboration with the Remote Sensing Laboratory of Postgraduate

[4] DZB00402700026H018001 Mission 4027, Frame 18 and DZB00402600042H023001 Mission 4026, Frame 23.

School of the Navy in Monterey, USA, and are the subject of a separate study (Oltean and Abell 2012).[5]

All the imagery collected is used as a base layer on which the interpretation and transcription of archaeological information is built into digital archaeological maps and site plans utilising GIS software (ArcView 3.3 and ArcGIS 9.0). The GIS environment presents further advantages as it allows data integration with published archaeological information from the area in order to facilitate chronological interpretations of the identified remains and subsequent full analysis of the evolution of the settlement pattern in the study area (e.g. Fig. 18.4).

18.3 Problems with Individual Data Sets

Data collection at the right time, particularly in relation to the limited period of visibility of agricultural cropmarks, and the visual clarity of the often quite small features captured on this type of imagery are essential premises for the optimal recovery of archaeological information from the air (see also Chap. 15 by Beck and Philip, this volume). As the majority of aerial and satellite data currently archived throughout the world were collected for purposes other than archaeology, they have often been criticised as having been collected at inappropriate times for archaeological investigation. Indeed, as military intelligence gathering was the primary reason for the acquisition of aerial photographs during the Second World War, their dates of acquisition are, inevitably, heavily dependent on the evolution of military operations rather than the cropmark season, and by no means, all were acquired in conditions ideal for the recovery of archaeological information (see Chap. 9 by Oltean, this volume, for a more detailed discussion). Thus, in southern Dobrogea, the German military block coverages of 30 April 1944, the 1940 and 1943 photomosaics and even the USAF coverage of 31 May 1944 were all taken too early in the growing season to best document archaeological remains through cropmark evidence. The same applies to the CORONA imagery collected during the spring of 1966.

Visual clarity of both Second World War aerial photographs and CORONA imagery to enable feature recognition may also raise problems for the interpreter. The age of the material and inconsistencies in its archival treatment have resulted in a variable state of preservation of the Second World War photographs and films, and physical and chemical deterioration of negatives, along with the presence of occasional annotations, may conceal archaeological features sometimes making it difficult to use the imagery (Figs. 18.7a and 9.3; for a more detailed discussion, see Chap. 9 by Oltean, this volume). The CORONA imagery is better preserved physically, but even at its best has slightly poorer spatial resolution than the Second World War vertical aerial photographs and can be affected to a greater degree by cloud coverage obscuring the ground. Fortunately, the data acquisition season of the

[5] Coverages were acquired from the 8th of July 2002 (005677175010_01), two from the 25th of June 2004 (005677173010_01; 005677174010_01; 005677178010_01) and one each from the 18th of July 2004 (005677172010_01) and the 8th of June 2008 (005677176010_01).

CORONA satellite imagery makes it very compatible with the Second World War material, both in terms of visual identification of archaeological remains as soilmarks and the types of archaeological features recoverable. Accordingly, despite these impediments, both types of data sets proved to be extremely prolific in their depiction of extant or soilmarked archaeological features and particularly useful for larger monuments, such as ancient roads, linear defensive systems and tumular cemeteries. The latter in particular were sometimes too large and the more scattered tumuli too numerous to be documented efficiently by traditional reconnaissance producing single photographs from an altitude of 500 m. They were much more readily plotted from the vertical block coverage provided by historical archival aerial or satellite photographs.

One of the major benefits of observer-directed aerial reconnaissance is its flexibility, allowing recording to take place at the optimum time of the year for the recovery of archaeological sites and to focus on those areas considered likely to be most productive (c.f. Wilson 2005). In addition, the lower flying height provides extremely detailed, high-resolution imagery. Furthermore, direct engagement in the acquisition process in the air provides the researcher with a better appreciation of the wider landscape. The application of such aerial reconnaissance in southern Dobrogea provided some excellent results, with 140 sites recorded, both extant and as cropmarks. Unfortunately, as a survey methodology which has not changed significantly since its inception in the 1920s, aerial reconnaissance remains restricted in its successful application by a number of problems, both specific to the conditions for undertaking this research project (such as finding a suitable base or establishing when and where it was possible to fly) and more general, related to the leading principles of its traditional methodology. The optimum time for data collection can only be estimated based on prior knowledge of natural conditions or, ideally, on extensive experience in aerial reconnaissance in the area under study. Under the conditions usually in force for research projects abroad, the resources to remain in the area for a more extended period in order to test these assumptions more rigorously are very restricted. At the beginning of the project the area was entirely new, not just to the authors but to aerial reconnaissance itself, and estimations had to be based on previous experience of aerial reconnaissance elsewhere in Romania in Western Transylvania between 1998 and 2003 (e.g. Hanson and Oltean 2001). Allowing for differences in relation to the local climate and agricultural regime in this more easterly coastal location, it was possible to estimate that the ideal time to fly to maximise recovery of cropmarks was likely to be in early to mid-June. However, throughout the first reconnaissance seasons of 2005 and 2006, problems of weather and the general wetness of the seasons either prevented flying or, in the longer term, affected the appearance of cropmarks. Indeed, in 2006 the Danube experienced severe flooding throughout the spring which was still apparent at the time of reconnaissance. Even if sometimes the floods could be turned to advantage where they aided the appreciation of the topography in relation to site location, as for example in the case of the Iron Age and Roman settlement at Oltina-Capu Dealului, generally the consecutive wet seasons are likely to have impacted significantly on the number and the nature of archaeological sites identified (see below). Fortunately, subsequent reconnaissance in 2011 did not suffer from problems of weather conditions, but this example serves to emphasise the need to take a long-term approach when undertaking aerial survey in any region and further reinforces the benefit of accessing imagery of different dates, however acquired.

A major benefit of aerial reconnaissance over terrestrial survey is the spatial coverage that it allows. Location of airfield facilities, however, is always problematic, especially in countries like Romania where a sharp decline in the previously state-owned airclub system and the current embryonic development of alternative privately owned airfields severely restricts choice. The nearest available facility, the private airfield on the Black Sea coast at Tuzla, was fortunately located within the study area, but its non-central position created an inherent bias in reconnaissance coverage towards the eastern part. A further limitation on archaeological aerial reconnaissance in the study area is restricted air space: light aircraft are greatly constrained when trying to fly anywhere north of the Danube-Black Sea Canal because of military restrictions; nor is reconnaissance allowed within 5 km of the nuclear power station at Cernavodă, nor within 5 km of the Bulgarian border. Restrictions of this nature are a serious impediment to scholarly attempts to study ancient landscapes, as most frequently modern boundaries do not correspond to those in force in the past and, as a result, studies based on information which is imbalanced in terms of spatial coverage are still shaping our understanding of the past.

More generally, however, the fundamental problem with traditional archaeological aerial reconnaissance, which has come under heavy criticism in recent years in terms of its cost-effectiveness and productivity rate (e.g. Palmer 2005), is the serendipitous and highly selective nature of the process. Surveyors fly routes selected on the basis of the probability of site visibility and record only what they perceive as significant when in the air. Observer subjectivity during flights further biases data recovery, as it is far too easy to miss sites simply by virtue of flying over the wrong field or looking out of the opposite side of the aircraft at a crucial moment, as confirmed by the evidence of archaeological sites being recorded only on the periphery of photographs taken with the intention of documenting other targets (Cowley 2002). Indeed, it is a commonplace amongst practising archaeological aerial surveyors that more is often visible on photographs when examined later than was appreciated at the time by the photographer, as might reasonably be expected given the greater time to peruse every detail and the facility to enhance visibility by digitally manipulating the image. By contrast, one of the benefits of archival vertical imagery is the block coverage that it provides, allowing time for a more systematic examination of the area under study (e.g. Fig. 18.2).

18.4 Data Integration Benefits

The integration of aerial reconnaissance data with historical and more recent archival material has made a major contribution to achievement of the aims of the project. The Second World War aerial and Cold War satellite military coverages were highly selective in relation to the strategic relevance of target areas, leaving important gaps in the coverage of the whole study area (Fig. 18.1). However, this was to some extent compensated for by the amount of high-resolution satellite imagery freely available via Google Earth covering the greatest part of the study area. This extensive and fairly systematic coverage enabled, for the first time in Romania, the reconstruction of an archaeological landscape and the analysis and detailed mapping of a large number of sites.

A wide range of sites and monuments have been identified and mapped. Some were previously known (e.g. Oltina, Izvoarele, *Sacidava*), but perhaps insufficiently documented or with imprecise locational data. However, many other sites, including large open settlements like Hagieni (see below), were entirely new to the local archaeological record. Prehistoric fortified settlements identified through aerial reconnaissance along the Danube and its lakes include Gârliţa (Bronze Age), Coslugea and Strunga (late Iron Age) or Ivrinezul Mic (of uncertain date, see http://archweb.cimec.ro/scripts/ARH/RAN/sel.asp -last viewed August 2010 RAN 62734.02). Further enclosures discovered in the same way included examples in the eastern part of the study area at 23 August, probably occupied during the late Iron Age and the Roman periods, or of uncertain date at Mangalia and Albeşti. Various other examples were identified on Luftwaffe imagery of April 1944 slightly further inland at Vâlcele, Topraisar, east of Conacu, Cetatea (probably from the Iron Age), Coroana (possibly Hellenistic) and Văleni (of unknown date) (Fig. 18.5c). Other individual enclosures of Roman, Late Roman or Byzantine date, some whose morphology suggests a probable military function, were also recorded on Second World War aerial photographs at Pietreni (Fig. 18.5a; associated with a linear group of barrows), Negureni (Fig. 18.5b), Căscioarele and Ovidiu (Oltean and Hanson 2007b, Fig. 5). A separate example located by aerial reconnaissance on the western shore of Lake Bugeac may be of funerary rather than military character (Oltean and Hanson 2007b, Fig. 4 *contra* Crăciun 2009).

Unenclosed settlements remain poorly represented in the study area. Rather than a cultural feature, this is most likely the result of problems affecting individual data sets, such as ground resolution or the stage of crop development (see discussion above). Open settlements were present within the area, as indicated by a large example discovered by aerial reconnaissance in 2005, and augmented by further survey in 2006 and 2011, located midway between Hagieni and Vama Veche (Fig. 18.4). Although previous information provides no indication of the presence of a settlement in this location, cropmarks revealed the layout of extensive, internally subdivided, stone-walled buildings buried beneath modern cereal crops, indicating the presence of a substantial aggregated ancient settlement of probable Roman date (Oltean and Hanson 2007b, Fig. 2; Oltean and Abell 2012: 294-298). Further examples of such

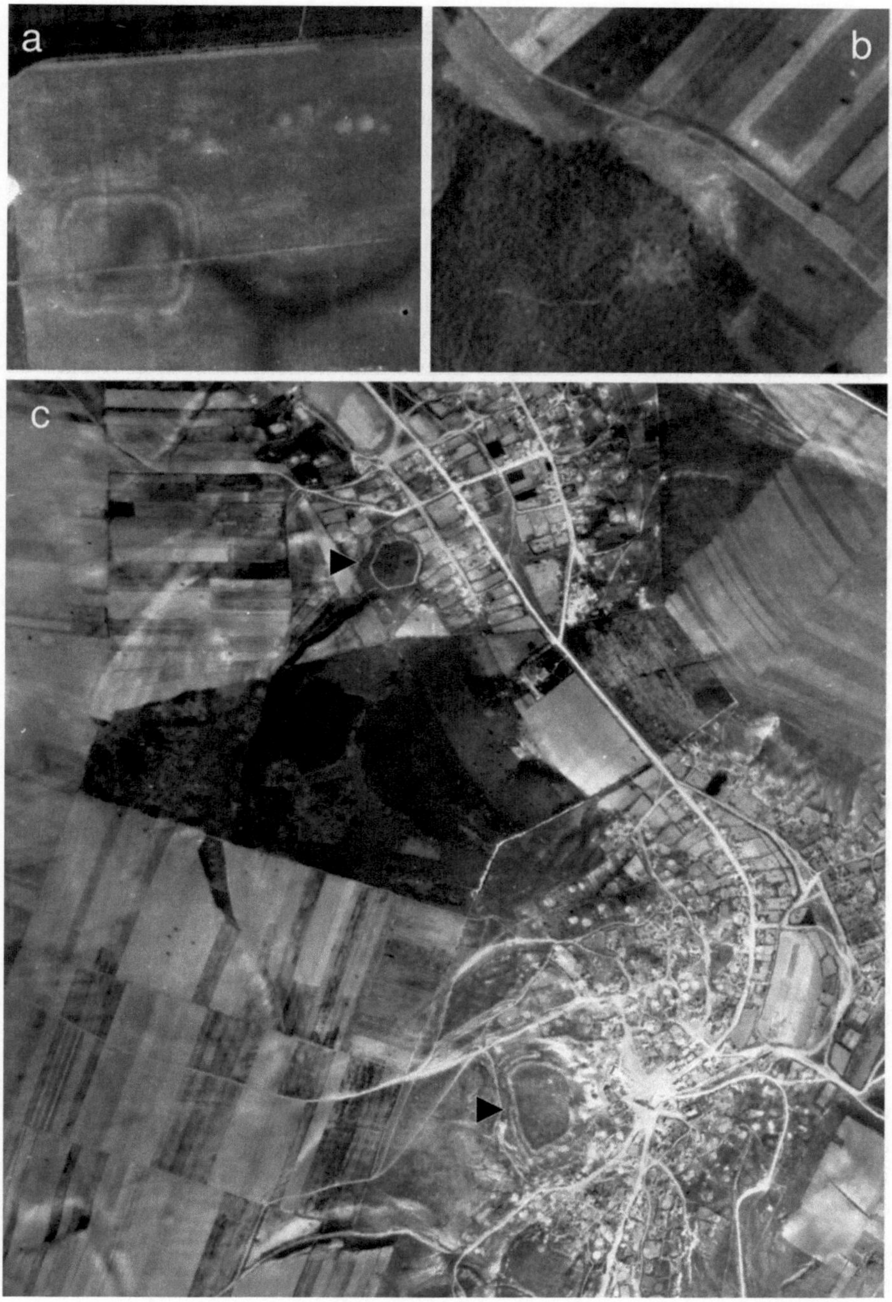

Fig. 18.5 Fortified sites of prehistoric, Roman, Late Roman and Medieval date from Luftwaffe aerial photographs: (**a**) northeast of Pietreni (GX 22249 frame 58), (**b**) east of Negureni (GX 22249 frame 68) and (**c**) at Văleni and Cetatea (GX 22250 frame 168) (All images: Licensor NCAP/aerial.rcahms.gov.uk)

settlements were identified in the southeastern part of the study area and along the Bulgarian border. Five kilometres away to the west, traces of lighter-coloured debris may indicate the location of a group of flattened stone buildings laid out around several roads. Further west, the remains of at least two other stone-built settlements have been confirmed on Second World War aerial photographs, one overlooking the dry canyon valley of the Limanu river at Coroana (Fig. 18.6) and the other 4 km downstream towards Vâlcele. Though no previous records exist for the latter, the former site may be identical with a previously reported Roman settlement (Lista Monumentelor Istorice 2004: CT-I-s-B-02637 Coroana-Oierie). In both cases, a number of households, internal roads and animal pens are visible because the slight traces of the stone-walled structures were still extant at the time of photography in April 1944. CORONA imagery attests that the settlement at Coroana was still extant in April 1966 (Fig. 18.6b), but that coverage did not reach as far as Vâlcele. Given their proximity to the Bulgarian border (2.4 and 1.8 km, respectively), neither of these examples could be reached by modern aerial reconnaissance.

However, it was the wider landscape approach facilitated by the aerial coverage, and particularly the historical imagery, that best documented other activities from the past in the study area. For instance, traces of previous cultivation in the form of field systems and rig and furrow have been documented for the first time in this part of Romania. Such extensive remains of rig and furrow were revealed by Luftwaffe aerial photographs of 1944 over much of the western uplands and particularly in the area of the modern state boundary overlapping it. So too the ancient road network, whose extent is more readily mapped from the extensive vertical coverage provided by the historical imagery (see below).

Of particular significance was the impact of the survey on our understanding of the extent and distribution of funerary barrows (tumuli) of prehistoric and Roman date in the region. Though some are still extant and have been recorded on modern topographic maps or, indeed, on some earlier maps of the nineteenth century[6] and a number have been sampled by excavation (see recent overview by Lungu 2007), what became immediately apparent from the very first reconnaissance flight was that the whole landscape is littered with thousands of barrows. For every extant example still visible, several more were revealed as cropmarks or soilmarks. Thus, the overwhelming majority had already been flattened by the plough and as a result had remained unrecognised by previous studies. They are distributed either as individual monuments or as clusters along Roman roads or in various dominant positions (Figs. 18.2 and 18.4). The use of such burial tumuli has a long tradition in the area from the sixth century BC to the fourth century AD. This research project has identified and mapped 8,758 such barrows so far. This is the first time they have been quantified, so it is not possible to indicate the percentage increase in knowledge.

---

[6] For example, *Sketch of the routes from Kustenjeh to Chernavoda and Rassova with the Karasu Lakes by Capt T. Spratt R.N. C.B. 24 July 1854. Made during a reconnaissance in company with Lieut. Col. the Hon. A. Gordon and Lieut. Col. J. Desaint de l'Etat Major*. (G236:4/6) – see http://www.nmm.ac.uk/collections/explore/object.cfm?ID=G236%3A4%2F6

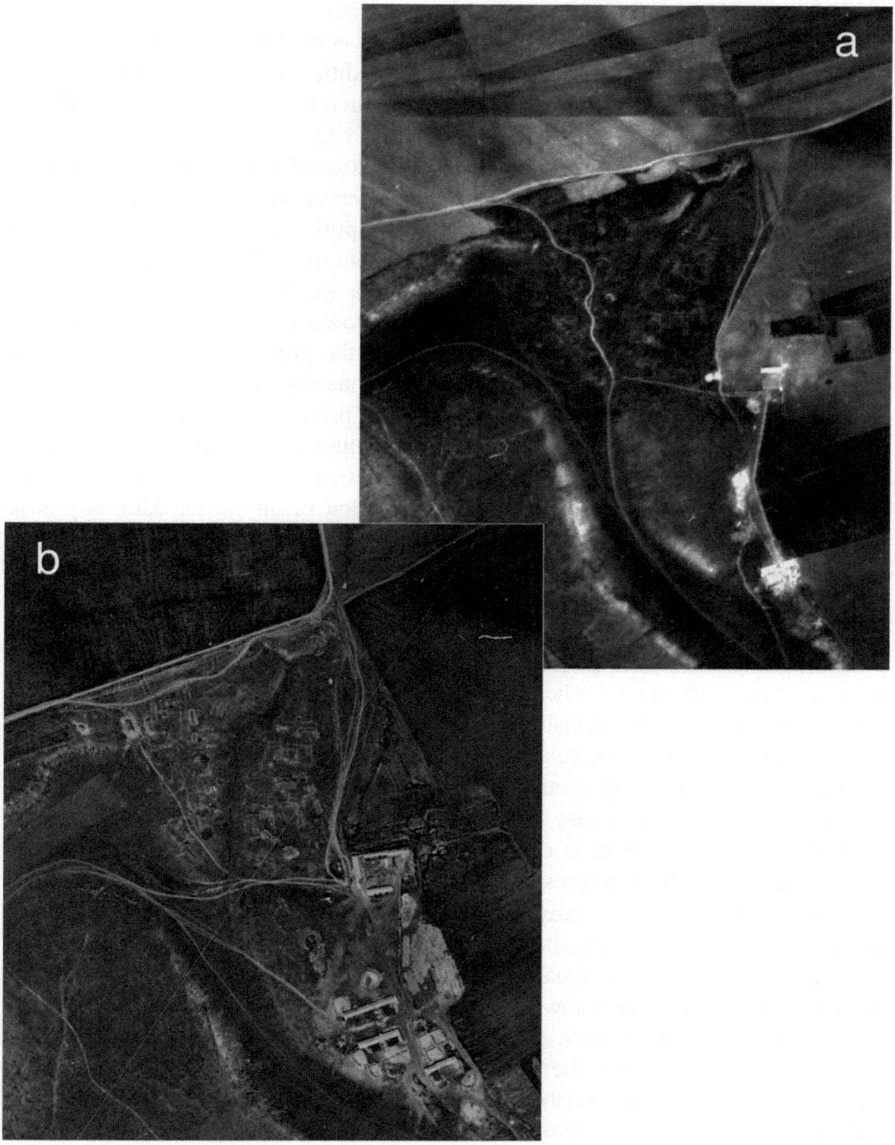

Fig. 18.6 Extant remains of stone walls from an unenclosed settlement of uncertain date east of Coroana as visible on (**a**) Luftwaffe aerial photograph GX 22248 frame 79 in April 1944 (Licensor NCAP/aerial.rcahms.gov.uk) and (**b**) on CORONA satellite photograph in April 1966

Oblique photographs allowed the better appreciation of their character, landscape location and state of preservation (e.g. Oltean and Hanson 2007b), but their widespread distribution and large numbers would have made futile any attempt to map all of them on the basis of such photographs. Fortunately, the more extensive coverage provided by the military intelligence archival material (both TARA and CORONA) and, indeed, the recent satellite imagery consulted has provided an ideal background for mapping their full extent. As an additional bonus, the sets of historical imagery utilised provided coverage of areas such as Constanţa (ancient *Tomis*) and Mangalia (ancient *Callatis*) before they had been affected by the aggressive post-war development and expansion which destroyed numerous archaeological remains. Thus, extensive traces of a tumular cemetery apparent on both CORONA and Luftwaffe material indicate that the ancient necropolis extended inland for several kilometres away from *Callatis* (Fig. 18.2) in a pattern similar to the better known cemetery outside ancient *Histria* in northern Dobrogea (Alexandrescu 1966). The modern town of Mangalia has expanded greatly towards the north and west since 1966, damaging these archaeological remains in the process (compare Figs. 18.2 and 18.3); yet, despite this and the numerous archaeological excavations since the 1930s, no overall site plan of the area has been produced until now (Fig. 18.4). Similarly, USAF aerial photographs of May 1944 outside modern Constanţa allowed the recovery of further remains of a similar nature from the ancient tumular necropolis of *Tomis*, though in this case urban expansion of an earlier date may have already destroyed significant parts of it.

The spatial distribution of the barrows identified is additionally significant as it is currently providing the quantitative basis for determining ancient trends in land occupation, distinguishing areas of more aggregated settlements from those with a predominantly individual pattern, and identifying the communication networks between these settlements.[7] The quantitative data recovered in relation to funerary activity allows us to distinguish several possible levels of aggregation in ancient settlement types and to appreciate their respective spatial distribution. By far, the predominant pattern of aggregation (86.5%) is that of single or small groups of barrows (≥5) evenly spread across the entire area. However, larger clusters of barrows suggest their interpretation as cemeteries belonging to ancient aggregated settlements, even where the structures of the settlement are otherwise unattested. Notable examples include Izvoru Mare, Fântâna Mare, Valea Dacilor, Cuza Vodâ and Bârâganu, with cemeteries perhaps large enough to belong to urban communities. All are located in the central part of the study area at a distance of at least 15 km from known major ancient towns, which lends further support to their probable identification as central places. Although more research on the ground is needed to confirm these observations, the analysis enabled by combined aerial photographic and satellite evidence is nevertheless the first consistent attempt to reconstruct the Hellenistic and Roman settlement pattern of ancient Dobrogea on the basis of archaeological, rather than epigraphic evidence.

Linear arrangements of barrows may suggest their placement along ancient roads, which can add to the stretches of major or secondary Roman roads directly identified by reconnaissance or on archival aerial and satellite imagery. Accordingly, in an analysis of a wide area between 2 Mai and Limanu, south of the former Greek

[7] A more detailed study by the first named author will appear shortly in *Antiquity*.

colony and Roman and Byzantine town at *Callatis* (Mangalia), the distribution of tumuli gives a clearer picture than ever before of the settlement pattern and road network in the vicinity of the town (Fig. 18.4). Moreover, throughout the study area these barrows revealed a tendency to be aligned not only along Roman roads but also with the banks of the Danube, the shores of Black Sea and in other highly visible positions (topographical characteristics attested elsewhere in European prehistory – see, for example, Tilley 1994; Bradley 1998). Such spatial distributions may further contribute towards our understanding of evolutionary patterns in the construction of funerary monuments.

The landscape approach espoused in this study and facilitated by the integrated analysis of a range of aerial and satellite imagery has significantly enhanced our understanding of the variable character of human occupation over time within its natural and archaeological context. To give two specific examples: at Cascioarele, a rectangular enclosure with Late Roman characteristics, overlooking a stretch of ancient road, partly overlaps a group of barrows from an earlier period; at Ovidiu, thanks to the Second World War USAF coverage of May 1944, the known late Roman small *burgus* can now be placed in a wider context alongside the newly discovered promontory enclosure, large rectangular enclosure and group of barrows, both of the latter now completely destroyed by post-war development (Oltean and Hanson 2007b, Fig. 5).

This integrated aerial approach has also provided a better insight and understanding of large-scale monuments, facilitating a major reassessment of the linear fortifications,[8] known collectively as the Valu lui Traian, which extend across Dobrogea from Cernavodâ (*Axiopolis*) on the Danube to the Black Sea coast at Constanţa (*Tomis*). Though considered to be Trajanic in popular tradition (hence the place name Valu lui Traian or Trajan's Rampart), the system has been assumed by Romanian archaeologists to be tenth to eleventh century AD in date since the late 1950s, on the basis of a Slavic inscription of AD 943 recovered from a fort on the Stone Wall to the south of Mircea Vodă and as a result does not feature in modern considerations of Roman frontiers. There has been minimal excavation and very little other work on the system since the 1960s. There are clearly three different lines of fortifications, two earthworks, the so-called Large and Small Earthen Walls, and a Stone Wall. The latter and the Large Earthen Wall mirror each other closely for almost half of their length. Both have a series of attached rectangular fortifications sufficiently strongly reminiscent of Roman military works to suggest that an original Roman date is much more probable than an early medieval one, but with later reuse. The fortification lines, including their attached forts, were extensively mapped some 100 years ago by P. Polonic for G. Tocilescu (1900) and by C. Schuchhardt, the latter later making use of aerial photographs from the First World War (Schuchhardt 1918; Bogdan-Cătăniciu 2006), the earliest example of the use of aerial photography for archaeology in Romania. Still largely extant, particularly in the western third, these

[8] A more detailed analysis of the remains and full reinterpretation of the function and history of these linear barriers, based on the significant new information obtained from historical photography augmented by further field data, is in press (Hanson and Oltean 2012).

Fig. 18.7 The ancient linear fortifications across Dobrogea on World War Two photographs: *above*: at Medgidia (Luftwaffe GX aerial photomosaic SO 40 228, April 1940); *below*: at Constanța (extract of RAF coverage, MAPRW 60 PR 460 frame 4015, May 1944) (Both images: Licensor NCAP/aerial.rcahms.gov.uk)

Fig. 18.8 (**a**) The line of the linear fortifications through Medgidia visible in Fig. 18.7 (*above*) now heavily built over and bisected by the canal, as shown on Google Earth in 2010 (© 2012 Google Earth; © 2012 GeoEye); (**b**) The line of the linear fortifications on the outskirts of Constanța visible in Fig. 18.7 (*below*) now entirely built over, as shown on Google Earth in 2011 (© 2012 Google Earth; © 2012 GeoEye; © 2012 TeleAtlas; © 2012 Basarsoft)

linear monuments have, however, suffered damage and destruction in places over time either from intensive agriculture or as a result of modern development, especially in the eastern and central sectors around towns such as Constanța and Medgidia. Examination of the 1940 Luftwaffe photomosaic of Medgidia (Fig. 18.7a) reveals not only the lines of both the so-called Large Earthen Wall and the Stone Wall as they enter the town from the west but a fort within a previously unrecognised massive enclosure attached to the Earthen Wall on the eastern side of the town, both the latter now lost under urban expansion and industrial development (Fig. 18.8a). Similarly, close examination of 1944 aerial coverage on the outskirts of Constanța (Fig. 18.7b) reveals the existence of two forts, one apparently attached to the Large Earthen Wall the other to the Stone Wall, immediately to the east of the intersection of the three walls; yet only one fort was recorded by both Schuchhardt and Tocilescu and consequently appears on modern maps of the frontier. The smaller fort had already been partially destroyed by a railway line and encroaching settlement by 1944, and both have since disappeared entirely under buildings or industrial development (Fig. 18.8b). In the same coverage of 1944 and especially in the CORONA imagery of 21 March 1966, a series of previously unrecognised features adjacent to these linear fortifications were apparent, including additional enclosures and barrows, along with extensive traces of ancient settlement associated with the forts and roads (see also Bogdan-Cătăniciu 2006).

Finally, the historical archival imagery can on occasions assist in establishing the likely date of certain sites identified during recent aerial reconnaissance and avoid their chronological misinterpretation. Thus, a group of extant earthwork remains of a small settlement north of Albeşti/Cotu Văii, which might easily be mistaken for a medieval or even earlier settlement, can be seen to still have been operating as a farmstead as recorded on the photography from 1944, and ditch-defined compounds with sunken-floored structures, such as those recorded outside Movilița, were still inhabited at that time.

18.5 Conclusions

As detailed above, the value of integrating a range of imagery, both historical and recent, lies not just in that it has enabled extensive detailed mapping of large archaeological landscapes in Romania for the first time, but also that it has allowed the recording of archaeological features substantially damaged or permanently destroyed by more recent development across wide areas. Thus, the methodology employed was successful in identifying thousands of archaeological features and sites in the area, most of them no longer extant, and has enabled the reconstruction of large parts of the archaeological landscape.

Each of the different types of imagery has made a contribution to building as clear of a picture as possible of the archaeological landscape, involving the identification of ancient roads, field systems, cemeteries, settlements and fortifications (e.g. Fig. 18.4). Also, the archival photography has helped clarify the complex interrelationships of the three neglected linear barriers between Constanța and Medgidia, with their associated forts and settlements, against a background of ancient roads and funerary monuments. Combining the imagery in this way has facilitated the production of the most complete large-scale mapping of an archaeological landscape so far undertaken in Romania. Constructing the data as layers within a GIS environment allows assessment of the evolution of the settlement pattern from the pre-Roman to the Roman period, though all sites are being mapped regardless of whether their date is considered to fall within the parameters of the primary research project. In the longer term, the GIS created will enable the Romanian authorities to exercise a more efficient management and protection regime for the sites in the region.

The utilisation of archival imagery, both the vertical photography from Second World War and CORONA satellite photography, has enabled the better identification and recognition of archaeological features before they had been largely ploughed out by intensive agriculture or, indeed, facilitated their discovery and recovery before they were destroyed by modern expansion of built-up urban areas or by industrial and infrastructural development. This is particularly important in a region with no history of sustained aerial reconnaissance for archaeology which can be drawn upon. It has also provided important coverage of areas now inaccessible to aerial reconnaissance for political or security reasons, such as along the Bulgarian border and north of the Danube-Black Sea Canal. Finally, the extensive and fairly systematic coverage provided by the archival imagery has enabled assessment of traces of past human activity largely overlooked by previous ground-based research, such as field systems and traces of cultivation, and has facilitated appreciation of the extent of road systems and the widespread distribution of funerary monuments. Thus, it greatly enhances the ability to map the wider archaeological landscape and provides a fuller appreciation of its development over time.

In areas where archaeological aerial reconnaissance has been undertaken intensively since the end of the Second World War, aerial survey can be shown to have had a huge impact on the discovery of archaeological sites and the mapping of archaeological landscapes (e.g. British Academy 2001). Yet even there, historical aerial photography, originally acquired for non-archaeological purposes, can contribute to our understanding by identifying previously unknown or unrecorded sites of all periods (e.g. Chap. 7 by Young, this volume). Such historical imagery is more widely available than is broadly appreciated, whether through international archives, such as TARA and the United States Geological Survey, or less well-known national collections (see the brief summary with references in Chap. 1 by Hanson and Oltean, this volume). Thus, the use of historical aerial and satellite photography to enhance and augment discoveries from archaeological aerial reconnaissance clearly has considerable wider potential significance in many other regions across Europe and more widely, as the geographical range of papers in this volume illustrate, and is particularly important in those regions which are, or have been until recently, unable to undertake their own such reconnaissance.

Bibliography

Alexandrescu, P. (1966). Necropola tumulară. Săpături 1955–1961. In S. Dimitriu (Ed.), *Histria. Monografie arheologică, II* (pp. 133–194). Bucharest: Editura Academiei R.S.R.

Batty, R. (2007). *Rome and the Nomads. The Pontic-Danubian realm in antiquity*. Oxford: University Press.

Bogdan-Cătăniciu, I. (2006). Cercetări aerofotografice pe valurile din Dobrogea. Aspecte demografice. In M. Mănucu-Adameşteanu (Ed.), *A la recherche d'une colonie. Actes du colleque international "40 ans de recherche archéologique à Orgamè/Argamum (Bucarest-Tulcea-Jurilovca, 3–5 octobre 2005)* (pp. 407–428). Bucharest: Editions AGIR.

Bradley, R. (1998). *The significance of monuments: On the shaping of human experience in Neolithic and Bronze Age Europe*. London/New York: Routledge.

British Academy. (2001). *Aerial survey for archaeology. Report of a British Academy working party 1999*. Compiled by Robert Bewley. London: British Academy.

Cowley, D. C. (2002). A case study in the analysis of patterns of aerial reconnaissance in a Lowland area of Southwest Scotland. *Archaeological Prospection, 9*, 255–65.

Crăciun, C. (2009). Structuri antice descoperite prin fotointerpretarea imaginilor aeriene. *Pontica, 41*, 357–392.

Hanson, W. S. (2007). Site discovery: Remote sensing approaches, aerial. In D. M. Pearsall (Ed.), *Encyclopedia of archaeology* (pp. 1907–1912). New York/London: Academic.

Hanson, W. S., & Oltean, I. A. (2001). Recent aerial survey in Western Transylvania: Problems and potential. In R. H. Bewley & W. Raczkowski (Eds.), *Aerial archaeology – Developing future practi*ce (Nato Science Series, pp. 109–115, 353–355). Amsterdam: IOS Press.

Hanson, W. S., & Oltean, I. A. (2003). The identification of Roman buildings from the air: Recent discoveries in Western Transylvania. *Archaeological Prospection, 10*, 101–17.

Hanson, W. S., & Oltean, I. A. (2012). The 'Valu lui Traian': A Roman frontier rehabilitated. *Journal of Roman Archaeology, 25*.

Lungu, V. (2007). Necropoles Greques du Pont Gauche: Istros, Orgame, Tomis, Callatis. In D. V. Grammenos & E. K. Petropoulos (Eds.), *Ancient Greek colonies in the Black Sea* (British Archaeological Reports International Series 1675, Vol. 2, pp. 337–382). Oxford: Archaeopress.

Oltean, I. A. (2007). *Dacia: Landscape, colonization, Romanization*. London: Routledge.

Oltean, I. A., & Abell, L. L. (2012). High-resolution satellite imagery and the detection of buried archaeological features in ploughed landscapes. In R. Lasaponara & N. Masini (Eds), *Satellite Remote Sensing, A New Tool for Archaeology* (pp. 291–305). New York: Springer.

Oltean, I. A., & Hanson, W. S. (2001). Military *vici* in Roman Dacia: An aerial perspective. *Acta Musei Napocensis, 38*(1), 123–34.

Oltean, I. A., & Hanson, W. S. (2007a). Cropmark formation in 'difficult' soils: Case studies from Romania. In J. Mills & R. Palmer (Eds.), *Populating clay landscapes* (pp. 75–88). Stroud: Tempus.

Oltean, I. A., & Hanson, W. S. (2007b). Villa settlements in Roman Transylvania. *Journal of Roman Archaeol, 20*, 1–25.

Palmer, R. (2005). If they used their own photographs they wouldn't take them like that. In K. Brophy & D. Cowley (Eds.), *From the air. Understanding aerial archaeology* (pp. 94–116). Stroud: Tempus.

Schuchhardt, C. (1918). *Die sogenannten Trajanswalle in der Dobrudscha*. Berlin: Reimer.

Tilley, C. (1994). *A phenomenology of landscape*. Oxford/Providence: Berg.

Tocilescu, G. G. (1900). *Fouilles et recherches archaeologiques en Roumanie*. Bucharest: Ispasesco and Bratanesco.

Wilson, D. R. (2005). Bias in aerial reconnaissance. In K. Brophy & D. Cowley (Eds.), *From the air. Understanding aerial archaeology* (pp. 64–72). Stroud: Tempus.

Index

W.S. Hanson and I.A. Oltean (eds.), *Archaeology from Historical Aerial and Satellite Archives*, DOI 10.1007/978-1-4614-4505-0,

MIX
Papier aus verantwortungsvollen Quellen
Paper from responsible sources
FSC® C105338

If you have any concerns about our products,
you can contact us on
ProductSafety@springernature.com

In case Publisher is established outside the EU,
the EU authorized representative is:
Springer Nature Customer Service Center GmbH
Europaplatz 3, 69115 Heidelberg, Germany

Printed by Libri Plureos GmbH
in Hamburg, Germany